Springer Collected Works in Mathematics

For further volumes:
http://www.springer.com/series/11104

Photo taken 1951

Kiyoshi Oka, 1901–1978

Kiyoshi Oka

Collected Papers

Translated from the French by Raghavan Narasimhan
With Commentaries by Henri Cartan
Edited by Reinhold Remmert

Reprint of the 1984 Edition

Springer

Author
Kiyoshi Oka (1901 Osaka,
Japan – 1978 Nara, Japan)
Nara Women's University
Japan

Editor
Reinhold Remmert
Mathematical Institute
University of Münster
Germany

Comments
Henri Cartan (1904 – 2008)

Translator
Raghavan Narasimhan
Mathematical Department
University of Chicago
USA

ISSN 2194-9875
ISBN 978-3-662-43412-3 (Softcover)
978-3-540-13240-0 (Hardcover)
DOI 10.1007/978-3-642-15083-8
Springer Heidelberg New York Dordrecht London

Library of Congress Control Number: 2012954381

Revised translation of the French edition "Sur les Fonctions Analytiques de Plusieurs Variables" par Kiyoshi Oka. Originally published by Iwanami Shoten Publishers, Tokyo (1961).

Printed on acid-free paper

Springer is part of Springer Science+Business Media (www.springer.com)

今ノ一当ハ
昨ノ百不当ノ
力ナリ

A success of today draws its strength from a hundred failures of yesterday. And indeed, one has no cause to speak of today's success without yesterday's hundred failures – just as, for example, even a voyage of a thousand leagues begins with one first step.

Calligraphy (first sentence only) by Heisuke Hironaka

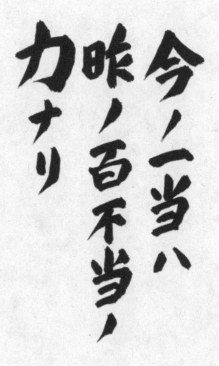

Curriculum vitae of Kiyoshi Oka

1901, April 19	Born in Osaka, Japan
1922–1925	Imperial University of Kyoto: student of physics, from 1923 on student of mathematics
1925	Graduation
1925–1929	Lecturer, Faculty of Science, Imperial University of Kyoto
1929–1932	Assistant Professor, Faculty of Science, Imperial University of Kyoto
1929–1932	Sabbatical stay in Paris, aquaintance with G. Julia
1932–1938	Assistant Professor, Faculty of Science, Hiroshima University
1938–1940	First retirement to Kimitôge in Wakayama
1940	Doctor of Science at Kyoto
1941–1942	Research fellow at Hokkaido University
1942–1949	Second retirement to Kimitôge in Wakayama, supported by the Hûju-kai Foundation (chairman: Professor T. Takagi)
1949–1964	Professor, Nara University for Women
1951	Medal of the Academy of Science, Japan
1954	Prize of Culture, Asahi-bunka-shô
1960	Medal of Culture, Bunka-kunshô
1969–1978	Professor, Industrial University of Kyoto
1978, March 1	Died in Nara, Japan

XI

Soit $(x_1, x_2, \cdots, x_n)$ l'espace projective, ou l'espace cylindrique (c'est à dire produit des plans analytiques x_i, $i = 1, 2, \cdots, n$)
Soit $\mathfrak{R}$ un domaine compact au sens large (c'est à dire, D peut contenir des points critiques).

~~Considérons~~ Soit Σ une surface caractéristique Σ sur $\mathfrak{R}$. Considérons le problème de construire une fonction f, uniforme sur $\mathfrak{R}$, holomorphe en dehors de Σ et avoir tout point de Σ pour pôle. Quelle est la condition nécessaire et suffisante?

~~P.~~ Considérons pour la simplicité $n = 2$.

Considérons une classe de fonctions $g(P)$ uniformes et continues ~~sur~~ dans des domaines sur $\mathfrak{R}$, P étant le point de $\mathfrak{R}$, telle que, soit P_0 un point ~~quelconque~~ où $g(P_0)$ est nulle, mais d'haieur quelcongeu, on puisse trouver une fonction holomorphe $\varphi(P)$ au voisinage de P_0 de façon que

$$g(P) = \varphi(P)\,\lambda(P),$$

$\lambda(P)$ étant une fonction uniforme, continue et non-nulle; nous l'appellerons classe (A).

Sait (B) la classe consistant des fonctions dont chaqune peut représenter par une rapport de deux fonctions de classe (A).

On peut toujour ~~représenter~~ trouver, pour la surface caractéristique Σ ~~par le pôle simple d'~~ une fonction $g(P)$ de classe (B) au voisinage de Σ, qui possède sur Σ des pôles simples. Soit N_1 le nombre de point où les surfaces de zéros de $g(P)$ rencontrent Σ et soit N_2 le nombre de point où les surfaces des pôles de $g(P)$ differentes de Σ rencontrent Σ. Posons

$$E = N_1 - N_2,.$$

~~et nous~~ Ce nombre E est indépendent de $g(P)$. Nous appellirons E le *nombre extérieur* de Σ.

Soit P_0 un point de ~~$\mathfrak{R}$~~ Σ, où on ne peut représenter Σ localemt par le zéro simple d'une seule fonction holomorphe; nous l'appel point *non cousinien*; on peut toujour le représenter si l'on admet des fonctions méromorphes; soit n le nombre de encontre du pôle d'une telle fonction au Σ au point Σ ; nous appellerons n l'ordre du point non cousinien.

Soit N la somme de tous les n sur Σ. Alors, la condition nécessaire et suffisante est donnée par

$$E \geqq N.$$

XII

Soit $(x_1, x_2, \cdots, x_n)$ l'espace fini des n variables complexes.

<u>Lemme</u>. Considérons une équation différentielle linéaire aux dérivées partielles d'ordre m de la forme

$$\frac{\partial^m u}{\partial x_1^m} = a_1 \frac{\partial^m u}{\partial x_1^{m-1} \partial x_2} + \cdots ,$$

u étant la fonction unconnue, $a_1, a_2, \cdots$ étant des fonctions entières <u>d'ordre fini</u>.

Considérons une condition de Cauchy sur $x_1 = 0$, telle que

$$u(0, x_2, \cdots, x_n) = \varphi_1 (x_2, \cdots, x_n),$$

$$- - - - - - - -$$

$$\frac{\partial^{m-1}}{\partial x_1^{m-1}} u(0, x_2 \cdots, x_n) = \varphi_m (x_2, \cdots, x_n),$$

$\varphi_i(x)$ $(i = 1, 2, \cdots, m)$ étant des fonctions entières <u>d'ordre fini</u>.

Alors, la solution est une fonction entière.

From a handwritten communication of Kiyoshi Oka to Henri Cartan

Table of Contents

Vorwort

„Wenn die Könige baun, haben die Kärrner zu tun." Kiyoshi OKA war ein König. Sein Reich war die Funktionentheorie mehrerer komplexer Veränderlichen. Er löste Probleme, die als unangreifbar galten; er entwickelte Methoden, deren Kühnheit die Mitwelt bewunderte. OKA gab der komplexen Analysis neues Leben. Seine Ideen wirken fort, weiterentwickelt von Mathematikern, die selbst Könige sind.

OKA hat sein Werk in französischer Sprache in zehn Mémoires niedergelegt. Das Studium der Originaltexte ist schwer. Die Mathematik Okas ist nicht, wie JACOBI einmal formulierte, eine Wissenschaft, bei der sich alles von selbst versteht. OKA war, um mit KRONECKER zu sprechen, „König und Kärrner zugleich"; sein diesem Band vorangestellter Spruch war sein Motto. Okas Mathematik bedarf der Interpretation. GOETHE spricht einmal von der Dumpfheit des Genies, das Dinge schaut, ohne dem Geschauten sofort den klaren Ausdruck geben zu können. Klarheit wird erst allmählich durch spätere Arbeit gewonnen.

Bereits 1960 edierte Y. AKIZUKI einen Band *Sur les Fonctions Analytiques de Plusieurs Variables* mit den ersten neun Okaschen Arbeiten in ihrer Originalfassung (Iwanami Shoten, Tokyo); er schreibt in seiner Einleitung: „Je suis heureux d'avoir pu ainsi participer à la réédition de ces travaux qui représentent une si importante contribution au développement de notre Science." Ich schätze mich ebenso glücklich, vom Springer-Verlag mit der Herausgabe des hier vorgelegten Bandes betraut worden zu sein. Aufgenommen sind die zehn bekannten Artikel sowie zwei weitere kurze Noten von OKA. Bis auf die Note *Sur les Fonctions Analytiques de Plusieurs Variables*, Kōdai Math. Sem. Rep. Nos. 5–6, Dec., 1949, dürfte es sich um alle gegenwärtig zugänglichen mathematischen Arbeiten Okas handeln.

R. NARASIMHAN hat sich der großen Mühe unterzogen, die Übersetzung ins Englische zu besorgen. H. CARTAN, der wie kein anderer das Okasche Œuvre kennt, hat das Gesamtwerk und jede einzelne Arbeit kritisch kommentiert. Beiden Kollegen gebührt Dank für ihre selbstlose Bereitschaft. Wir hoffen, der mathematischen Welt das Werk eines Mannes näherzubringen, der einen so großen Einfluß auf unsere und die jüngere Generation gehabt hat.

Münster (Westfalen), 18. September 1983 — R. REMMERT

Sur l'Œuvre de Kiyoshi Oka

L'œuvre mathématique de Kiyoshi Oka (1901–1978) s'échelonne sur presque trente ans, de 1934 à 1962. Elle est tout entière consacrée aux fonctions analytiques de plusieurs variables complexes. L'intérêt d'Oka pour ce sujet remonte peut-être au séjour qu'il fit à Paris en 1929, séjour qui lui donna l'occasion de rencontrer Gaston Julia. Est-ce pour cette raison qu'Oka écrivit tous ses mémoires en français, à vrai dire dans une langue française un peu particulière qui lui est bien personnelle, mais à laquelle avec un peu d'exercice on finit par s'habituer? Cette coutume de publier en français, il l'a léguée à ses élèves et aux élèves de ses élèves.

La publication, en 1934, de la monographie de Behnke-Thullen faisant le point sur l'état de la théorie des fonctions analytiques de plusieurs variables complexes à un moment crucial de son développement, et mettant en évidence les principaux problèmes ouverts à cette époque, semble avoir joué un rôle déterminant dans l'orientation des recherches d'Oka: il se fixa pour tâche de résoudre ces problèmes difficiles, tâche quasi-surhumaine. On peut dire qu'il y réussit, surmontant l'un après l'autre les obstacles redoutables qui se trouvaient sur sa route.

Mais il faut avouer que les aspects techniques de ses démonstrations et le mode de présentation de ses résultats rendent difficile la tâche du lecteur, et que ce n'est qu'au prix d'un réel effort que l'on parvient à saisir la portée de ses résultats, qui est considérable. C'est pourquoi il est peut-être encore utile aujourd'hui, en hommage au grand créateur que fut Kiyoshi Oka, de présenter l'ensemble de son œuvre.

En 1961, la maison d'édition japonaise Iwanami Shoten avait édité un volume publié par les soins du professeur Yasuo Akizuki, sous le titre: «Sur les fonctions analytiques de plusieurs variables par Kiyoshi Oka». Ce volume réunissait, sans commentaires, les neuf mémoires, numérotés de I à IX par l'auteur lui-même, parus entre 1936 et 1953. On en trouvera la liste ci-dessous. Curieusement, ce volume publié du vivant d'Oka et visiblement avec son consentement, ne contenait pas un article publié dès 1934 dans le Journal of Science of Hiroshima University (4, 1934, p. 93–98) sous le titre:

«Note sur les familles de fonctions analytiques multiformes, etc.».

Il semble que cet article était une première ébauche d'un sujet qu'Oka devait reprendre plus tard dans une publication postérieure au volume de 1961, sous le titre:

«Sur les fonctions analytiques de plusieurs variables
X – Une mode nouvelle engendrant les domaines pseudoconvexes»
(Japanese Journal of Mathematics, XXXII, 1962, p. 1–12).

Le volume de 1961 ne contenait pas non plus une Note parue en 1941 aux

Proceedings Imperial Academy Tokyo (vol. 17, p. 7–10), qui en fait annonçait les résultats du mémoire VI.

Enfin, je possède un bref manuscrit de la main d'Oka, remplissant deux pages grand format, sous les numéros XI et XII, mais ne comportant aucun titre.

Comme les titres des articles d'Oka ne permettent pas toujours d'avoir une idée de leur contenu véritable, je tenterai d'exposer, pour chacun d'eux, un résumé des principaux résultats, en utilisant la terminologie en usage aujourd'hui, ce qui, je l'espère, facilitera la tâche du lecteur.

Henri CARTAN (août 1982)

A Note on Oka's Terminology

OKA uses a few expressions which are not those used today, or which are unusual. The following list of some of them and their modern equivalents might facilitate reading Oka's papers.

1. *Finite space*, *finite part of space*, *without points at infinity*.
   It would seem that the basic entity in Oka's mind is projective space. The above expressions all refer to what we would call $\mathbb{C}^n$.
2. *Finite domain*, *domain without points at infinity*.
   As with the expressions in 1, these refer to domains in $\mathbb{C}^n$ or to domains spread over $\mathbb{C}^n$, depending on the context.
3. *In the interior*, *completely interior*.
   These refer to sets which are relatively compact.
4. *Characteristic surface*.
   Analytic set of codimension 1, hypersurface.
5. *Characteristic variety*.
   Analytic set, analytic variety.
6. *Characteristic plane*.
   Usually just a lower dimensional affine subspace of $\mathbb{C}^n$; sometimes parallel to the coordinate planes.
7. *Circle*, *sphere*, *hypersphere*.
   Disc, ball. OKA refers to the boundary of the disc as the circumference of the circle.
8. *Pole*.
   In the early papers, OKA seems to exclude points of indeterminacy.

This is usually clear from the context.

R. NARASIMHAN

I. Rationally Convex Domains

Domaines convexes par rapport aux fonctions rationelles

Journal of Science of the Hiroshima University **6** (1936), p. 245–255

Introduction. Despite recent progress in the theory of analytic functions of several variables, several important matters remain more or less obscure, most notably: the kind of domains for which the theorem of Runge or those of Mr. P. COUSIN remain valid, the relationship between the convexity of Mr. F. HARTOGS and that of Messrs. H. CARTAN and P. THULLEN[1]; and there are intimate relations between these matters. The present memoir and those which will follow are meant to treat these problems.

Now, I have noticed that one can sometimes reduce the difficulty of these problems by raising suitably the dimension of the spaces in which one works. In the present memoir, realising this general idea in a special case, I shall establish a principle which, so to speak, reduces the study of the domains of the title to cylindrical domains of higher dimension. (For the concrete form of this principle, see Problem I of No. 1.)

Once the principle is established, one can deduce that the theorem of Mr. P. COUSIN concerning prescribed poles remains valid for the domains of the title. (For the exact form of this result, see Theorem I of No. 5.) The converse is also true. I shall actually prove these theorems simultaneously by a process of induction. Using the above principle, one also recovers immediately the Runge theorem for these domains, which was expounded by Mr. A. WEIL[2].

Thus, I shall be concerned, in the present memoir, with the interior of domains which are convex with respect to rational functions; this will enable me at the same time to investigate, under less restrictive hypotheses than hitherto, some lemmas which are indispensable to me.[3]

1. Definitions. In the space $((x))$ of n complex variables $x_1, \dots, x_n$, let us consider the region[4] Δ defined by

$$(\Delta) \qquad x_i \in X_i, \quad R_j((x)) \in Y_j \quad (i=1,2,\dots,n; j=1,2,\dots,\nu),$$

where X_i, Y_j are univalent (schlicht) bounded domains[5] in the plane, and $R_j((x))$ are rational functions. For simplicity, we shall say that any region which can be described in such a way *belongs to the class* (Ω_0). Given a region

[1] See the book of Messrs. H. BEHNKE and P. THULLEN: *Theorie der Funktionen mehrerer komplexer Veränderlichen*, in particular on pages 54, 68, 79.

[2] *Sur les séries de polynomes de deux variables complexes.* C.R. Acad. Sci., Paris 1932. *L'intégrale de Cauchy et les fonctions de plusieurs variables.* Math. Annalen, 1935.

[3] As for the generalisation of the theorem of Mr. P. COUSIN, I think that one can do this using the integral of Mr. A. WEIL cited above.

[4],[5] In what follows, an open set will be called domain or region to distinguish between when it is known to be connected and when it is not; it is understood that the regions (domains) of this memoir are, without exception, univalent.

of (Ω_0) in the space $((x))$, we shall call the smallest number of rational functions of the n variables x_i, the functions x_i being excluded, which are needed to define the region, *the order of the region.* The order of the region Δ above is $\leqq \nu$; every region of (Ω_0) of order 0 is a cylindrical domain.

It is convenient to give here, in concrete form, the problems to be dealt with.

Problem I. Let us introduce new complex variables $y_1, y_2, \ldots, y_\nu$ and consider, in the space $((x, y))$, the variety Σ defined by

$$(\Sigma) \qquad y_j = R_j((x)), \quad ((x)) \in \Delta, \quad (j = 1, 2, \ldots, \nu).$$

The boundary of the variety Σ is entirely situated on the boundary of the cylindrical domain

$$(C) \qquad x_i \in X_i, \quad y_j \in Y_j \quad (i = 1, 2, \ldots, n; j = 1, 2, \ldots, \nu).$$

Let $f((x))$ be a holomorphic function of the n variables x_i in Δ; [6] if we regard it as a function of $n+\nu$ variables x_i, y_j, it is holomorphic at every point $M = ((x^0, y^0))$ on Σ since $((x^0))$ is necessarily a point of Δ.

Under these circumstances, *construct a holomorphic function on* (C') *having the value* $f(M)$ *for any point* M *on the portion of* Σ *lying in* (C'), (C') being a given domain in the interior of (C).

We shall call this *Problem I of order* ν, where ν is half the difference between the dimension of the space $((x, y))$ and that of the variety Σ. In this case, the order starts with the value 1.

Problem II. We prescribe poles $(\mathfrak{p})$ in the region Δ by the usual method, that is to say: we make correspond to any point P of Δ a polycylinder (γ) and a function $g((x))$ which is meromorphic on (γ) in such a way that on any part common to several of the (γ), the functions $g((x))$ attached to them are mutually equivalent under subtraction (the difference is holomorphic); we define the poles $(\mathfrak{p})$ locally as those of the functions $g((x))$. Here is the problem:

Find a meromorphic function with the poles $(\mathfrak{p})$ *on a region* Δ' *given à priori in the interior of* Δ.

This will be called *Problem II of order* μ, μ being the order of Δ. We know, thanks to Mr. P. Cousin[7], that Problem II of order zero is solvable.

2. Reduction of Problem I. We shall first prove the following:

If Problems I and II are solvable for all orders $< \nu$, *Problem I is solvable for order* ν.

---

[6] When the region Δ consists of several connected components, $f((x))$ can, naturally, be composed of distinct analytic functions.

[7] Acta, 1895.

We suppose that we can solve Problems I and II for any order $<\nu$. We take Problem I of order ν in the exact form in which it is given in the preceding section. In the space $(x_1, x_2, \dots, x_n, y_\nu)$, consider the region

$$(D) \qquad x_i \in X_i, \quad y_\nu \in Y_\nu, \quad R_k((x)) \in Y_k \quad (i=1,2,\dots,n;\ k=1,2,\dots,\nu-1),$$

and the variety

$$(S) \qquad y_\nu = R_\nu((x)), \quad ((x)) \in \Delta.$$

We find that the boundary of the variety S is supported by that of D.

To each point of the region D, we attach a function $g((x), y_\nu)$ of the $n+1$ variables x_i, y_ν in such a way that:

1°. For any point P outside the variety S,

$$g((x, y_\nu) = 0.$$

2°. For any point M on S,

$$g((x), y_\nu) = \frac{f((x))}{Q((x)) \cdot y_\nu - P((x))},$$

where $P((x))$ and $Q((x))$ are polynomials in x_i, relatively prime to each other (in the algebraic sense) and such that $R_\nu((x)) = P((x))/Q((x))$.

We are going to restrict each of the functions g to a polycylinder (γ). For any point P outside of S, it is sufficient to take a (γ) with center P which does not contain any point of S. Let then M be any point of S; there are two things to which we must pay attention: 1°. We must choose a polycylinder around M on which the function f is holomorphic. This is possible since f is holomorphic on the region Δ, and, on the other hand, if $((x^0), y_\nu^0)$ are the coordinates of M, $((x^0))$ belongs to Δ. 2°. Let T be the part of the variety $Q((x))\, y_\nu - P((x)) = 0$ in the interior of D; for any point $M' = ((x'), y'_\nu)$ on T, there are two cases possible: either M' is a point of S which implies that $((x'))$ belongs to Δ, or $((x'))$ is a point of indeterminacy of $R_\nu((x))$, when $((x'))$ can never fall inside Δ.

It is necessary to avoid points M' with the second property. This can always be achieved, because, in view of what we have seen, these points are at a positive distance from M. For any point M on S, we attach a polycylinder (γ) with center M satisfying these conditions.

The functions $g((x), y_\nu)$ defined thus are meromorphic in their own polycylinders (γ) and satisfy mutually the equivalence condition. These functions therefore define poles $(\mathfrak{p})$ in D. These poles all lie on S. Now, the region D of the space $((x), y_\nu)$ belongs evidently to (Ω_0) and its order is no bigger than $\nu-1$. Hence, by hypothesis, there exists a function $G((x), y_\nu)$, meromorphic in D' and having $(\mathfrak{p})$ as poles, D' being a region completely interior to D, whose exact form will be given later.

Let us set

$$\varphi((x), y_\nu) = G((x), y_\nu)\,[Q((x))\, y_\nu - P((x))].$$

It is verified immediately that the function φ so obtained is holomorphic in D', and that, in addition,

$$\varphi[(x), R_\nu((x))] = f((x)),$$

for any point on S lying in D'.[8)]

For D' we take the region[*)]

$$(D') \qquad x_i \in X'_i, \quad y_\nu \in Y'_\nu, \quad R_k((x)) \in Y'_k$$
$$(i=1,2,\dots,n; k=1,2,\dots,\nu-1),$$

X'_i, Y'_j being $(n+\nu)$ domains completely interior to X_i, Y_j respectively, and this, in such a way that the cylindrical domain

$$x_i \in X'_i, \quad y_j \in Y'_j \quad (i=1,2,\dots,n; j=1,2,\dots,\nu)$$

contains the given domain (C'), together with its boundary.

We want to construct a function F of the $(n+\nu)$ variables x_i, y_j, holomorphic in (C'), such that

$$F[x_1,\dots,x_n, R_1((x)),\dots,R_{\nu-1}((x)), y_\nu] = \varphi(x_1,\dots,x_n, y_\nu)$$

on the part of the variety $y_k = R_k((x))$ $(k=1,2,\dots,\nu-1)$ inside (C'). Such a function $F((x,y))$ certainly exists by our hypothesis; for this is just a Problem I of order $\nu-1$.

Now, for any point $M = ((x,R))$ on the variety Σ inside (C'), we have

$$F[x_1,\dots,x_n, R_1((x)),\dots,R_\nu((x))] = \varphi[x_1,\dots,x_n, R_\nu((x))] = f(x_1,\dots,x_n).$$

The function $F((x,y))$ is therefore the desired solution.

Let us remark, by the way, that *Problem I of order 1 is solvable.*

3. Reduction of Problem II. Conclusion. We now consider Problem II. To apply the classical method of Mr. P. Cousin to the present case, one must overcome the following difficulty. One has to construct a function having a given jump in our regions; more precisely:

Problem A. Consider the region Δ of No. 1. In the domain X_1 of the x_1-plane, we draw a simple rectifiable Jordan arc L with extremities a and b. Let T be the set of points $((x))$ satisfying $x_1 \in L$, $((x)) \in \Delta$. Let Δ' be an arbitrary region completely interior to Δ, T' the part of T contained in Δ'.

Under these circumstances, given a function $f((x))$ of the n variables x_i, well-defined and holomorphic in the neighbourhood of any point of T, *find a*

---

8) I owe to Mr. H. Cartan the idea of using the theorem of Mr. Cousin in this way. See: *Sur les fonctions de deux variables complexes*. Bull. Sci. math. 1930.

*) *Erratum, taken from Memoir IV.* The open set D' is not necessarily completely interior to D. This is a phenomenon which arises because of points of indeterminacy. To avoid this, we proceed as follows: 1°. The reasoning of Memoir I is correct for polynomials. 2°. One can therefore verify the theorem in No. 4 and Theorem I in complete analogy with No. 5 of Memoir II; note that Theorem I of Memoir II is obvious for rational functions. 3°. One can then establish Theorem II without modifying the proof given in No. 5.

function $\varphi((x))$ *of the n variables* x_i*, well-defined and holomorphic at every point of* Δ' *except on* T'*, and admitting a jump on* T' *of the following kind:*

1°. If one crosses T' excepting the points $x_1 = a, b$, $\varphi((x))$ can be analytically continued, and this in such a way that

$$\varphi((x)) - \psi((x)) = \pm f((x)),$$

$\psi((x))$ being the function obtained by continuation, the sign + being valid for $((x))$ lying to the left of T'; by this we mean that the point x_1 lies to the left of the curve L described from a to b.

2°. In the neighbourhood of any point on $x_1 = b$ in Δ', the analytic continuation of the function

$$\varphi((x)) - \frac{f((x))}{2\pi i} \log(b - x_1)$$

remains holomorphic. For $x_1 = a$, one has only to replace the logarithm by $-\log(a - x_1)$.

We shall call this Problem A of order μ when μ is the order of Δ. We shall see that:

If it is possible to solve Problem I of any order $\leqq \nu$*, one can solve Problem A of any order* $\leqq \nu$.

In fact, we shall prove that Problem A formulated above necessarily has solutions under the hypothesis that any Problem I of order $\leqq \nu$ is solvable. This is sufficient, for although the order of the problem may be less than ν, we can put any Problem A of order ν in this form. For the sake of clarity, we suppose that the given region Δ' is of the form

$$(\Delta') \qquad x_i \in X'_i, \quad R_j((x)) \in Y'_j \quad (i = 1, 2, \ldots, n; j = 1, 2, \ldots, \nu).$$

To the region Δ, we make correspond the cylindrical domain (C) and the variety Σ in the space $((x, y))$ by the method of No. 1. Let (C'), Σ' be those which correspond to Δ'.

The function $f((x))$ being well-defined and holomorphic in the neighbourhood of any point of T', we can find, given any region completely interior to Δ, a domain G containing L in such a way that $f((x))$ is holomorphic at any common point of the given region and the domain $x_1 \in G$. We suppose, for the sake of simplicity, that this is the case for Δ itself; this does not make it less general, because Δ' and its boundary are contained in Δ. We also suppose that

$$L \subset G' \subset G \subset X'_1,$$

where G' is again a given domain completely interior to G; this again can be achieved without loss of generality.

Under these conditions, we want to find a function $F((x, y))$ of the $n + \nu$ variables x_i, y_j, holomorphic on the common part of (C') and $x_1 \in G'$, and having the value $f((x))$ for any point $((x, y))$ of Σ' belonging to $x_1 \in G'$. This being a Problem I of order ν, $F((x, y))$ exists by hypothesis.

Let us now consider the integral

$$\Phi((x,y))=\frac{1}{2\pi i}\int_L \frac{F(t,x_2,\dots,x_n,y_1,\dots,y_\nu)}{t-x_1}\,dt$$

taken following the curve L in the x_1-plane from a to b. The region where the function $F((x,y))$ is defined is just the cylindrical domain

$$x_1\in G',\quad x_k\in X'_k,\quad y_j\in Y'_j\quad (k=2,3,\dots,n;j=1,2,\dots,\nu),$$

where G' contains L. The properties of the integral $\Phi((x,y))$ are therefore well known, thanks to Mr. P. COUSIN [9]. They are as follows.

Let U' be the set of points such that $x_1\in L$, $((x,y))\in(C')$; the integral $\Phi((x,y))$ has a meaning and represents a holomorphic function at every point of (C') except on U' where the function has a certain property; to state this property, it is sufficient to modify conditions 1°, 2° to be satisfied by the required function $\varphi((x))$ by replacing

$$\Delta',\quad T',\quad \varphi((x)),\quad \psi((x)),\quad f((x))$$

by

$$(C'),\quad U',\quad \Phi((x,y)),\quad \Psi((x,y)),\quad F((x,y))$$

respectively.

Now, when $((x))$ runs over Δ', the point M of the $((x,y))$-space having coordinates $x=x_i$, $y_j=R_j((x))$ describes the variety Σ'. Let us therefore set

$$\varphi((x))=\Phi(M).$$

The function $\varphi((x))$ is then well-defined on Δ' except on T' and represents a holomorphic function of the n variables x_i at each of these points. Since, in addition,

$$f((x))=F(M)$$

near U' for any point M on Σ', the required properties of $\varphi((x))$ follow immediately from the properties of $\Phi((x,y))\cdot\varphi((x))$ is therefore a solution of the problem. Q.E.D.

It remains for us to show that we can solve any Problem II of order $\leqq\nu$ under the assumption that any Problem A of order $\leqq\nu$ has solutions. But now, the classical method being applicable almost literally [10], we shall content ourselves with referring the reader to this [11].

We thus have the proposition that if every Problem I of order $\leqq\nu$ is solvable, one can solve any Problem II of order $\leqq\nu$.

---

[9]) See e.g. OSGOOD: Lehrbuch der Funktionstheorie II_1 (1929), §22 of Chap. III.

[10]) All the modifications necessary are simply of the following form: instead of considering "a cylindrical domain (d), $x_p\in(a_p)$, $x_q\in A_q$ $(p=1,2,\dots,\lambda;\ q=\lambda+1,\lambda+2,\dots,n)$, in the given cylindrical domain (D), $x_i\in A_i$, $(i=1,2,\dots,n)$", one should consider, in the present case "the part common to (d) and the given region Δ, (D) being now an arbitrary cylindrical domain containing Δ in its interior".

[11]) OSGOOD: §23, 24 of Chap. III.

From the two auxiliary propositions, we obtain:

Conclusion. *Problems I, II of No. 1 are always solvable.*

4. Remarks on Expansions. In what follows, we shall be concerned with completing the preceding proposition. To do this, we shall start by using some known facts about expansions of functions.

1°. Let us first consider in the space $((x))$ a domain Δ of the form

$$(\Delta) \qquad |x_i| < r_0, \quad |R_j((x))| < 1, \quad (i=1,2,\dots,n; j=1,2,\dots,\nu),$$

where the $R_j((x))$ are rational functions of the x_i, r_0 being a positive constant. If the set of points defined by these inequalities is composed of several connected components, we understand by Δ one of these components. For such a domain, one knows by the work of Mr. A. WEIL that every holomorphic function can be expanded in a series of rational functions[12]. This is an immediate consequence of the principle we have just established.

In fact, given a holomorphic function $f((x))$ in the domain Δ, we can find, for any $0<r<r_0$, $0<\rho<1$, a new function $F((x, y))$ holomorphic in the polycylinder.

$$(C') \qquad |x_i| < r, \quad |y_j| < \rho, \quad (i=1,2,\dots,n; j=1,2,\dots,\nu),$$

such that for any point M of (C') having coordinates of the form $x_i = x_i$, $y_j = R_j((x))$, we have

$$F(M) = f((x)).$$

The function $F((x, y))$ can be expanded in a Taylor series about the origin, valid at least on (C'). From this, if we set $y_j = R_j((x))$ in the expansion, we obtain a series of rational functions converging uniformly to $f((x))$ in the interior of Δ',

$$|x_i| < r, \quad |R_j((x))| < \rho, \quad ((x)) \in \Delta, \quad (i=1,2,\dots,n; j=1,2,\dots,\nu).$$

More precisely, each term of the series is a polynomial in x_i, $R_j((x))$. From this, one obtains an expansion of the same nature for Δ itself.

2°. Let us now consider a bounded domain D in the space $((x))$ which is convex with respect to a class $\mathfrak{R}$ of rational functions of the n variables x_i, holomorphic in D.[13] D is then necessarily univalent. Let D' be an arbitrary domain completely interior to D, and let δ be the "minimal distance" of D' relative to D, $\delta > 0$. We consider in D the set of points Σ whose "boundary distance" relative to D is equal to $\delta/2$. To each point M on Σ, there is always at least one function f of $\mathfrak{R}$ such that

$$|f(M)| > 1,$$

[12]) Work cited above.

[13]) Mrssrs. H. CARTAN and P. THULLEN: Regularitäts- und Konvergenzbereiche. Math. Annalen, 1932.

while, on the other hand,

$$|f| < 1 \quad \text{on} \quad D';$$

let (γ) be a polycylinder around M, interior to D, and on which $|f| > 1$. Thus, every point M of the closed set Σ is the centre of a (γ) having the above property. We can therefore cover Σ by a finite number of these (γ) by the Borel-Lebesgue lemma.

Let $f_1, f_2, \ldots, f_\nu$ be the functions of $\mathfrak{R}$ corresponding to these (γ). Consider, in D, the set of points satisfying

$$|f_j((x))| < 1, \quad (j = 1, 2, \ldots, \nu).$$

Among the connected components of this set, there is one which contains D', which we denote by Δ. The boundary of Δ is still contained in D because the domain Δ cannot extend beyond Σ.

Thus, to any domain D' completely interior to D, we can make correspond a domain Δ of the above kind. This remains true even for an unbounded domain so long as the domain contains only points of finite distance as is immediately verified. Thus, in view of the preceding considerations, we have the following proposition:

Let D be a domain in the space $((x))$ [in the proper sense, i.e. in the finite part of the space] which is convex with respect to a class $\mathfrak{R}$ of rational functions of the n variables x_i, holomorphic in D. Every function holomorphic in D can be expanded in a series of rational functions converging uniformly in the interior of D, and this, in such a way that the general term of the series is a polynomial in the x_i and functions of $\mathfrak{R}$.

5. Theorem I. *Let D be a domain having the character described above. Given poles* $(\mathfrak{p})$ *in D as in No. 1, there exists always a meromorphic function in D having exactly the poles* $(\mathfrak{p})$.

In fact, we construct in D a sequence of domains

$$\Delta_1, \Delta_2, \ldots, \Delta_n, \ldots$$

tending to D, Δ_n having the same nature as the domain Δ of the preceding section. We know that this can be achieved, irrespective of whether D is bounded or not. To simplify our language, we shall suppose that Δ_n is completely interior to Δ_{n+1}, n being arbitrary.

We consider next a sequence of functions

$$\Phi_1((x)), \quad \Phi_2((x)), \ldots, \Phi_n((x)), \ldots,$$

where $\Phi_n((x))$ stands for a meromorphic function of the n variables x_i, having the poles $(\mathfrak{p})$ on Δ_n, Δ_n being any domain in the above sequence. Such a function $\Phi_n((x))$ certainly exists since, to find it, it suffices to solve a Problem II on Δ_{n+1}, a domain which evidently belongs to (Ω_0).

Now, the difference

$$\Omega_n((x)) = \Phi_{n+1}((x)) - \Phi_n((x))$$

is a holomorphic function on Δ_n for every n. We can therefore expand $\Omega_n((x))$ on Δ_n in a series of functions holomorphic on the given domain D in view of the preceding proposition.

After what we have seen so far, we can complete the proof exactly as for cylindrical domains[14]. We shall not repeat this. Q.E.D.

Theorem II. *Under the conditions of Problem I formulated in No. 1, we can find a holomorphic function of the* $n+\nu$ *variables* x_i, y_j *on the cylindrical domain* (C) *having the value* $f(M)$ *for any point* M *on the variety* Σ.

We shall say, in fact, that Problem I or II given in No. 1 is *completely* solvable if it is solvable for $(C')=(C)$ or $\Delta'=\Delta$ respectively.

Consider Problem II. It is checked immediately that all the connected components of the region Δ satisfy the conditions imposed on the domain D in Theorem I. Problem II is therefore completely solvable for any connected component, and consequently for Δ itself, by definition.

As for Problem I, the reasoning given in No. 2 remains applicable to the present situation without modification. From this, in veiw of the preceding, it follows that if Problem I of any order smaller than ν is completely solvable, so is any Problem I of order ν for any $\nu>1$, and, for $\nu=1$, it is always solvable. One can therefore solve Problem I completely. Q.E.D.

Commentaire de H. Cartan

Une des idées d'OKA consiste à réaliser certains domaines de $\mathbb{C}^n$ comme sous-variétés analytiques dans des polydisques de dimension plus grande. Il le fait ici pour les domaines définis, pour $x=(x_i)\in\mathbb{C}^n$, par

$$(\Delta) \qquad x_i\in X_i, \quad R_j(x)\in Y_j \quad (1\leqq i\leqq n,\ 1\leqq j\leqq \nu),$$

où les X_i et les Y_j sont des domaines de $\mathbb{C}$, et où les R_j sont des fonctions *rationnelles*. Un tel domaine Δ est isomorphe à la sous-variété Σ du produit

$$C=(\prod_i X_i)\times(\prod_j Y_j)$$

définie par les équations $y_j=R_j(x)$.

L'auteur prouve les deux théorèmes suivants:

Théorème I. Le premier problème de Cousin est résoluble dans un tel domaine Δ.

Théorème II. Toute fonction holomorphe sur Σ est induite par une fonction holomorphe dans C.

---

[14]) OSGOOD, §24 of Chap. III.

La méthode de démonstration est la suivante. On appelle *problème I d'ordre* ν le problème qui consiste à trouver, pour tout ouvert C' relativement compact de C, une fonction holomorphe dans C' qui induise, en tout point de $\Sigma \cap C'$, une fonction f donnée, holomorphe dans Σ. On appelle *problème II d'ordre* ν le problème qui consiste à trouver, pour tout ouvert C' relativement compact de C, une fonction méromorphe dans $\Sigma \cap C'$ qui admette des parties principales données dans Σ. (Dans ces formulations, ν désigne le nombre des fonctions R_j.) Alors OKA prouve:

(i) si les problèmes I et II ont une solution pour tout entier $\nu < k$, le problème I a une solution pour $\nu = k$.

(ii) si le problème I a une solution pour tout entier $\nu \leqq k$, il en est de même du problème II.

La conclusion est évidemment que les problèmes I et II ont toujours une solution.

Pour prouver (ii), OKA se sert de la méthode de COUSIN, qui utilise l'intégrale de CAUCHY.

Il reste ensuite à faire un passage à la limite pour obtenir les théorèmes I et II (le lecteur notera que c'est la solution du problème II qui conduit au théorème I, et vice-versa).

Le théorème I avait été annoncé sans démonstration, au moins pour $n=2$, par H. CARTAN dès 1934 (Comptes Rendus Ac. Sciences Paris, 199, p. 925–927).

Le théorème II entraîne aussitôt que toute fonction holomorphe dans Δ est développable en série de fractions rationnelles qui converge uniformément sur tout compact de Δ; plus précisément, ces fractions rationnelles sont des polynômes en les x_i et les $R_j(x)$. On en déduit: dans tout domaine convexe par rapport à une classe K de fractions rationnelles, toute fonction holomorphe est développable en série de polynômes en les coordonnées et les fonctions de K.

II. Domains of Holomorphy

Domaines d'holomorphie

Journal of Science of the Hiroshima University 7 (1937), p. 115–130

Introduction. In the preceding memoir [1], I dealt with the subject of domains convex with respect ro rational functions. I shall now examine the same question relative to holomorphic functions; and this will be done by applying the same idea, that is to say, by going over to larger spaces.

Given a bounded univalent region in the space of several complex variables which is convex relative to a finite number of holomorphic functions, we construct from this a variety Σ in a larger space by the procedure adopted previously. It is at this variety Σ and the nature of its convexity that we shall look more precisely.

As a consequence we shall find, with the help of the theorems established in the preceding memoir, that the variety Σ is, in some sense, convex with respect to polynomials. (See Theorem I in No. 4.) In our opinion, this is a fundamental fact about domains of holomorphy. From this, in view of a well known theorem of Messrs. H. CARTAN and P. THULLEN [2], one can easily give an affirmative solution to one of the unsolved problems [3] of the theory of the title; namely, that the theorem of Mr. P. COUSIN concerning functions with given poles remains valid for univalent, bounded domains of holomorphy.

1. Generalities [4]. Consider the space $((x))$ generated by n complex variables $x_1, x_2, \ldots, x_n$. We shall first give some definitions. *A region is said to belong to the class* (P_0) if it can be defined as the set of points satisfying

$$|x_i| < r_i, \quad |P_j((x))| < 1, \quad (i=1,2,\ldots,n;\ j=1,2,\ldots,\nu),$$

the r_i being positive constants and the $P_j((x))$ polynomials in the x_i.

We say that a *closed set F belongs to the class* (P_1), if we can find, in the class (P_0), a decreasing sequence of regions having F as its limit.

Given a bounded set E, consider $\mathfrak{E}$, the set common to all sets of class (P_1) which contain E. We shall call this *the smallest set of class* (P_1) *containing* E, *because* $\mathfrak{E}$ *belongs also to the class* (P_1).

In fact, $\mathfrak{E}$ is necessarily a closed set. Let (C) be a polycylinder containing the given set E and its points of accumulation; let F be a set in the class (P_1) which is contained in (C) and which contains E, but otherwise arbitrary; $\mathfrak{E}$

[1] This journal, 6 (1936).

[2] Paper cited previously.

[3] See the book of Messrs. H. BEHNKE and P. THULLEN, 68, cited previously.

[4] In what follows, an open set will be called domain or region to distinguish between when it is known to be connected and not; to simplify language, we shall suppose that regions (domains) are always univalent and bounded, except when the contrary is stated.

is then the set common to all these F. For any positive number ρ, let us form the set A consisting of all points of the polycylinder (C) and its boundary for which the distance to $\mathfrak{E}$ is $\geqq \rho$; A is necessarily a closed set.

Let M be an arbitrary point of A; since M is an exterior point for a certain F, one can find a polynomial $P((x))$ such that we have:

$$|P((x))| > 1$$

in a sufficiently small polycylinder (γ) of centre M, and

$$|P((x))| < 1$$

on $\mathfrak{E}$; then, by means of the BOREL-LEBESGUE lemma, one can cover A by a finite number of such polycylinders (γ); let $P_j((x))$, $(j = 1, 2, \ldots, \nu)$, be the corresponding polynomials. Consider a region D of the form

$$((x)) \in (C), \qquad |P_j((x))| < 1, \qquad (j = 1, 2, \ldots, \nu).$$

The region D, which clearly belongs to the class (P_0) covers $\mathfrak{E}$ without containing any point having distance $> \rho$ from $\mathfrak{E}$; and this holds whatever be ρ. The set $\mathfrak{E}$ belongs therefore to the class (P_1). Q.E.D.

Let E be an arbitrary bounded set in the space $((x))$, and $\mathfrak{E}$ the smallest set of class (P_1) containing E. Let (σ_t) be a family of characteristic surfaces[5] *of the form*

$$(\sigma_t) \qquad f((x), t) = 0, \qquad ((x)) \in U, \qquad 0 \leqq t \leqq 1,$$

U being a domain in the space $((x))$ and $f((x), t)$ a single-valued function of the variables x_i and t defined for $((x)) \in U$, $0 \leqq t \leqq 1$, holomorphic at every point of this set, and such that for any value of t in the interval, $f((x), t)$ does not vanish identically.

These surfaces can never be situated in such a manner as to have the following properties:

1°. for each characteristic surface of the family, its boundary does not enter into a neighbourhood V of $\mathfrak{E}$;

2°. none of the characteristic surfaces passes through a point of E or one of its points of accumulation;

3°. the characteristic surface (σ_0) passes through a point of $\mathfrak{E}$, while (σ_1) remains outside V.*)

---

5) For characteristic surfaces (charakteristische Flächenstücke), see the book of Messrs. BEHNKE and THULLEN, cited previously, p. 24.

*) *Erratum taken from Memoir IV.* One should add to the proposition the condition that $f((x), t)$ be a function which is single valued and continuous on $((x)) \in \overline{U}$, $0 \leqq t \leqq 1$, where $\overline{U}$ denotes the closed domain consisting of U and its boundary; we are consequently assuming that U is bounded.

Suppose, in fact, that a family (σ_t) of this nature exists. We can then find a domain T in the t-plane, containing the segment $(0,1)$ and belonging to the class (P_0) in the plane in such a way that all the above circumstances remain unchanged when we pass from the segment $(0,1)$ to the domain T, at least if we shrink a little the given regions U and V. We may also suppose that the region V belongs to (P_0) in the space $((x))$, since $\mathfrak{E}$ is a set in (P_1). The region (V,T) so constructed belongs to (P_0) in the space $((x),t)$, and one can find on it a meromorphic function $G((x),t)$ of the variables x_i and t having as its poles $1/f((x),t)$, thanks to Theorem I of the preceding memoir.

Let (σ_α) be the last characteristic which passes through points of $\mathfrak{E}$, when t traces the segment from 0 to 1 and M one of the points of $\mathfrak{E}$ on this characteristic. When t tends to α along the segment $(\alpha,1)$, the value $G(M,t)$ becomes arbitrarily large since the function necessarily has the point (M,α) as a pole; on the other hand, the function $G((x),t)$, being regular at every point of the set $[E,(0,1)]$ as well as its points of accumulation, is bounded on this set; one can therefore choose a value β near α on the segment $(\alpha,1)$ in such a way that

$$\max|G(E,\beta)|<|G(M,\beta)|,$$

the first term meaning the upper bound of $|G((x),\beta)|$ on E.

Now, as the function $G((x),\beta)$ is regular at every point of the closed set $\mathfrak{E}$ belonging to the class (P_1) in the space $((x))$, we can expand it in a series of polynomials in the neighbourhood of $\mathfrak{E}$ because of our studies in the preceding memoir[6]; from this follows the existence of a polynomial $\Phi((x))$ in the x_i such that

$$\max|\Phi(E)|<|\Phi(M)|,$$

which is a contradiction since $\mathfrak{E}$ is minimal. A family of characteristics of this nature cannot therefore exist. Q.E.D.

Remark. We can replace polynomials by rational functions in the above considerations and define classes (R_0) and (R_1) and the smallest set in the class (R_1). One can also check, without changing our method of argument, that this set also has the above property.

2. Class (H_0). Functions $R_j(x_1)$. Let us consider in a region $\mathfrak{G}$ of the space $((x))$ [assumed univalent and bounded] ν holomorphic functions $f_j((x))$ of the variables x_i, with the help of which we form a set Δ of points as follows:

$$(\Delta)\qquad |x_i|\leqq r_i,\quad |f_j((x))|\leqq 1\quad (i=1,2,\dots,n;\ j=1,2,\dots,\nu),\quad ((x))\in\mathfrak{G},$$

the r_i being positive constants. We assume also that no point of accumulation of Δ lies on the boundary of $\mathfrak{G}$, in other words that Δ is a closed set. We shall say that this closed set Δ, which corresponds to the region Δ of the preceding memoir, belongs to *the class* (H_0).

[6]) See No. 4, where we expounded this proposition on a domain; but since we did not use the connectedness hypothesis in the proof, the proposition remains valid for regions.

As in the previous case, we introduce ν new complex variables y_j and construct, in the $((x, y))$-space, a closed set Σ of the form

$$(\Sigma) \qquad y_j = f_j((x)), \quad ((x)) \in \Delta, \quad (j = 1, 2, \ldots, \nu);$$

one finds then that the whole boundary of the variety Σ lies on that of the polycylinder

$$|x_i| < r_i, \quad |y_j| < 1.$$

In this new space $((x, y))$ we consider the smallest set $\mathfrak{A}$ in the class (P_1) containing Σ, and we shall apply the general proposition we have just proved to this set $\mathfrak{A}$; in the first place, it is obvious that $\mathfrak{A}$ is contained in the closed polycylinder $|x_i| \leqq r_i$, $|y_j| \leqq 1$.

Let B *be the image of* $\mathfrak{A}$ *projected onto the space* $((x))$,[7] and $\mathfrak{C}$ that of projection on the x_1-plane. $\mathfrak{A}$ being a closed set, so are $\mathfrak{B}$ and $\mathfrak{C}$. We form *the section of* $\mathfrak{B}$ *by the characteristic plane* $x_1 = x_1'$ *(constant)* which we consider as a set of points in the space $(x_2, x_3, \ldots, x_n)$ and denote by Bx_1'.[8] Bx_1' is a closed set. Let Gx_1' be the section of $\mathfrak{G}$; Gx_1' is an open set. In the x_1-plane, let us mark the set Ω consisting of those points x_1' for each of which Bx_1' has at least one point $(x_2, x_3, \ldots, x_n)$ outside of Gx_1'. $\mathfrak{B}$ being a closed set and $\mathfrak{G}$ an open set, the set Ω, if it is not empty, is necessarily closed. Let (ω) be the region of the x_1-plane having the points of Ω and the point at infinity as exterior or boundary points, but containing all other points.

In the region (ω) so obtained, we shall define real functions $R_j(x_1)$ of the variable x_1 $(j = 1, 2, \ldots, \nu)$ as follows: If x_1 is not in $\mathfrak{C}$, we set

$$R_j(x_1) = 0.$$

Let then x_1^0 be a point of $\mathfrak{C}$ in (ω). Let $(x_1^0, \xi_2, \ldots, \xi_n;\ \eta_1, \eta_2, \ldots, \eta_\nu)$ be a point of $\mathfrak{A}$ for which $x_1 = x_1^0$, but otherwise arbitrary. Since x_1^0 is a point of (ω), $(x_1^0, \xi_2, \ldots, \xi_n)$ lies necessarily on $\mathfrak{G}$ by very definition of (ω). Consequently, the functions $f_j((x))$ being well-defined at this point, we can set

$$R_j(x_1^0) = \max |\eta_j - f_j(x_1^0, \xi_2, \ldots, \xi_n)|, \quad (x_1^0, \xi_2, \ldots, \xi_n; \eta_1, \eta_2, \ldots, \eta_\nu) \in \mathfrak{A}.$$

We shall show that the functions $R_j(x_1)$ have the following property:

$R_j(x_1)$ *are logarithmically subharmonic functions*[9]. Let us take the function $R_1(x_1)$, to fix our ideas. We start by showing that $\log R_1(x_1)$ is upper semi-

---

7) This means that to each point $((x'))$ of $\mathfrak{B}$ there corresponds at least one point $((x, y))$ of $\mathfrak{A}$ such that $((x)) = ((x'))$, and conversely.

8) This means that if $(x_2', x_3', \ldots, x_n')$ is a point of Bx_1', the point $((x'))$ belongs necessarily to $\mathfrak{B}$; and in the contrary case, $((x'))$ does not belong to $\mathfrak{B}$.

9) That is to say $\log R_j(x_1)$ are subharmonic functions, the logarithms being, naturally, real. We call subharmonic a function $\varphi(z)$ of the complex variable which is real, upper semi-continuous, bounded above in the interior of its domain of existence and such that if we draw a circle in the interior of the domain of existence, the arithmetic mean of $\varphi(z)$ taken in the sense of Mr. Lebesgue on the circumference is bigger than or equal to the value at the centre. We shall count the constant $-\infty$ among these functions. For subharmonic functions, see the book of Mr. G. Julia: Principes géométriques d'analyse, t.II.

continuous; it is sufficient to check this for $R_1(x_1)$ itself. To do this, it will be enough to affirm the semi continuity on $\mathfrak{C}'$, $\mathfrak{C}'$ being the portion of $\mathfrak{C}$ contained in the region (ω). Let, therefore, x_1^0 be an arbitrary point of $\mathfrak{C}'$, and let

$$x_1^{(1)}, x_1^{(2)}, \dots, x_1^{(p)}, \dots$$

be a sequence of points of $\mathfrak{C}'$ tending towards $x_1^{(0)}$ and such that

$$\lim R_1(x_1^{(p)}) = \alpha,$$

where α is a constant; the sequence is otherwise arbitrary. It is sufficient to prove that

$$\alpha \leqq R_1(x_1^{(0)}).$$

Now, to each $x_1^{(p)}$ in the sequence, there corresponds at least one point $M_p = ((x^{(p)}, y^{(p)}))$ of $\mathfrak{A}$ such that

$$R_1(x_1^{(p)}) = |y_1^{(p)} - f_1((x^{(p)}))|;$$

and we have a new sequence of points

$$M_1, M_2, \dots, M_p, \dots;$$

let $M_0 = ((\xi, \eta))$ be one of its limit points. We then have $\xi_1 = x_1^{(0)}$; in addition, $\mathfrak{A}$ being closed, M_0 is part of $\mathfrak{A}$. Letting $x_1^{(p)}$ tend towards $x_1^{(0)}$, it follows from this that

$$\alpha = |\eta_1 - f_1(x_1^{(0)}, \xi_2, \dots, \xi_n) \leqq R_1(x_1^{(0)});$$

$R_1(x_1)$ is therefore upper semi-continuous.

We must now look at harmonic majorants of the function $\log R_1(x_1)$, the logarithm being real. Suppose that we have a circle (γ),

$$|x_1 - x_1^{(0)}| < \rho$$

in the interior[10)] of the domain of existence such that the arithmetic mean of the function $\log R_1(x_1)$ taken over the circumference in the sense of Mr. LEBESGUE is smaller than its value at the centre. It will be sufficient to show that this hypothesis leads to a contradiction. Now, under the above hypothesis, by using the Poisson integral, one can easily find a holomorphic function $\Psi(x_1)$ of the variable x_1 in the circle, whose absolute value is single-valued, continuous and non-zero on $|x_1 - x_1^{(0)}| \leqq \rho$, and such that

$$R_1(x_1)|\Psi(x_1)| < 1$$

for any x_1 on the circumference, while, at the centre,

$$R_1(x_1^{(0)})|\Psi(x_1^{(0)})| = 1.$$

---

10) We say that a condition is fulfilled in the interior of a certain region if it is fulfilled on every closed set contained in the region.

With this function $\Psi(x_1)$, let us form the family of characteristic surfaces

$$(\sigma_t) \qquad [y_1 - f_1((x))] \cdot \Psi(x_1) = e^{i\theta}(1+t),$$
$$|x_1 - x_1^{(0)}| < \rho, \quad ((x)) \in \mathfrak{G}, \quad 0 \leqq t < +\infty,$$

θ being a certain argument which we shall determine later, and $i = \sqrt{-1}$. We see that both sides of the equation are holomorphic functions of the $n+1$ variables x_i and t at the points indicated. For this family (σ_t), we shall now examine successively the conditions formulated in No. 1.

1°. Let us look at the boundaries of the characteristics; there are two kinds of boundary points, one over the boundary of $\mathfrak{G}$ and the other over $|x_1 - x_1^{(0)}| = \rho$; let us begin with the first kind. The closed circle $|x_1 - x_1^{(0)}| \leqq \rho$ has been traced in the region (ω), the part of $\mathfrak{B}$ in this circle is contained in $\mathfrak{G}$, $\mathfrak{B}$ being the image of $\mathfrak{A}$ under projection on the space $((x))$; consequently, if we take a region V containing $\mathfrak{A}$ and sufficiently close to $\mathfrak{A}$, the boundary of $\mathfrak{G}$ will lie entirely exterior to V; the same is true of the boundary points of (σ_t) under consideration. As for the second kind of boundary points, since we have

$$|\eta_1 - f_1((\xi))| \, |\Psi(\xi_1)| < 1$$

for every point $((\xi, \eta))$ of $\mathfrak{A}$ over $|x_1 - x_1^{(0)}| = \rho$, if we choose V sufficiently close to $\mathfrak{A}$, we can ensure that all the boundary points in question remain outside V.

2°. For any t in the interval $0 \leqq t < +\infty$, the characteristic (σ_t) never passes through a point of Σ since Σ lies entirely on the surface $y_1 - f_1((x)) = 0$.

3°. Since $R_1(x_1^{(0)})$ is necessarily positive because of our hypothesis, the centre $x_1^{(0)}$ lies on $\mathfrak{C}$. Consequently, we can find a point $((x^0, y^0))$ of $\mathfrak{A}$ for which we have

$$|y_1^{(0)} - f_1((x_1^{(0)}))| = R_1(x_1^{(0)}),$$

and therefore

$$|y_1^{(0)} - f_1((x_1^{(0)}))| \, |\Psi(x_1^{(0)})| = 1;$$

we can therefore choose the argument θ so that the characteristic (σ_0) passes through the point $((x^0, y^0))$ of $\mathfrak{A}$. On the other hand, in the equation for (σ_t), we may regard the function $f_1((x))$ as being bounded on the region under consideration, since this can always be achieved by taking, if necessary, a region a little smaller than $\mathfrak{G}$. Then, from a certain value of the parameter on, none of the characteristics (σ_t) will enter V.

We can thus construct a family of characteristics fulfilling all the conditions formulated earlier. This contradicts the general proposition, and the function $R_1(x_1)$ is therefore logarithmically convex, and so also are the other functions. Q.E.D.

Let us decompose the region (ω) into connected components

$$(\omega_1), (\omega_2), \ldots, (\omega_i), \ldots,$$

on each of which the functions $R_j(x_1)$ are logarithmically subharmonic. *Suppose that one of these domains, (ω_1) for example, contains a point exterior to $\mathfrak{C}$.*

Then, *all the functions* $R_j(x_1)$ *vanish* in the neighbourhood of this point by very definition, and consequently vanish *identically on* (ω_1) because they are logarithmically subharmonic.

Let us also remark that *if, for a point* x_1^0 *of* $\mathfrak{C}$ *contained in* (ω), *all* ν *functions* R_j *are zero, we necessarily have*

$$Ex_1^0 = Ax_1^0,$$

where Ex_1^0 *and* Ax_1^0 *represent respectively the sections of* Σ *and* $\mathfrak{A}$ *by* $x_1 = x_1^0$. In fact, if this is not the case, the set Ax_1^0 would contain at least one point $(x_2^0, x_3^0, \dots, x_n^0, y_1^0, \dots, y_\nu^0)$ exterior to Ex_1^0. Now, x_1^0 being in (ω), the point $((x^0))$ belongs to $\mathfrak{G}$. On the other hand, this point is outside Δ because if $((x^0))$ were a point of Δ, the point $(x_2^0, \dots, y_\nu^0)$ would belong to Ex_1^0 because all the functions R_j vanish at x_1^0. Under these conditions, it follows from the definition of Δ that at least one of the functions, let us say $f_1((x))$, has an absolute value bigger than 1 at $((x^0))$. Since $R_1(x_1^0) = 0$, we have $y_1^0 = f_1((x^0))$ and consequently $|y_1^0| > 1$. This is absurd, because, as we remarked above, $\mathfrak{A}$ is necessarily contained in $|x_i| \leqq r_i$, $|y_j| \leqq 1$; we must therefore have $Ex_1^0 = Ax_1^0$.

3. Preliminary Proposition. In what follows, we are going to look at the boundary of (ω), and to do this, we shall first prove the following proposition.

Let $\varphi(x)$ *be a subharmonic function of the complex variable* x. *If we let the point* x *run along a simple Jordan arc* L *in the interior of the domain of existence and terminating at a point* ξ, *we always have as limit*

$$\overline{\lim}\, \varphi(x) = \varphi(\xi), \quad (x \neq \xi).^{11)}$$

We draw two concentric circles (γ) and (C) around ξ in the domain of existence of $\varphi(x)$; of these (γ) is inside (C). Let us suppose that when the point x describes the curve L starting at ξ, this point meets the circumference γ for the last time at the point b, and the outside circumference C for the first time at the point B. Let l be the arc of the curve from b to B, and D the domain having l and C as boundary. This domain is simply connected; let us map it conformally on $|y| < 1$ in such a way that the point ξ corresponds to the origin in the y-plane. This transformation will be denoted by

$$x = F(y).$$

To simplify the notation, let us suppose that (C) is the unit circle and let (γ) be of radius ρ. The function $\dfrac{F(y)}{y}$ is holomorphic and non-zero in the unit circle and takes the value $F'(0)$ at the origin. In addition, the absolute value of

[11]) We have supposed here that the curve is a simple JORDAN curve. But this is just for the sake of simplicity. The proposition remains valid for any continuous curve which converges uniquely to ξ. The proof of this can be given similarly, using well-known theorems of Messrs. FATOU and LEBESGUE.

the function remains single-valued and continuous up to the circumference and does not vanish there either. Consequently, the arithmetic mean of the real function $\log|F(y)|$ on $|y|=1$ equals $\log|F'(0)|$. If we denote by $2\pi\alpha$ the length of the arc on $|y|=1$ corresponding to the arc l in the x-plane, we obtain from this

$$\rho^{\alpha}<|F'(0)|.$$

On the other hand, by the theorem of Mr. KOEBE, we have

$$\rho \geqq k|F'(0)|,$$

k being a certain positive constant; we therefore have

$$\rho^{(1-\alpha)} \geqq k;$$

we find therefore that *when ρ tends to* 0, *α tends necessarily to* 1.

Since the function $\varphi(x)$ is upper semi-continuous on $|x| \leqq 1$, it is bounded above on this set; let M be its upper bound on this set, and let N be its supremum on the arc of the curve L from B to the origin, the origin being excluded. Let us consider the function

$$\varphi[F(y)];$$

it is subharmonic in $|y|<1$ and single-valued and upper semi-continuous on $|y| \leqq 1$. If we compare its mean on the circumference with its value at the centre, we obtain

$$\alpha N+(1-\alpha) M \geqq \varphi(0).$$

If we let y tend to the origin, we obtain

$$N \geqq \varphi(0),$$

from which it follows that

$$\overline{\lim}\, \varphi(x) \geqq \varphi(0),$$

in which the limit is taken along the curve $L (x \neq 0)$. In the relation thus obtained, only equality can hold because $\varphi(x)$ is upper semi-continuous. Q.E.D.

Let us also remark that *any function which is logarithmically subharmonic is itself subharmonic*[12]. The above proposition can therefore be applied to both types of functions.

4. Theorem I. *Given a set Δ in the class (H_0) in the space $((x))$, we construct from it the variety Σ in the larger space $((x, y))$ by the method of No. 2. Then Σ belongs to the class (P_1) in the new space.*

In other words, we shall show that

$$\Sigma=\mathfrak{A};$$

[12] This is easily verified by comparing majorant functions, and the essential point is this: if $u(x)$ is harmonic in x, $e^{u(x)}$ will be subharmonic.

and this will be done by a process of induction with respect to λ, λ being half the number of dimensions of the initial space $((x))$. We begin by establishing that the theorem remains valid for $\lambda = n$ $(n \geqq 2)$ if we admit the hypothesis that this is true for $\lambda < n$,

In view of what we have seen in No. 2, it is sufficient for this purpose to show that the region (ω) contains every finite point in the x_1-plane. Let us therefore suppose that this is not so. Then each connected component (ω_i) would have its own boundary points at finite distance. Among these components there is at least one which contains points exterior to $\mathfrak{C}$; let (ω_1) be a domain of this nature and ξ one of its finite boundary points.

I assert that E_ξ is not empty (E_ξ being the section of Σ by $x_1 = \xi$). In fact, if ξ is an interior point of $\mathfrak{C}$, then it is the limit of a sequence of points belonging to $\mathfrak{C}$ and to (ω_1); for every x_1 of this sequence, since $E x_1 = \mathfrak{A} x_1$ (sections by x_1 = constant x_1) and since $\mathfrak{A} x_1$ necessarily contains at least one point, $E x_1$ is not empty; consequently, Σ being a closed set, E_ξ is also non-empty. If ξ is a boundary point of $\mathfrak{C}$ and E_ξ were empty, the same would be true for all x_1 in a neighbourhood of ξ since Σ is closed. Then, in view of what we have seen about the expansion of functions in series of polynomials, we can easily form a polynomial in x_i, y_j in such a way as to contradict the minimality of $\mathfrak{A}$. [Translator's note: It is sufficient to expand $\dfrac{1}{x_1 - x_1^0}$ in a series of polynomials on $\mathfrak{A}$ where x_1^0 is outside $\mathfrak{C}$ but sufficiently close to ξ.] E_ξ is therefore always non-empty.

Since the point ξ is in Ω, the set $[\xi, A_\xi]$ has, by definition, at least one point outside the region $[\mathfrak{G}, |y_j| < \infty]$, $(j = 1, 2, \ldots, \nu)$, while this region contains Σ. We can therefore find a point M of A exterior to E_ξ. Now, we recognize without difficulty that the set E_ξ in the space $(x_2, x_3, \ldots, x_n; y_1, y_2, \ldots, y_\nu)$ is of the same form as the set Σ in question. Consequently, by our induction hypothesis, E_ξ belongs to the class (P_1) in this space. We can therefore find a region D_α of the form

$$(D_\alpha) \qquad |P_i| < \alpha \quad (i = 1, 2, \ldots, m),$$

P_i being a polynomial in x_i, y_j $(i = 2, 3, \ldots, n;\ j = 1, 2, \ldots, \nu)$ and α a positive constant in such a way that D_α contains E_ξ without containing the point M.

The D_α consisting of all points satisfying these inequalities will vary continuously with the parameter α. Let us consider another positive value of the parameter such that

$$\beta < \alpha,$$

and a circle (γ) of centre ξ in the x_1-plane. By taking D_α and D_β sufficiently close to E_ξ, and (γ) sufficiently small, we can make them satisfy the following conditions:

$$[(\gamma), D_\alpha] \subset [\mathfrak{G}, |y_j| < \infty], \qquad (j = 1, 2, \ldots, \nu),$$
$$[(\gamma), E x_1] \subset [(\gamma), D_\beta];$$

in this, the first relation comes from the fact that the term on the right contains (ξ, E_ξ) which is non-empty. Under these circumstances, let us form the function $\varphi(x_1)$ on the circle (γ) in such a way that

$$\varphi(x_1) = \max[\beta, \max|P_i(Ax_1)|], \qquad (i=1,2,\ldots,m)$$

on $\mathfrak{C}$, while, outside $\mathfrak{C}$ we have

$$\varphi(x_1) = \beta.$$

We shall show *that* $\varphi(x_1)$ *is logarithmically subharmonic*. In fact, $\varphi(x_1)$ is upper semi-continuous because in the definition of the function, Ax_1 is the section of a closed set and the bounds involved are simply upper bounds.

This being the case, suppose that $\varphi(x_1)$ is not logarithmically subharmonic; one can then find in the circle (γ) another closed circle

$$|x_1 - x_1^0| \leqq \rho$$

such that the mean of $\log\varphi(x_1)$ on the circumference is smaller than $\log\varphi(x_1^0)$. From this one can construct, using the Poisson integral, a holomorphic function $\Psi(x_1)$ of x_1 in the circle $|x_1 - x_1^0| < \rho$, having an absolute value which is single-valued, continuous and non-zero upto the circumference in such a way that the value of

$$\varphi(x_1)|\Psi(x_1)|$$

is smaller than 1 on the circumference and equal to 1 at the centre x_1^0. Since $\varphi(x_1)$ is always at least equal to β, we have

$$|\Psi(x_1)| < \frac{1}{\beta}$$

on the circumference and consequently also in the circle. Since, by hypothesis, $\varphi(x_1^0) > \beta$, $\varphi(x_1^0)$ is represented by a quantity different from β in the definition. Let us assume that

$$\varphi(x_1^0) = \max|P_1(Ax_1^0)|.$$

Under these conditions, we form the following family of characteristics

$$(\sigma_t) \qquad P_1\Psi_1(x_1) = e^{i\theta}(1+t),$$

for

$$|x_1 - x_1^0| < \rho, \qquad 0 \leqq t < +\infty,$$

θ being an argument which will be determined later and $i = \sqrt{-1}$. For any fixed value of t in the interval, the characteristic (σ_t) has no boundary points at finite distance except on the set $|x_1 - x_1^0| = \rho$; on this set $\varphi(x_1)|\Psi(x_1)|$ being smaller than 1, we have, in particular, for such boundary points on $\mathfrak{A}$:

$$|P_1\Psi| < 1.$$

This shows that none of the characteristics of the family has a boundary point in the neighbourhood of $\mathfrak{A}$. None of the characteristics of the family passes through a point of Σ because we have simultaneously

$$\max|P_1(Ex_1)| < \beta, \qquad |\Psi(x_1)| < \frac{1}{\beta}$$

at points on the closed circle. On the other hand, the characteristic (σ_0) will meet $\mathfrak{A}$ if we choose the argument θ suitably because, for $x_1 = x_1^0$, we have

$$\max |P_1(Ax_1^0)|\,|\Psi(x_1^0)| = 1.$$

We also see that for any sufficiently large value of the parameter (σ_t) lies entirely outside a certain neighbourhood of $\mathfrak{A}$. This, we have seen, contradicts the proposition of No. 1 because of the minimality of $\mathfrak{A}$. $\varphi(x_1)$ is therefore logarithmically subharmonic in the circle (γ).

Consider now the set (e) of points of the circle (γ) satisfying

$$\varphi(x_1) \geqq \alpha;$$

since $\varphi(x_1)$ is upper semi-continuous, every point of accumulation of the set (e) contained in the circle belongs again to (e). We take a point a of the domain (ω_1) in (γ) and consider the pencil of circles whose Poncelet circles are the point a and its image by reflection in the circumference γ. [Translator's note: In this instance, this is the pencil obtained as the image of the family of concentric circles with centre ξ under a fractional linear transformation which leaves the circle invariant and takes the centre to a.] If we start at the point a and let the circumference of the circles of this pencil increase towards γ, we reach a circumference C which meets the set (e) for the first time because the point a is outside (e) since $\varphi(a) = \beta$, the point ξ lies on (e) and the set (e) is closed in the interior of (γ).

For every point x_1 of the circle (C), we have, by definition

$$\varphi(x_1) < \alpha;$$

recalling the way $\varphi(x_1)$ is defined, this means that Ax_1 is contained in D_α, hence in $[Gx_1, ((y))$ arbitrary$]$ where Gx_1 is the section of $\mathfrak{G}$ by $x_1 = \text{constant}$. In other words, x_1 belongs to (ω) and consequently to (ω_1). We then have $Ax_1 = Ex_1$; Ax_1 is therefore contained in D_β; in other words

$$\varphi(x_1) = \beta,$$

and this identically on the circle (C).

Since $\varphi(x_1)$ is logarithmically subharmonic, this equality will continue to hold upto the circumference because of the lemma of the preceding section. This is a contradiction, one which results from the assumption made at the beginning. We must therefore have $\Sigma = \mathfrak{A}$.

It remains only to prove the theorem for $\lambda = 1$. In this case, if we define again the ν functions $R_j(x_1)$ in the same way as in No. 2 (where we were concerned with these functions with the understanding that $\lambda \geqq 2$), it is obvious that these functions are defined for all finite values x_1; it follows from this that all these functions vanish identically, and hence that $\Sigma = \mathfrak{A}$. Q.E.D.

5. We shall apply the theorem to the prblem of finding meromorphic functions having given poles. Let us take again, in the space $((x))$, a closed set in the class (H_0),

$$|x_i| \leqq r_i, \quad |f_j((x))| \leqq 1, \quad ((x)) \in \mathfrak{G}, \quad (i = 1, 2, \ldots, n;\ j = 1, 2, \ldots, \nu),$$

to which corresponds, in the space $((x, y))$, the variety Σ

$$y_j = f_j((x)), \quad ((x)) \in \Delta, \quad (j = 1, 2, \ldots, \nu),$$

belonging to the class (P_1) in the new space.

1°. Let $F((x))$ be a holomorphic function of the n variables x_i in a neighbourhood of Δ. If we regard this as a function of the $n+\nu$ variables x_i, y_j, it is holomorphic in a neighbourhood of Σ. Consequently, in view of our studies in No. 4 of the preceding memoir, we can expand the function in a series of polynomials in a neighbourhood of Σ. Substituting the ν relations $y_j = f_j((x))$ in the expansion, we find that, in a neighbourhood of Δ, the function $F((x))$ can be expanded in a uniformly convergent series whose general term is a polynomial in x_i and $f_j((x))$.

2°. Given poles $(\mathfrak{p})$[13] in a neighbourhood of Δ in the usual way, we consider $(\mathfrak{p})$ as being poles distributed in a certain neighbourhood of Σ, as is always possible. One can find a meromorphic function $\Phi((x, y))$ of the variables x_i, y_j in a neighbourhood of Σ having $(\mathfrak{p})$ as its poles, by virtue of Theorem I of the preceding memoir. Now, at a pole (or point of indeterminacy) $((x^0, y^0))$ of this function, it can be represented in the form

$$\Phi(x, y) = g((x)) + h((x, y)),$$

where $g((x))$ is a meromorphic function which depends only on the x_i and $h((x, y))$ is a holomorphic function. By setting $y_j = f_j((x))$ in the function $\Phi((x, y))$ we therefore obtain a meromorphic function of the variables x_i having the poles $(\mathfrak{p})$ in a neighbourhood of Δ.

3°. Let D be a univalent domain of holomorphy (bounded or not) in the finite part of the space $((x))$. Thanks to the work of Messrs. H. Cartan and P. Thullen, we know that the domain D is convex with respect to the class $\mathfrak{K}$ of all holomorphic functions on D. From this, one establishes that for any given domain D' completely interior to D, one can construct, by repeating the argument expounded in the preceding memoir, a closed set Δ of class (H_0) containing D', and this in such a way that Δ is defined by functions from the class $\mathfrak{K}$.

In view of the results we have so far, we obtain the following theorem; we shall content ourselves with stating it immediately since we can prove it just as for cylindrical domains.

Theorem II. *On a univalent domain of holomorphy which contains only points at finite distance, we can always find a meromorphic function having given poles (and points of indeterminacy).*

Commentaire de H. Cartan

Dans le Mémoire I, Oka a démontré que le premier problème de Cousin est résoluble dans tout domaine de $\mathbb{C}^n$ convexe par rapport aux fonctions ra-

[13]) This expression does not exclude points of indeterminacy from being present in $(\mathfrak{p})$. (See No. 1 of the preceding memoir.)

tionnelles. Ici il prouve davantage, en démontrant que *le premier problème de Cousin est résoluble dans tout domaine d'holomorphie* (théorème II du présent Mémoire). Ce théorème II est déduit du

Théorème I. On désigne par Δ un sous-ensemble compact de $\mathbb{C}^n$ défini par

$$|x_i| \leqq r_i, \qquad |f_j(x)| \leqq 1 \qquad (1 \leqq i \leqq n,\ 1 \leqq j \leqq \nu),$$

où les f_j sont holomorphes dans un ouvert $\mathfrak{G}$ contenant Δ. On considère, dans $\mathbb{C}^{n+\nu}$, la sous-variété Σ définie par

$$(\Sigma) \qquad x \in \Delta, \qquad y_j = f_j(x).$$

Alors Σ appartient à la classe (P_1) de $\mathbb{C}^{n+\nu}$, i.e. Σ est intersection d'une suite décroissante d'ouverts définis par des inégalités (strictes) portant sur des *polynômes.*

La démonstration du théorème I utilise des fonctions dont le logarithme est sous-harmonique, et procède par récurrence sur n, en se servant des résultats du Mémoire I.

Pour déduire le théorème II du théorème I, on procède ainsi: si D est un domaine d'holomorphie, tout domaine D' dont l'adhérence est compacte et contenue dans D est contenu dans un Δ du type précédent.

Remarques du commentateur. On sait aujourd'hui que le théorème II résulte immédiatement de deux faits: (1) cohérence du faisceau des germes de fonctions holomorphes (résultat prouvé par Oka dans son Mémoire VII); (2) application du théorème B à ce faisceau.

Quant au théorème I, on a plus généralement le résultat suivant: dans un polycylindre *compact* $D \subset \mathbb{C}^N$, soit Σ un sous-ensemble analytique fermé; alors Σ est intersection d'une suite décroissante d'ouverts dont chacun est défini par des inégalités en nombre fini de la forme $|P(x)| < 1$, où P est un polynôme. Ceci peut se démontrer comme suit: d'après le théorème B, il existe, pour tout point $a \in D - \Sigma$, une f holomorphe dans D telle que $f(x) = 0$ pour tout $x \in \Sigma$, $|f(a)| > 1$. Comme f est limite uniforme de polynômes, il existe un polynôme P tel que $|P(x)| \leqq 1/2$ pour tout $x \in \Sigma$, $|P(a)| > 1$. Etant donné un voisinage ouvert V de Σ dans D, la compacité de $D - V$ entraîne l'existence d'un nombre fini de polynômes P_k tels que l'ensemble des $x \in D$ tels que $|P_k(x)| < 1$ pour tout k soit contenu dans V et contienne Σ.

III. The Second Cousin Problem

Deuxième problème de Cousin

Journal of Science of the Hiroshima University **9** (1939), p. 7–19

Introduction[1]. P. Cousin was the first to pose the problem of finding meromorphic functions with given poles and points of indeterminacy, and that of finding holomorphic functions with given zeros. We shall call these, with Mr. H. Cartan[2], the first and second Cousin problem respectively. These two problems are intimately related but differ more profoundly than may appear at first sight. For example, the first problem is always solvable on univalent domains of holomorphy without points at infinity as we have seen in the preceding memoir, but one cannot expect the same result to hold for the second problem; for there effectively exist domains whose character is contrary to this[3]. In the present memoir, we shall attempt to extract that which characterizes the second problem.

Here is an example from which the author started. Consider, in the space of three real variables x, y and z, the cylinder D and the segment L given by

$$(D) \qquad x^2+y^2<1, \quad |z|<2,$$

$$(L) \qquad x=y=0, \quad |z|\leqq 1;$$

there are infinitely many continuous functions $F(x,y,z)$, well-defined on D, which vanish at every point of L, without having zeros outside. But a purely geometric method such as this is not the method defining zeros in the second Cousin problem. Let us therefore add the following condition:

$$F(x,y,0)=(x+iy)\,\lambda(x,y) \qquad \text{for } x^2+y^2<\rho^2,$$

i being the imaginary unit, λ a function which is continuous and non-vanishing, and ρ a positive number, be it ever so small; and one finds that such an F no longer exists.

---

[1]) For the bibliography of the problem of this paper, we shall content ourselves with referring the reader to the following:
H. Behnke and P. Thullen. Theorie der Funktionen mehrerer komplexer Veränderlichen, 1934.
H. Behnke and K. Stein: Analytische Funktionen mehrerer Veränderlichen zu vorgegebenen Null- und Polstellenflächen. Jbr. Deutsch. Math.-Vereinig., 1937.

[2]) H. Cartan: Les problèmes de Poincaré et de Cousin pour les fonctions de plusieurs variables complexes. C.R. Acad. Sci., Paris, 1934.
In consideration of this paper, the author must confess that Theorem I expounded in Memoir I should be attributed to him.

[3]) One will find a concrete example in the following paper:
T.H. Gronwall: On the expressibility of uniform functions of several complex variables as quotient of two functions of entire character. Amer. Math. Soc. Trans., 1917.
We shall demonstrate a very simple new example in what follows.

This example, particularly interesting when one associates it with the nature of the second Cousin problem mentioned above, forces us to extend this problem to the domain of continuous functions. This can be done without difficulty by defining equivalence of continuous functions exactly as for holomorphic functions. The condition that the problem so generalised be always solvable is obviously topological. Leaving aside the case in which the given zeros fill out some portion of space, we shall study this problem for cylindrical domains which are univalent and at finite distance, and we shall find that for this, it is necessary and sufficient that with at most one exception, all the projections of the given domain on the coordinate planes be simply connected.

If we define the zeros analytically by the usual method in a domain D which is univalent in the finite space of several complex variables, we are presented simultaneously with the second Cousin problem and the generalised problem. Here is the relation between them: Let D be a domain of holomorphy; if non-analytic solutions exist, so do analytic solutions. This is the result that we wanted[4)].

In view of the preceding, we shall find that the second Cousin problem is always solvable on a domain of holomorphy D, if we can map it topologically onto a cylindrical domain of the kind indicated above.

Non-analytic Solutions

1. Definitions. We shall extend the second Cousin problem to the domain of non-analytic functions.

Consider a domain D in the space of several real variables and two continuous functions f_1 and f_2 on it[5)]. If these functions satisfy a relation of the form $f_1 = f_2 \lambda$ on D, λ being a continuous non-vanishing function, they will be called *equivalent* on the domain[6)]. Two functions, continuous at a point P of finite space will be called equivalent at P if we can find a hypersphere of centre P on which the functions are equivalent.

Consider a mechanism of the following type on D. To each point P of the domain, there corresponds a hypersphere (γ) centred at P and a continuous function f on (γ), and this, in such a way that for each pair of contiguous hyperspheres, the corresponding functions are equivalent on the part common to the two hyperspheres. We shall call f the function attached to P. Under these conditions we ask to find a continuous function F on D equivalent at every point P of the domain to the function attached to P. We shall say that the function F, if it exists, has the given zeros.

---

4) To establish this result, we shall make use of certain results from the two preceding memoirs:
I. Rationally Convex Domains. This journal, 1936.
II. Domains of holomorphy. This journal, 1937.

5) In the present memoir domains will be understood to be univalent and in the finite part of space unless the contrary is explicitly indicated.

6) We shall apply the same definition to two functions continuous on an arbitrary set of points in finite space.

Now, the problem so generalised is not necessarily solvable even on a sphere[7]. To avoid this behaviour, *we abandon, once for all, all cases in which the given zeros fill out a whole portion of the space of the variables.*

Let us make a remark regarding this convention. Let F_1, F_2 be two continuous functions on the domain D which are equivalent at each point of D. These functions are not, in general, equivalent on the domain[8]; but they are certainly equivalent if the set of points on which they vanish has no interior points. In fact, F_1 and F_2 being equivalent at a point P of D, one finds a relation $F_1 = F_2 \lambda$ in the neighbourhood of P, λ being continuous and non-vanishing. Now, because of the hypothesis, λ is uniquely determined in the neighbourhood of P and this for every P in D. The global equivalence follows from this.

Let us consider zeros $(\mathfrak{z})$ in a domain D, and let us map D onto another domain in the same space by a bijective, bicontinuous transformation. The mechanism defining $(\mathfrak{z})$ can be transported to D', and generates new zeros in D', say $(\mathfrak{z}')$. Under these circumstances, if our problem is solvable for D and $(\mathfrak{z})$, it is certainly also solvable for D' and $(\mathfrak{z}')$, and conversely. Thus, *the generalised problem is invariant under topological transformations.*

2. An Example. Let us consider, in the space of the two complex variables x, y, a cylindrical domain (Γ, Γ') of the form

$$r < |x| < 1, \qquad r' < |y| < 1,$$

where the sum $r + r'$ of the radii will be assumed to be greater than 1. The characteristic surface $y = x - 1$ decomposes, in (Γ, Γ'), into two pieces, one situated above the real axes and the other below them. Let (σ) be the first piece.

Suppose that we had a continuous function $F(x, y)$ in (Γ, Γ') equivalent to $f(x, y) = y - x + 1$ in the neighbourhood of (σ) and equivalent to 1 in the exterior of (σ). In the annulus Γ, let us consider points ξ_1 and ξ_2, of which ξ_1 is on the real axis and positive, and ξ_2 is on the real axis and negative. Let us draw in the annulus Γ' a circumference C' about the origin. Since $F(\xi_1, y)$ does not vanish on C', we can calculate the variation of the argument of F without ambiguity when y describes C' once in the positive sense; we shall denote this

---

7) In fact, let us consider, in the space of three real variables x, y, z, two spheres of centre $(0, 0, 1)$ and $(0, 0, -1)$ and of radius 1. We can find an infinity of functions, continuous and real on all of finite space, which vanish on these two spheres without vanishing anywhere in the exterior of the two. Let $\psi(x, y, z)$ be one of these functions. We attach to each point (x, y, z) of space a sphere (γ) around this point of radius ρ and a function $f(x, y, z)$ in such a way that

$$f(x, y, z) = \psi(x, y, z) \qquad \text{for } |z| \geqq 1,$$
$$f(x, y, z) = (x + iy)\,\psi(x, y, z) \qquad \text{for } |z| \leqq 1,$$

i being the imaginary unit. If the radius ρ is sufficiently small, the equivalence condition is satisfied, but there is no continuous function having the given zeros in the sphere of centre $((0))$ and radius 3.

8) The two functions $\psi(x, y, 1)$ and $(x + iy)\,\psi(x, y, 1)$ are equivalent at every point of the circle $x^2 + y^2 < 2$; but they are not on the whole circle.

by $V(F,\xi_1)$.[9] We shall use this notation in the general case. We can similarly calculate $V(F,\xi_2)$. Let x' be a point of Γ situated below the real axis, but otherwise arbitrary; $F(x',y)$ is never zero in Γ'; from this it follows that

$$V(F,\xi_1)=V(F,\xi_2).$$

On the other hand, let A be the set of points of Γ situated on or above the real axis. Since the functions $F(x,y)$ and $f(x,y)$ are equivalent at every point of (A,Γ'), there is a function $\lambda(x,y)$, continuous and non-vanishing on (A,Γ') such that one has, identically, $F=f\lambda$. For this λ, we have

$$V(\lambda,\xi_1)=V(\lambda,\xi_2).$$

As for f, since it equals $y-x+1$, we have

$$V(f,\xi_1)=2\pi,\qquad V(f,\xi_2)=0$$

$$\therefore\quad V(F,\xi_1)=V(F,\xi_2)+2\pi.$$

This contradicts the equality established earlier; therefore F does not exist. We have thus seen that *the generalised problem is not necessarily solvable in the domain* (Γ,Γ').

Remark. We have found, at the same time, that *the second Cousin problem is not always solvable in a domain of holomorphy.*

3. The Generalised Problem in a Cylindrical Domain. We shall start by studying the generalised problem in cylindrical domains to keep things from getting complicated from the beginning.

Consider a cylindrical domain $(X_1,X_2,\ldots,X_n)$ in the space of n complex variables $x_1,x_2,\ldots,x_n$, X_i being a domain in the x_i-plane. We shall denote this simply by $((X))$. From the preceding example, it follows immediately that if *at least two of the* X_i *are multiply connected, the generalised problem is not always solvable on* $((X))$. We shall therefore suppose, in what follows, that *all the* X_i *are simply connected, excepting at most one of them.* We now ask if the problem is solvable on $((X))$.

1°. On the plane x_1, we draw a rectangle R which we decompose into equal small rectangles (ω_i) by two systems of straight lines parallel to the sides of R. Let A_1 be a closed domain consisting of one or more of the (ω_i) and their boundaries, and A_2 a closed domain consisting of one of the (ω_i), contiguous to A_1, and its boundary. We shall denote the common part of the boundaries of A_1 and A_2 by L, and the closed domain consisting of A_1 and A_2 by A. Let us take a domain B in the space $(x_2,x_3,\ldots,x_n)$.

In what follows, it will be understood that the point $(x_2,\ldots,x_n)$ lies in the domain B. Let f_1,f_2 be two continuous functions of the n variables x_i in neighbourhoods of A_1 and A_2 respectively[10], which are globally equivalent in a neighbourhood of L. Under these conditions, one can find a continuous

[9] This is obviously finite.
[10] That is, in certain domains containing A_1,A_2.

function of the n variables x_i, defined in a neighbourhood of A, and equivalent to at least one of the two given functions f_1, f_2 at any point of A, *if L does not coincide with the entire boundary of the rectangle A_2 and if B is simply connected.*

In fact, the functions f_1 and f_2 being globally equivalent, satisfy an equation $f_1 = f_2 \lambda$, where λ is a function of the n variables x_i, continuous and non-vanishing in a neighbourhood of L. We propose to find a function μ of the n variables, continuous and non-vanishing in a neighbourhood of the closed rectangle A_2 and equal to λ in a neighbourhood of L.

This would in general be impossible if L coincided with the boundary of A_2 or if B were multiply connected[11]. But, in the case actually before us, this is immediate because any determination of $\log \lambda$ is single valued in a neighbourhood of L. Let us set

$$F((x)) = f_1((x))$$

in a neighbourhood of A_1 and

$$F((x)) = f_2((x))\,\mu((x))$$

in a neighbourhood of A_2. The function F has, obviously, the required properties.
Q.E.D.

4. 2°. Given zeros (3) in the cylindrical domain $((X))$, there exists, on any domain Δ completely interior to $((X))$,[12] a continuous function having the zeros (3) on Δ.

This is an immediate consequence of the preceding proposition because all the X_i are supposed to be simply connected with at most one exception, and in addition, the zeros are supposed not to fill out a whole portion of space.

3°. We can always find a continuous function having given zeros on the cylindrical domain $((X))$.

It is the following kind of extension of non-vanishing continuous functions which will be important in the proof of this proposition: Let $\lambda(z)$ be a function of the variable z which is continuous and non-vanishing in $1 \leqq |z| \leqq 2$; extend $\lambda(z)$ to $0 < |z| < \infty$ without ceasing to be continuous and non-vanishing. This can be done by extending the function along each half-line from the origin without changing the original values.

Let us consider in the interior of each X_i a domain X_i' bounded by one or several simple closed JORDAN curves (without common points) such that each of these curves, if it is surrounded by another one of them, contains in its interior a point not belonging to X_i. We can construct a cylindrical domain $((X'))$ so as to contain a domain completely interior to $((X))$ given à priori. We thus obtain a sequence $((X^p))$ of cylindrical domains of the kind indicated ($p = 1, 2, \ldots$) tending towards $((X))$; we can assume that each is completely interior to the following.

---

[11]) The first case is obvious; for the second, one can find an example immediately using the results of the preceding section.

[12]) This means that Δ is contained, together with its boundary, in $((X))$.

Let $(\mathfrak{z})$ be given zeros in $((X))$. By virtue of the preceding proposition, there exists, for each p, a continuous function $F_p((x))$ having $(\mathfrak{z})$ as its zeros in a neighbourhood of the closed domain $((X^p))$. We must construct a continuous function $F((x))$ having $(\mathfrak{z})$ as its zeros on the domain $((X))$, which we shall do successively as follows.

For the closed domain $((X^1))$, let us set

$$F((x)) = F_1((x)).$$

Let us pass to $((X^2))$. Since the two functions F_1 and F_2 are equivalent on $((X^1))$, they satisfy a relation $F_1 = F_2\lambda$, λ being a function which is continuous and non-vanishing on the closed set $((X^1))$. We can certainly extend this function λ to all of $((X))$ without losing these properties. Let μ be the extension; let us set

$$F((x)) = F_2((x))\,\mu((x)).$$

The new $F((x))$ is a continuous extension of the old one to $((X^2))$ and has the given zeros in a neighbourhood of the closed domain $((X^2))$. Repeating the same procedure with $((X^3)), ((X^4)), \ldots$, we obtain the required function. Q.E.D.

Lemma I. *Let $((X))$ be a cylindrical domain in the space $((x))$. For the generalised problem to be solvable on $((X))$, always assuming that the given zeros never fill out a whole portion of the space $((x))$, it is necessary and sufficient that all the X_i, with at most one exception, be simply connected.*

Analytic Solutions

5. Definitions. From now on we shall study the second Cousin problem in domains of holomorphy with the help of what we have just seen.

Consider a domain D in the space of the n complex variables x_i. Zeros $(\mathfrak{z})$ in D will be called *analytic* if to every point of D there corresponds a holomorphic function (of the variables x_i) having $(\mathfrak{z})$ as its zeros in the neighbourhood of that point. The second Cousin problem consists in finding a holomorphic function having given analytic zeros in D; solutions of this problem will sometimes be called analytic to distinguish them from solutions of the generalised problem.

Let us introduce the following auxiliary notion. Given zeros $(\mathfrak{z})$ in D, we shall say that they can be *swept out* if, to each point P of D there corresponds a hypersphere (γ) and a continuous function $f((x),t)$ on $[(\gamma), 0 \leqq t \leqq 1]$ such that:

1°. $f((x),0)$ has $(\mathfrak{z})$ as its zeros and $f((x),1)$ is non-vanishing.

2°. $f((x),t)$ is not identically zero on a whole portion of $(2n+1)$ dimensions.

3°. For each pair (γ_1), (γ_2) of contiguous hyperspheres, the corresponding functions $f_1((x),t)$ and $f_2((x),t)$ are equivalent on $[(\delta), 0 \leqq t \leqq 1]$, (δ) being the part common to (γ_1) and (γ_2).

The second condition is essential, since, without it, all zeros could be swept out[13]. Let $F_1((x),t), F_2((x),t)$ be continuous functions on $[\Delta, 0 \leqq t \leqq 1]$, where Δ is an arbitrary domain in D, which are equivalent to at least one of the functions $f((x),t)$ when $((x))$ is in the neighbourhood of any point of Δ, and t on the closed interval. Then, because of the same second condition, it follows that F_1 and F_2 are globally equivalent on $[\Delta, 0 \leqq t \leqq 1]$.

If there exists a continuous function having given zeros in D, these zeros can necessarily be swept out.

In fact, let $F((x))$ be the solution. Let us set

$$f((x),t) = (1-t)\,F((x)) + t;$$

we find that the three conditions are fulfilled.

It is the converse of this statement with which we shall be occupied in what follows.

6. A Preliminary Proposition. The example given in the introduction leads us to a proposition which we shall use frequently in what follows.

Let D be a domain in the space $((x))$ and $\lambda((x))$ a function continuous and non-vanishing in D. We shall say ghat $\lambda((x))$ *satisfies condition* (α) in the domain if there exists a function $\lambda((x),t)$ on $(D, 0 \leqq t \leqq 1)$, continuous and non-vanishing on this set, and such that

$$\lambda((x),0) = \lambda((x)), \quad \lambda((x),1) = 1.$$

If $\log \lambda((x))$ is single valued in D, condition (α) is necessarily satisfied; to see this, it is enough to set

$$\lambda((x),t) = e^{(1-t)\log \lambda((x))}.$$

Suppose, conversely, that condition (α) is fulfilled for the function $\lambda((x))$ in the domain D; let $\lambda((x),t)$ be one of the functions realising this. Let us draw a closed curve C in D. When $((x))$ describes C in a definite sense, and t is fixed in $0 \leqq t \leqq 1$, the variation of the argument of $\lambda((x),t)$ can be calculated unambiguously since $\lambda((x),t)$ is never zero. We shall denote the variation by $V(t)$. This is a continuous function of the variable t on $0 \leqq t \leqq 1$; the values it can take are discrete; it is therefore constant, and, in particular, $V(0) = V(1)$.

Now, for $t=1$, we have $\lambda = 1$ identically. $V(1)$ is therefore equal to zero; hence $V(0)$ is also zero. This is true whatever be C. Hence, any determination of $\log \lambda((x))$ is single-valued on D.

We have thus established the following proposition: *Every determination of* $\log \lambda((x))$ *is single-valued in D if and only if* $\lambda((x))$ *fulfills condition* (α) *on D.*

7. Zeros Which Can be Swept Out. 1°. Let us consider once again the configuration of No. 3 in the x_1-plane, consisting of the rectangles R, (ω_i), the

[13]) In fact, let $f((x))$ be any function attached to a point (defining the zeros) and let us set $f((x),t) = (1-3t)f((x)), \ =0, \ =(3t-2)$ according as $0 \leqq 3t \leqq 1$, $1 \leqq 3t \leqq 2$, $2 \leqq 3t \leqq 3$. The functions $f((x),t)$ so defined satisfy the first and the third condition.

closed domains A, A_1, A_2 and the common part L of the boundaries of A_1 and A_2. Let us consider also the closed set Δ in finite $((x))$ space.

In what follows, it will be understood that $((x))$ remains in a neighbourhood of Δ and t in the interval $(0,1)$. Let f_1 be a continuous function of the variables x_i, t $(i=1,2,\ldots,n)$ in a neighbourhood of A_1, *analytic for* $t=0$ *and identically* 1 *for* $t=1$. Let f_2 be a function having these properties in a neighbourhood of A_2. Suppose that f_1 and f_2 are globally equivalent in a neighbourhood of L as functions of $n+1$ variables. We shall prove the following:

We can find a continuous function of the $n+1$ variables having the properties indicated above in a neighbourhood of A and equivalent to at least one of the given functions at every point $((x))$ for any t in $(0,1)$ if L does not coincide with the entire boundary of the rectangle A_2, and if, in addition, Δ *belongs to the class* (P_1).

In fact, as f_1 and f_2 are globally equivalent, we have a relation

$$f_1((x),t)=f_2((x),t)\,\lambda((x),t),$$

where λ is a function which is continuous and non-vanishing in a neighbourhood of L, from which it follows that $\lambda((x),0)$ is analytic and $\lambda((x),1)=1$. Hence, by the preceding proposition, $\log\lambda((x),0)$ is single valued, the determination being arbitrary. Thus $\log\lambda((x),0)$ is a holomorphic function in a neighbourhood of L. Δ is a set in the class (P_1). In view of the results of No. 3 in Memoir I, we can therefore find two holomorphic functions $\varphi_1((x)), \varphi_2((x))$ in neighbourhoods of A_1, A_2 respectively, such that in a neighbourhood of L we have the relation

$$\varphi_1((x))-\varphi_2((x))=-\log\lambda((x),0).$$

Let us set

$$f_1((x),t)\,e^{(1-t)(\varphi_1+2k\pi i)}=\mu((x),t)\,f_2((x),t)\,e^{(1-t)\varphi_2},$$

k being a suitable integer depending on the domains. The function μ is continuous and non-zero in a neighbourhood of L and reduces to 1 for $t=0$ and $t=1$. We find a function $v((x),t)$ having the same properties as μ in a neighbourhood of the rectangle A_2, and identical with μ in a neighbourhood of L. By virtue of the preliminary proposition, every determination of $\log\mu((x),t)$ is single-valued, and therefore, v certainly exists. We set

$$F((x),t)=f_1((x),t)\,e^{(1-t)(\varphi_1+2k\pi i)}$$

or

$$F((x),t)=v((x),t)\,f_2((x),t)\,e^{(1-t)\varphi_2}$$

according as x_1 is situated on A_1 or on A_2. The function F satisfies the condition imposed.

8. 2°. Let us consider a *domain of holomorphy* D in the space $((x))$. We construct a sequence of closed sets:

$$(S)\qquad \Delta_1,\Delta_2,\ldots,\Delta_p,\ldots,$$

tending to D from the interior such that each Δ_p is of the form

$$|x_i|\leqq r_{p,i},\qquad |g_{p,j}((x))|\leqq 1,\qquad (i=1,2,\ldots,n;\, j=1,2,\ldots,p')$$

where the $r_{p,i}$ are positive constants and the $g_{p,j}((x))$ functions holomorphic on D. Thanks to the work of Messrs. H. CARTAN and P. THULLEN, (S) always exists. We shall also suppose that Δ_p consists of points interior to Δ_{p+1}.

Let us consider in the domain D zeros (σ) *which are analytic and can be swept out*. There then exists, for every point P of the domain, a hypersphere (γ) of centre P, a holomorphic function $\varphi((x))$ having (σ) as its zeros on (γ) and a function $f((x),t)$ satisfying the three conditions formulated in No. 5 on $[(\gamma),(0,1)]$. Now, we can always suppose that f is analytic for $t=0$ and reduces to 1 for $t=1$. In fact, $\varphi((x))$ and $f((x),0)$ are equivalent in (γ) and $f((x),1)$ is never zero. Hence, if we put

$$\varphi((x))=\frac{f((x),0)}{f((x),1)}\lambda((x)),$$

λ is continuous and non-vanishing on (γ); it satisfies condition (α) on (γ) because (γ) is simply connected. Let $\lambda((x),t)$ be a function realising condition (α), and let us set

$$\varphi((x),t)=\frac{f((x),t)}{f((x),1)}\lambda((x),t).$$

The function $\varphi((x),t)$ has the properties required, and we shall denote it again by $f((x),t)$.

We shall prove the following: To each Δ_p of the sequence S there corresponds a continuous function $F_p((x),t)$ in the neighbourhood of Δ_p which is equivalent to at least one of the functions $f((x),t)$ at every point of Δ_p, t being in $(0,1)$, and such that it is holomorphic for $t=0$ and reduces to 1 for $t=1$.

When the Δ_p belong to (P_1), this is an immediate consequence of the preceding proposition.

Let us consider the general case. In view of Theorem I in Memoir II, the closed set Σ,

$$y_j=g_{p,j}((x)),\quad ((x))\in\Delta_p,\quad (j=1,2,\ldots,p'),$$

always belongs to the class (P_1) in the space $((x,y))$. We pose the present problem in the space $((x,y))$ by regarding the function $f((x),t)$ as being given in a neighbourhood of Σ. This being the case above, there exists a function $\Phi((x,y),t)$ in a neighbourhood of Σ and having the properties indicated above. Substituting $y_j=g_{p,j}((x))$ in Φ, we obtain the required function.

9. 3°. From the sequence of functions obtained thus

$$F_1((x),t), F_2((x),t),\ldots, F_p((x),t),\ldots,$$

we shall construct a holomorphic function $\Phi((x))$ having the zeros (σ) in the domain D.

Let ε_p, $(p=1,2,\ldots)$, be a sequence of positive numbers such that the series $\Sigma\,\varepsilon_p$ converges. We first set

$$\Phi_1((x))=F_1((x),0);$$

Φ_1 is a holomorphic function having the zeros (σ) in a neighbourhood of Δ_1.

The functions $F_1((x),t)$, $F_2((x),t)$ are globally equivalent in a neighbourhood of Δ_1 because they have the same zeros which do not fill out any portion of

$((x), t)$ space; hence, if we set

$$F_1((x), t) = F_2((x), t)\,\lambda_1((x), t),$$

the function λ_1 is continuous and non-vanishing in a neighbourhood of Δ_1, t being in $(0, 1)$. It is holomorphic for $t=0$ and reduces to 1 for $t=1$. Hence $\log \lambda_1((x), 0)$ is holomorphic in a neighbourhood of Δ_1, the determination being arbitrary. In view of the results expounded in No. 5 of Memoir II, we can therefore find a holomorphic function $\varphi_1((x))$ in the domain, satisfying, in a neighbourhood of Δ_1, the condition

$$|\varphi_1((x)) - \log \lambda_1((x), 0)| < \varepsilon_1.$$

We set

$$\Phi_2((x)) = F_2((x), 0)\, e^{\varphi_1}.$$

Φ_2 is a holomorphic function having the zeros (σ) in a neighbourhood of Δ_2; we see that

$$e^{-\varepsilon_1} < \left|\frac{\Phi_2((x))}{\Phi_1((x))}\right| < e^{\varepsilon_1}$$

in a neighbourhood of Δ_1. Moreover, the logarithms of Φ_2/F_3 with $t=0$ are holomorphic in a neighbourhood of Δ_2.

From this last property of Φ_2, we form a function $\Phi_3((x))$ holomorphic in a neighbourhood of Δ_3 and satisfying the condition

$$e^{-\varepsilon_2} < \left|\frac{\Phi_3((x))}{\Phi_2((x))}\right| < e^{\varepsilon_2}.$$

in a neighbourhood of Δ_2, and such that the logarithms of Φ_3/F_4 are holomorphic in a neighbourhood of Δ_3 when $t=0$; and so on. Every limit of the sequence of functions so obtained gives us the required function. We have thus established the following proposition.

Lemma II. *Given analytic zeros which can be swept out on a univalent domain of holomorphy without points at infinity, we can find a holomorphic function having the given zeros on the domain.*

Remark. Let us mention the corresponding proposition for the generalised problem. *Given zeros which can be swept out on a univalent domain in the finite space of several complex variables, we can find a continuous function having the given zeros on the domain* [14].

One can prove this in exactly the same way, and in fact even more simply.

10. Theorem I. *Given analytic zeros in a univalent domain of holomorphy without points at infinity, if there exists a continuous solution, an analytic solution also exists.*

This is an immediate consequence of Lemma II since zeros admitting a continuous solution can necessarily be swept out.

---

[14]) Whether these zeros fill out a portion of space or not.

Theorem II. *Let Γ be a univalent cylindrical domain in the finite space of several complex variables, all of whose projections on the coordinate planes, with at most one exception, are simply connected.*

The second Cousin problem is always solvable on a univalent domain of holomorphy in the same finite space if it can be mapped onto Γ by a topological transformation.

For, by Lemma I, the generalised problem is always solvable on Γ; consequently the same is true on the given domain of holomorphy.

Commentaire de H. Cartan

Oka ayant résolu dans son Mémoire II le premier problème de Cousin pour tout domaine d'holomorphie, s'attaque ici au deuxième problème de Cousin. Rappelons d'abord en quoi il consiste.

Chaque fonction holomorphe f dans un domaine D, non identiquement nulle, définit un diviseur positif $\Sigma(f)$, qui consiste en la donnée d'une famille localement finie d'hypersurfaces analytiques dont chacune est munie d'un ordre de multiplicité (entier >0). Réciproquement, tout diviseur positif peut être *localement* défini par une fonction holomorphe $\not\equiv 0$, et si deux fonctions holomorphes f_1 et f_2 définissent le même diviseur on a $f_2 = \varphi f_1$, où φ est holomorphe partout $\neq 0$ (φ est uniquement déterminèe par f_1 et f_2). Le deuxième problème de Cousin est le suivant: étant donné, dans un domaine $D \subset \mathbb{C}^n$, un diviseur positif Σ, il s'agit de trouver une fonction f holomorphe dans D dont le diviseur $\Sigma(f)$ soit égal à Σ.

Cousin (Acta Mathematica, 19, 1895, p. 1–62) avait montré que si D est le produit de n domaines $D_i \subset \mathbb{C}$, ce problème possède une solution quel que soit Σ, pourvu toutefois que les D_i soient simplement connexes sauf au plus l'un d'entre eux; cette dernière condition, qui avait en réalité échappé à l'attention de Cousin, avait été signalée par Gronwall (Trans. A.M.S., 18, 1917, p. 50–64). Il s'ensuit que le deuxième problène de Cousin pour un domaine D n'a pas toujours de solution, même si D est un domaine d'holomorphie.

Dans son Mémoire III, Oka montre que si D est un domaine d'holomorphie, la condition à laquelle doit satisfaire un diviseur Σ pour que Σ soit le diviseur d'une fonction holomorphe dans D est de nature *purement topologique.* Voici comment on peut présenter le résultat d'Oka. Il introduit essentiellement la notion de *diviseur continu:* un tel diviseur est défini *localement* par la donnée d'une fonction continue (à valeurs complexes) dont l'ensemble des zéros a un intérieur vide; deux telles fonctions f_1 et f_2 définissent le même diviseur continu si l'on a $f_2 = \varphi f_1$, où φ est une fonction continue partout $\neq 0$ (φ est alors uniquement déterminée par f_1 et f_2). Par opposition à cette notion de diviseur continu, nous appellerons *diviseur analytique* le diviseur défini par la donnée locale de fonctions holomorphes.

Le résultat fondamental du Mémoire III d'Oka peut alors se formuler comme suit: si, dans un domaine d'holomorphie D, un diviseur analytique Σ est le diviseur d'une fonction *continue* dans D, c'est aussi le diviseur d'une fonction *holomorphe* dans D. Pour prouver ce résultat, Oka introduit la notion de diviseur *balayable:* essentiellement, un diviseur Σ dans D est balayable s'il existe dans $D \times [0,1]$ un diviseur continu dont la restriction à $D \times \{0\}$ (identifié à D) soit Σ, et dont la restriction à $D \times \{1\}$ soit vide. Alors Oka prouve: 1° si un diviseur Σ dans D est le diviseur d'une fonction continue, Σ est balayable (ceci est facile); 2° si un diviseur analytique Σ dans D est balayable, c'est le diviseur d'une fonction holomorphe dans D. La démonstration de ce deuxième point utilise les résultats des Mémoires I et II, et nécessite un passage à la limite.

Rappelons qu'après ce résultat d'Oka, il restait à expliciter les *obstructions topologiques* associées à un diviseur analytique Σ dans D. Karl Stein fut le premier à expliciter de telles obstructions (Math. Ann. 117, 1941, p. 727–757, et Math. Ann. 123, 1951, p. 201–222). Dans le deuxième article de Stein le deuxième problème de Cousin est considéré, plus généralement, dans une variété de Stein, et plus seulement dans un domaine d'holomorphie. C'est Jean-Pierre Serre (Colloque sur les fonctions de plusieurs variables Bruxelles 1953, p. 57–68) qui a donné la solution finale du deuxième problème de Cousin pour une variété de Stein X: pour qu'un diviseur analytique Σ soit le diviseur d'une fonction holomorphe dans X, il faut et il suffit que l'élément du groupe de cohomologie $H^2(X; Z)$ défini par Σ soit nul; ce résultat est même valable si X n'est pas une variété de Stein, pourvu que $H^1(X, \mathcal{O}) = 0$ (cohomologie à coefficients dans le faisceau $\mathcal{O}$ des germes de fonctions holomorphes sur X).

IV. Domains of Holomorphy and Rationally Convex Domains

Domaines d'holomorphie et domaines rationnellement convexes

Japanese Journal of Mathematics **17** (1941), p. 517–521

As we have said before, the subject of the present researches consists of the following problems: the problems of COUSIN, representations of holomorphic functions, and the classification of domains[1]. In the present memoir, we shall be concerned with the last problem[2].

1. We shall consider domains in the space $((x))$ described by the complex variables $x_1, x_2, \dots, x_n$. In the present research, we shall always suppose that they are univalent without points at infinity to keep matters from becoming complicated from the begiining. If we classify *domains of the type discovered by F. Hartogs* using the *method of Mr. H. Cartan*, and simplify the classification, we obtain the following types of domains:

(*I*) *Cylindrical domains.*
(*II*) *Rationally convex domains.*
(*III*) *Domains of holomorphy.*
(*IV*) *Pseudoconvex domains.*

A domain D will be called *rationally convex* if it is convex with respect to rational functions which are holomorphic on D, or if D can be approximated from the interior by domains of the above class.

Let us consider a bounded domain D. Let E be the set consisting of points of finite space $((x))$ which do not belong to D. If, in the neighbourhood of an arbitrary boundary point P of D, the set E satisfies the continuity theorem[3], and if this property remains invariant under all bijective pseudoconformal transformations in the neighbourhood of P, D will be called *pseudoconvex*. More generally, we shall call pseudoconvex any domain which can be approximated from the interior by domains of the above character.

These 4 fundamental types satisfy the following relations:

$$(I) \subset (II) \subseteqq (III) \subseteqq (IV).$$

---

[1] For these problems, see, above all G. JULIA, Sur les familles de fonctions analytiques de plusieurs variables, 1926 (Acta Mathematica).

[2] The 3 preceding memoirs are the following:
I. Rationally convex domains, 1936.
II. Domains of holomorphy, 1937.
III. The second COUSIN problem, 1939. (Journal of Science of the Hiroshima University.)

[3] This means that there exists a hypersphere S of centre P having the following property: Consider in S an arbitrary point $((a))$ and a circumference of the form $x_1 = a_1, x_2 = a_2, \dots, x_{n-1} = a_{n-1}$, $|x_n - a_n| = r$, r being an arbitrary radius. If this circumference is exterior to E but the point $((a))$ is not, we can find a positive number d such that to each point $(x_1^0, x_2^0, \dots, x_{n-1}^0)$ in $|x_k - a_k| < d$ $(k = 1, 2, \dots, n-1)$ there corresponds at least one point x_n^0 in $|x_n - a_n| < r$ such that the point $((x^0))$ belongs to E.

In other words, these types are nested in each other. The two last distinctions are but provisional[4]. It is possible that there exist only 2 types (I) and (II). *Here then are two problems:* $(II)\subsetneqq(III)$? and $(III)\subsetneqq(IV)$? The second of these will be treated later. In the present memoir, we shall treat the first.

2. Consider the di-annulus

$$(\Gamma,\Gamma') \qquad r<|x|<1, \quad r'<|y|<1$$

in the space of two complex variables x, y, and, at the same time, the characteristic plane

$$y=x-1.$$

The part of this characteristic in the di-annulus decomposes into two (continuous) pieces if the radii satisfy the relation

$$r+r'>1.$$

Let us consider this case. One of the pieces consists of points (x',y') such that x' as well as y' lie in the upper half-plane. We denote this by (σ).[5] Let us construct a function $G(x,y)$, meromorphic in (Γ,Γ'), having

$$\frac{1}{y-x+1}$$

as poles on (σ), and holomorphic outside (σ); $G(x,y)$ certainly exists, by COUSIN's theorem. With this G, we construct the set of points of the form

$$(A) \qquad r<|x|<1, \quad r'<|y|<1,$$

$$|G(x,y)|<M,$$

where M is a positive number which we shall determine later. The set A is open; it consists of domains of holomorphy.

We choose a positive number d as follows:

$$6d<1-r';$$

we draw in the x-plane a circumference C around the origin of radius half of $(1+r)$, and in the y-plane, the annulus

$$(\Gamma_0') \qquad r'-d<|y|<1-d.$$

Let (x',y') be any point on (σ). On the characteristic plane $x=x'$, we describe a circle $(\gamma_{x'})$ of centre y' and radius d. (γ_x) is uniquely determined, if it is defined, by x. Let E be the set of points consisting of all the circles $(\gamma_{x'})$. Let (x',y') be now a point belonging to the set

$$|x|=\frac{1+r}{2}, \quad r'-d\leqq|y|\leqq1-d$$

---

[4] In Memoir I, we looked for a method to pass from (II) to (I); and in Memoir II for the passage from (III) to (II).

[5] This is the configuration we used once before in the second COUSIN problem.

without belonging to E. The set consisting of all these (x', y') will be denoted by F; F is a continuum contained in (Γ, Γ'). At any point of F, $G(x, y)$ is holomorphic. Hence A contains F as long as M is sufficiently large. *Let A_0 be a domain of the form A and containing F.* We shall prove that *the domain of holomorphy A_0 is not rationally convex.*

Let us suppose that A_0 is rationally convex. Then, thanks to the theorem of Weil explained in No. 4 of Memoir I, we can represent the function $G(x, y)$ in the neighbourhood of F[6] as the limit of a sequence of holomorphic rational functions:

$$(S) \qquad f_1(x, y), f_2(x, y), \dots, f_n(x, y), \dots$$

converging uniformly.

Let ξ be a point on the circumference C such that (γ_ξ) is contained in the annulus Γ_0'; ξ certainly exists, because the distance between the two pieces of the boundary of Γ_0' is bigger than $4d$, and, on the other hand, the radius of (γ_ξ) equals d. Let η be the centre of this circle. Then, in the sequence

$$(S') \qquad f_1(\xi, y), f_2(\xi, y), \dots, f_n(\xi, y), \dots$$

one can find at least one function having poles in the circle (γ_ξ) in the y-plane. In fact, if not, every function of (S') would be holomorphic in (γ_ξ). Now, each of these functions is a priori holomorphic on the circumference since the circumference, considered in (x, y) space, belongs to F, and for this reason, the sequence (S') converges uniformly to $G(\xi, y)$ on the circumference. $G(\xi, y)$ would therefore be holomorphic in the circle, in contradicting its having the pole $1/(y-\xi+1)$ at the centre. Let then $f_m(\xi, y)$ be a function of (S') having poles in (γ_ξ); let us set

$$f_m(x, y) = \frac{\varphi(x, y)}{\psi(x, y)},$$

φ and ψ being two polynomials without common factors.

Let (x', y') be a point having the same properties as (ξ, η) but otherwise arbitrary. Consider the equation

$$\psi(x, y) = 0,$$

in which we consider x to be an independent parameter. Since $\psi(x, y)$ is non-zero on F, no root of this equation can leave the circle (γ_ξ) when x describes the arc of C from ξ to x'. Consequently, the equation $\psi(x', y) = 0$ has at least one root in the circle $(\gamma_{x'})$ of the y-plane, x' being arbitrary. The number of these roots is independent of x', and we shall denote it by λ.

Let us draw, in the y-plane, a circumference C' around the origin, of radius half of $(1+r')$. When x' describes the circumference C once positively, the circle $(\gamma_{x'})$ in the y-plane crosses C' once from the interior to the exterior of the circle (C'). Now, the radius of $(\gamma_{x'})$ equals d; the distance of the circumference C' to the boundary of Γ_0' is greater than $2d$; the circle $(\gamma_{x'})$ and its circumference are contained in Γ_0' at the moment it crosses C'.

[6]) We shall say that a certain phenomenon occurs in a neighbourhood of a set E if this is the case on a certain open set containing E and its points of accumulation.

This is very curious. For then, when x describes C once positively, λ roots of the algebraic equation $\psi=0$ in (C') will leave this circle; this is, of course, absurd. Thus, A_0 is not rationally convex.

Theorem. *A domain of holomorphy is not necessarily rationally convex, even if it is univalent (and without points at infinity).*

This is the reason for establishing *Theorem I* of Memoir II. We have seen at the same time that *the convexity of Mr. H. Cartan*[7] and *the theorem of Messrs. H. Cartan and P. Thullen*[8] (Hauptsatz über die gleichzeitige Fortsetzbarkeit) are indispensable, even for univalent domains. The theorem shows, in addition, that it is in general impossible to expand holomorphic functions on a univalent domain of holomorphy without points at infinity in a series of rational functions (holomorphic on the domain or not) which converges uniformly in the interior of the domain[9]. In other words, *Runge's theorem* is no longer valid for the most general univalent domains of holomorphy.

Commentaire de H. Cartan

OKA construit explicitement dans $\mathbb{C}^2$ un domaine d'holomorphie, noté A_0, qui n'est pas rationnellement convexe, parce que les fonctions rationnelles qui sont holomorphes dans A_0 ne permettent pas d'approcher, uniformément sur tout compact de A_0, une certaine fonction holomorphe dans A_0.

---

[7]) H. CARTAN: Sur les domaines d'existence des fonctions de plusieurs variables complexes, 1931 (Bull. Soc. Math. France).

[8]) H. CARTAN und P. THULLEN: Regularitäts- und Konvergenzbereiche, 1932 (Math. Annalen).

[9]) We say that a phenomenon occurs in the interior of a domain D if it does on any domain contained together with its boundary in D.

V. The Cauchy Integral

L'intégrale de Cauchy

Japanese Journal of Mathematics **17** (1941), p. 523–531

In the present memoir, we shall be concerned with the problem of representations of holomorphic functions in a univalent domain of holomorphy. Now, as we have seen, it is in general impossible to do this by means of series of rational functions. We shall therefore go over to Cauchy integrals, more precisely to the integral of Mr. A. WEIL.

1. The Weil Integral. We shall begin by describing the integral of Mr. WEIL by translating a part of his article[1)].

"Let $X_1, X_2, \dots, X_N$ be N holomorphic functions (non-constant) in a domain D (univalent and bounded) in the space of two complex variables x, y; let Δ_i be a closed (univalent) bounded domain in the X_i plane, whose boundary C_i consists of a finite number of analytic arcs. Let Δ be a connected component, or the sum of several connected components, of the set in D defined by the conditions:

$$X_i(x, y) \in \Delta_i.$$

We suppose that Δ is completely interior to D, and, denoting by S_i the set of boundary points of Δ for which $X_i \in C_i$, that no two of the S_i have common elements of more than two dimensions (it is always sufficient, in order to fulfil this condition, to make infinitely small displacements of the contours C_i as needed). Let σ_{ij} be the sum of the two dimensional elements of the intersection of S_i and S_j, the orientation being suitably defined by the sequence S_i, $S_i \times S_j$."

We shall say that Δ is of type (α), D being a bounded open set containing the closed set Δ, but otherwise arbitrary.

"Suppose, on the other hand, that we can make correspond to each $X_i(x, y)$, two functions $P_i(x, y; x_0, y_0)$, $Q_i(x, y; x_0, y_0)$ which are holomorphic when $(x, y) \in D$ and $(x_0, y_0) \in D$, and such that, we have

$$X_i(x, y) - X_i(x_0, y_0) = (x - x_0) P_i + (y - y_0) Q_i$$

identically."

We shall call this condition (β).

"Under these conditions, let $f(x, y)$ be a single-valued function holomorphic at every point of Δ. Consider the sum of integrals:

$$I = \frac{-1}{4\pi^2} \sum_{(i,j)} \int_{\sigma_{ij}} \frac{(P_i Q_j - P_j Q_i) f(x, y)}{[X_i(x, y) - X_i(x_0, y_0)][X_j(x, y) - X_j(x_0, y_0)]} dx\, dy,$$

[1)] L'intégrale de CAUCHY et les fonctions de plusieurs variables, 1935 (Math. Annalen).

the summation being over all pairs of indices (i,j); we have

$$I=f(x_0,y_0), \quad \text{or} \quad =0,$$

according as (x_0,y_0) is a point interior to Δ, or is exterior to Δ."

This is the result of Mr. WEIL. Put briefly, *if the "polyhedron" Δ is of type (α) and satisfies condition (β), then Δ possesses a Cauchy integral $I(x_0,y_0)$*. The integral is taken along the σ_{ij}, intersection of the "faces" S_i and S_j.

2. The New Condition (γ). We propose the following new *condition* (γ):

Suppose that we can make correspond to each $X_i(x,y)$ 2 functions $P_i(x,y;x_0,y_0)$, $Q_i(x,y;x_0,y_0)$, holomorphic when $(x,y)\in D$ and $(x_0,y_0)\in D$ and such that we have

$$(X_i-X_i^0)R=(x-x_0)P_i+(y-y_0)Q_i$$

identically, where X_i^0 stands for $X_i(x_0,y_0)$, and $R(x,y;x_0,y_0)$ is a new function, holomorphic when $(x,y)\in D$ and $(x_0,y_0)\in D$, independent of X_i and such that

$$R(x_0,y_0;x_0,y_0)=1.$$

We construct a CAUCHY integral for the "polyhedron" Δ of type (α) under this new hypothesis. Let us form the integral of the same form as I with the new functions P_i, Q_j:

$$J=\frac{-1}{4\pi^2}\sum_{(i,j)}\int_{\sigma_{ij}}\frac{(P_iQ_j-P_jQ_i)\,f(x,y)}{(X_i-X_i^0)(X_j-X_j^0)}\,dx\,dy.$$

One finds, *exactly as before*, that

$$J=f(x_0,y_0), \quad \text{or} \quad =0$$

according as (x_0,y_0) is an interior point of Δ, or is exterior to Δ.

The set of the integrals J contains the integrals I. We shall still call this integral J *the Weil integral.*

We shall study condition (γ).

3. We first make a remark on Theorem I of Memoir II.

Consider a closed set Δ belonging to the class (H_0) in the space of 2 complex variables of the form:

$$(\Delta) \qquad |x|\leqq r, \quad |y|\leqq r', \quad |f_j(x,y)|\leqq 1, \quad (x,y)\in G, \quad (j=1,2,\ldots,\nu),$$

where G is a bounded open set in the space (x,y), the f_j are holomorphic on G and r and r' are positive numbers. It is understood that Δ is contained in the interior of D so as to be a closed set. We raise the dimension of the space in which we work by introducing ν new complex variables $z_1,z_2,\ldots,z_\nu$, and consider a closed set Σ on a characteristic variety of dimension 4 in the space $[x,y,(z_i)]$ as follows:

$$(\Sigma) \qquad z_j=f_j(x,y), \quad (x,y)\in\Delta, \quad (j=1,2,\ldots,\nu).$$

Then Σ belongs to the class (P_1), that is to say, one can find a sequence of open sets, convex with respect to polynomials in x, y and z_i, which tends towards Σ from the exterior. This is Theorem I of Memoir II.

We are going to make the characteristic variety on which Σ lies *algebraic* by a sufficiently small deformation. Since Σ is of class (P_1), we can choose a suitable open set V containing Σ and expand the holomorphic functions

$$z_j - f_j(x, y) \quad (j=1, 2, \dots, \nu)$$

in a series of polynomials in x, y, z_j converging uniformly on V. We can therefore find polynomials $F_j[x, y, (z)]$ sufficiently close to the initial functions. We shall now look at the system of roots of the simultaneous equations

(1) $$F_j(x, y, z_1, z_2, \dots, z_\nu) = 0 \quad (j=1, 2, \dots, \nu),$$

in which we consider x and y as independent parameters.

We want to start with the functions $F_j[x, y, (z)]$ instead of the $z_j - f_j(x, y)$.

Let (x, y) be an arbitrary point of Δ. In the space $((z))$, we consider the polycylinder

(γ) $$|z_j - f_j(x, y)| < \rho, \quad (j=1, 2, \dots, \nu),$$

choosing ρ sufficiently small so that the open set V contains the set of points $[\Delta, (\gamma)]$ completely in its interior. We then take the F_j sufficiently close to the $(z_j - f_j)$ respectively for the *following three conditions* to be satisfied on the open set V:

1°. Let δ be a positive number, given à priori, so that $\delta < \rho$ and let p be any one of $1, 2, \dots, \nu$;

$$\text{if } z_p = f_p(x, y), \quad \text{we have } |F_p[x, y, (z)]| < \delta.$$

2°. Let, moreover, ρ' be a positive number given à priori so that $\rho' < \delta$;

$$\text{if } |F_p| \leqq \delta, \quad \text{we have } |z_p - f_p(x, y)| < \rho;$$

$$\text{if } |F_p| = \delta, \quad \text{we have } |z_p - f_p(x, y)| > \rho'.$$

3°. $$J = \frac{\partial(F_1, F_2, \dots, F_\nu)}{\partial(z_1, z_2, \dots, z_\nu)} \neq 0.$$

These are the 3 conditions. The first two conditions are certainly realisable; as for the third, we have only to remark that $J \to 1$.

Under these conditions, we shall prove the following concerning the system of roots

$$z_j = \varphi_j(x, y) \quad (j=1, 2, \dots, \nu)$$

of the simultaneous equations (1): *there is one, and only one, such system in* (γ); *moreover, the functions* $\varphi_j(x, y)$ *are holomorphic when* $(x, y) \in \Delta$.

We consider the auxiliary system of simultaneous equations

(2) $$F_j[x, y, (z)] = u_j, \quad (j=1, 2, \dots, \nu),$$

where the u_j are independent complex variables. Let us take an arbitrary point (x_0, y_0) in Δ, which we consider as being fixed. The only variables then are z_j and u_j; we denote the polycylinder (γ) corresponding to the point (x_0, y_0) by (γ_0), and $f_j(x_0, y_0)$ by z_j^0.

By condition 1°, to the centre $((z^0))$ of (γ_0), there corresponds a point of $|u_j|<\delta$ which we call $((u^0))$. Since $J \neq 0$, the simultaneous equations (2) have a unique solution

$$z_j = \Phi_j((u)), \qquad (j=1,2,\ldots,\nu),$$

near $((z^0))$ for any point $((u))$ in a neighbourhood of $((u^0))$; the functions $\Phi_j((u))$ are holomorphic.

Let us continue the functions $\Phi_j((u))$ analytically in the polycylinder $|u_j|<\delta$. In view of the second condition, the point with coordinates $\Phi_j((u))$ remains in (γ_0). Hence, by the third condition, J never vanishes. Consequently, we do not meet any singular point of any of the functions Φ_j; the $\Phi_j((u))$ are therefore holomorphic on $|u_j|<\delta$. Therefore, the pseudoconformal transformation $z_j = \Phi_j((u))$ $(j=1,2,\ldots,\nu)$ maps the polycylinder $|u_j|<\delta$ onto a univalent domain contained in (γ_0); we shall denote this domain by Z; Z contains the polycylinder $|z_j - z_j^0| < \rho'$ because of the second condition.

Suppose we had another system of roots of (2),

$$z_j = \Psi_j((u)) \qquad (j=1,2,\ldots,\nu),$$

such that the point $((\Psi))$ enters (γ_0) when $((u))$ describes the whole polycylinder $|u_j|<\delta$. From this hypothesis, it follows in the same way that $z_j = \Psi_j((u))$ $(j=1,2,\ldots,\nu)$ maps $|u_j|<\delta$ onto a univalent domain Z' containing $|z_j - z_j^0| < \rho'$. Thus, the domains Z and Z' both contain the polycylinder $|z - z_j^0| < \rho'$; if therefore we choose carefully the point $((z))$ in this polycylinder, there corresponds at least 2 points in $|u_j|<\delta$, each satisfying the simultaneous equations (2), which is obviously absurd. $z_j = \Phi_j((u))$ is thus the only system of roots which enters (γ_0) when $((u))$ describes $|u_j|<\delta$. In addition, as we have seen, this system remains in (γ_0).

If we set $u_j = 0$ $(j=1,2,\ldots,\nu)$, we find that the simultaneous equations (1) have a solution in (γ_0) for $(x, y) = (x_0, y_0)$, which we shall denote by

$$\Phi_j(x_0, y_0, (0)) = \varphi_j(x_0, y_0),$$

and there is only one such solution for an arbitrary point (x_0, y_0) of Δ. The functions $\varphi_j(x, y)$ are obviously single valued, and are also holomorphic at any point of Δ since $J \neq 0$.

We formulate what we have seen above as:

Complement to Theorem I of Memoir II. "Suppose given a number ρ such that $0 < \rho < \rho_0$, where ρ_0 is a suitable upper bound. We can find simultaneous *algebraic* equations

$$F_j(x, y; z_1, z_2, \ldots, z_\nu) = 0 \qquad (j=1,2,\ldots,\nu),$$

the F_j being polynomials, such that the equations have a unique system of roots

$$z_j = \varphi_j(x, y) \qquad (j=1,2,\ldots,\nu)$$

in

$$|z_j - f_j(x,y)| < \rho \qquad (j=1,2,\dots,\nu);$$

the functions $\varphi_j(x,y)$ are holomorphic for $(x,y)\in\Delta$."

We have seen, incidentally, that we may suppose that

$$J = \frac{\partial(F_1, F_2, \dots, F_\nu)}{\partial(z_1, z_2, \dots, z_\nu)} \neq 0$$

at every point of the set $(x,y)\in\Delta$, $|z_j - f_j(x,y)| < \rho$ $(j=1,2,\dots,\nu)$.

Let us make another remark, although it will not be needed in what follows. We may suppose, without loss of generality that the algebraic functions $\varphi_j(x,y)$ have neither poles nor points of indeterminacy in any finite portion of (x,y) space. For this, it suffices to take polynomials of the form

$$F_j[x,y,(z)] = \alpha_j z_j^{N_j} + \dots \qquad (j=1,2,\dots,\nu)$$

where the α_j are constants different from zero, and the other terms are of degree smaller than N_j with respect to the z_k $(k=1,2,\dots,\nu)$.

4. Let us consider a univalent domain of holomorphy D without points at infinity in the space (x,y). Because of the theorem of CARTAN-THULLEN, there exists a sequence of sets belonging to the class (H_0) which tends to D from the interior. Let Δ be an arbitrary set of this sequence, and having the form

$$(\Delta) \qquad |x| \leqq r, \quad |y| \leqq r', \quad |f_j(x,y)| \leqq 1,$$
$$(x,y)\in G \qquad (j=1,2,\dots,\nu),$$

the notation being the same as at the beginning of the preceding section. Now, in view of what we have just seen, we can suppose without loss of generality that $z_j = f_j(x,y)$ $(j=1,2,\dots,\nu)$ is a system of roots of simultaneous algebraic equations $F_j[x,y,(z)]=0$ $(j=1,2,\dots,\nu)$. We can moreover suppose that the F_j are polynomials such that

$$J = \frac{\partial(F_1, F_2, \dots, F_\nu)}{\partial(z_1, z_2, \dots, z_\nu)} \neq 0$$

in the set (V):

$$(x,y)\in G, \quad |z_j - f_j(x,y)| < \rho \qquad (j=1,2,\dots,\nu),$$

where ρ is a certain positive number.

Let $[x,y,(z)]$, $[x_0,y_0,(z^0)]$ be two arbitrary points in a finite portion of space. Since the F_j are polynomials, we can set

$$(1) \qquad F_j - F_j^0 = (x-x_0)A_j + (y-y_0)B_j + \sum (z_k - z_k^0) C_{jk}, \qquad (j,k=1,2,\dots,\nu),$$

where F_j^0 stands for $F_j[x_0,y_0,(z^0)]$; we shall often use similar abbreviations in what follows. A_j, B_j, C_{jk} are polynomials in the variables x, y, z_j, x_0, y_0, z_j^0. We look upon the $(z_k - z_k^0)$ as the only unknowns and solve the simultaneous linear equations

$$(2) \qquad \sum (z_k - z_k^0) C_{jk} + (x-x_0)A_j + (y-y_0)B_j = 0 \qquad (j,k=1,2,\dots,\nu),$$

and we have

$$(z_k - z_k^0) N = (x - x_0) a_k + (y - y_0) b_k,$$

N, a_k and b_k being polynomials in x, y, z_j, x_0, y_0, z_j^0, and, in particular,

$$N = |C_{jk}| \qquad (j, k = 1, 2, \ldots, \nu).$$ [2]

Let us look at the determinant $|C_{jk}|$: for $x = x_0$, $y = y_0$, $z_j = z_j^0$, we evidently have

$$C_{jk}^0 = \left(\frac{\partial F_j}{\partial z_k}\right)_0;$$

we therefore have

$$N_0 = J_0.$$

J_0 is a polynomial in x_0, y_0, z_j^0, non-vanishing in V.

Now, if we substitute $z_j = f_j(x, y)$, $z_j^0 = f_j(x_0, y_0)$ in (1), we have $F_j = F_j^0 = 0$. Hence, if we denote by N', a_k', b_k', J_0' the functions obtained by making this substitution in N, a_k, b_k, J_0 respectively, and if we set

$$\frac{N'}{J_0'} = R(x, y; x_0, y_0)$$

$$\frac{a_k'}{J_0'} = p_k, \qquad \frac{b_k'}{J_0'} = q_k,$$

we obtain

$$(3) \qquad (f_k - f_k^0) R = (x - x_0) p_k + (y - y_0) q_k \qquad (k = 1, 2, \ldots, \nu),$$

where R, p_k and q_k are functions of the $x, y; x_0, y_0$, holomorphic in (G, G) and

$$R(x_0, y_0; x_0, y_0) = 1.$$

Let now $\varphi(x, y)$ be a holomorphic function in a neighbourhood of Δ; because of Theorem I of Memoir II and WEIL's theorem, we can expand it in a series of polynomials in x, y and $f_j(x, y)$, converging uniformly in a neighbourhood of Δ. Hence, corresponding to any positive number ε, we can find a polynomial in x, y and $f_j(x, y)$, which we shall denote by $\psi(x, y)$, such that

$$|\varphi(x, y) - \psi(x, y)| < \varepsilon$$

in a neighbourhood of $\Delta \cdot \psi - \psi_0$ is of the form

$$\psi - \psi_0 = (x - x_0)\alpha + (y - y_0)\beta + \sum (f_j - f_j^0)\gamma_j,$$

α, β and γ_j being holomorphic functions of the variables $x, y; x_0, y_0$ in a neighbourhood of (Δ, Δ). It follows from this, if we substitute (3), that

$$(\psi - \psi_0) R = (x - x_0) P + (y - y_0) Q,$$

P and Q being holomorphic functions of all their variables in a neighbourhood of (Δ, Δ). We have thus obtained the following lemma.

---

[2] $|C_{jk}|$ is the determinant of $C_{j1}, C_{j2}, \ldots, C_{j\nu}$ when j runs successively through $1, 2, \ldots, \nu$.

Lemma. *Let D be a univalent domain of holomorphy without points at infinity in the space* (x,y), *and let* D_0 *be a univalent domain completely interior to D. We can find a holomorphic function* $R(x,y;x_0,y_0)$ *on* (D_0,D_0) *such that*

$$R(x_0,y_0;x_0,y_0)=1$$

and playing the following role: Given a positive number ε, *to any function* $f(x,y)$ *holomorphic on D there corresponds a holomorphic function* $\varphi(x,y)$ *on* D_0 *such that*

$$|f(x,y)-\varphi(x,y)|<\varepsilon,$$

and

$$(\varphi-\varphi_0)\cdot R=(x-x_0)P+(y-y_0)Q,$$

where φ_0 *stands for* $\varphi(x_0,y_0)$, *and P and Q are holomorphic functions of the variables* $x,y;x_0,y_0$ *in* (D_0,D_0).

This proposition continues to hold for an arbitrary number of variables, as can be seen in exactly the same way.

5. Let us return to the Weil integral. Consider a "polyhedron" Δ of type (α) as in No. 1. Δ is of the form

$$X_i(x,y)\in\Delta_i \quad (i=1,2,\ldots,N).$$

Δ can obviously be approximated from the outside by domains of holomorphy. Therefore, in view of the above lemma, *we can construct a "polyhedron"* Δ' *of type* (α) *and having property* (γ) *by making arbitrarily small deformations of* X_i *and of* Δ_i; *thus* Δ' *has the Weil integral given in No. 2.*

From this and the CARTAN-THULLEN theorem, we obtain the:

Theorem. *Given a univalent domain of holomorphy D without points at infinity in the space of 2 complex variables, we can represent holomorphic functions on D by the Weil integral in the interior of D.*[3)]

The author believes that this theorem too holds for any number of variables.

Commentaire de H. Cartan

Il s'agit de la formule intégrale d'ANDRÉ WEIL (Math. Annalen, 111, 1935, p. 178–182; Collected Papers, vol. I, p. 104–108). OKA considère ici le cas de deux variables x et y. La formule de WEIL exprime la valeur d'une $f(x,y)$ holomorphe en tout point intérieur à un polyèdre analytique défini par des relations $|X_i(x,y)|\in\Delta_i(\Delta_i$ compact dans $\mathbb{C})$ à l'aide des valeurs de f sur les arêtes (à 2 dimensions réelles) de ce polyèdre. Les X_i sont supposées holomor-

[3)] The form of the integral depends, of course, on domains completely interior to *D*.

phes dans un domaine $D \subset \mathbb{C}^2$, et l'on suppose que l'ensemble des points de D définis par les relations précédentes est compact. Malheureusement la formule de WEIL ne peut être écrite que si l'on suppose l'existence de fonctions P_i et Q_i holomorphes dans $D \times D$ et satisfaisant à

$$X_i(x, y) - X_i(x_0, y_0) = (x - x_0)\, P_i(x, y, x_0, y_0) + (y - y_0)\, Q_i(x, y, x_0, y_0).$$

On sait aujourd'hui que, si D est un domaine d'holomorphie, de telles fonctions P_i et Q_i existent, grâce à la théorie des faisceaux cohérents appliquée à $D \times D$ et à sa diagonale. Mais lorsque OKA écrivit son Mémoire en 1940, on l'ignorait, sauf dans certains cas particuliers (par exemple si les X_i sont des polynômes). C'est pourquoi OKA essaie de tourner la difficulté; pour cela, il prouve d'abord que la formule de WEIL subsiste sans changement sous l'hypothèse plus générale d'une relation de la forme

$$[X_i(x, y) - X_i(x_0, y_0)]\, R(x, y, x_0, y_0) = (x - x_0)\, P_i + (y - y_0)\, Q_i,$$

où la fonction R, indépendante de l'indice i, est holomorphe dans $D \times D$ et égale à 1 sur la diagonale. Et il prouve l'existence de telles relations, non pas pour les fonctions X_i données, mais pour des fonctions qui les approchent arbitrairement. Il s'ensuit que tout domaine d'holomorphie D est réunion croissante de polyèdres analytiques auxquels s'applique la formule d'ANDRÉ WEIL.

OKA affirme que ce résultat s'étend au cas de n variables.

VI. Pseudoconvex Domains

Domaines pseudoconvexes

Tôhoku Mathematical Journal **49** (1942), p. 15–52

Introduction. In 1906, F. HARTOGS discovered that domains of holomorphy are subject to a very curious restriction[1]; I believe that it was with this very discovery that the recent development of the theory of analytic functions of several variables was begun.

The same restriction was found at the foundations of different branches of the theory successively by E.E. LEVI, G. JULIA, W. SAXER and the author[2]. We shall call a domain restricted in this way *pseudoconvex*[3].

Convexity of this sort can be analysed locally. Now, in 1932, H. CARTAN and P. THULLEN found that domains of holomorphy are even, in a certain sense, globally convex[4]. We established several global theorems concerning domains of holomorphy because of this property[5].

We thus have several classes of pseudoconvex domains about which we know almost nothing, except on domains of holomorphy[6]. We are thus brought back to F. HARTOGS, to ask whether or not conversely every pseudoconvex domain is a domain of holomorphy. And if these 2 types of domains coincided, we would have a local criterion for domains of holomorphy. Among the different classes of pseudoconvex domains, those defined by G. JULIA have

---

[1]) F. HARTOGS: Einige Folgerungen aus der CAUCHYschen Integralformel bei Funktionen mehrerer Veränderlichen, 1906 (Münch. Berichte).

[2]) E.E. LEVI: Studii sui punti singolari essenziali delle funzioni analitiche di due o più variabili complesse, 1910 (Annali di Matematica).
G. JULIA: Sur les familles de fonctions analytiques de plusieurs variables, 1926 (Acta Mathematica).
W. SAXER: Sur les familles de fonctions meromorphes de plusieurs variables, 1931 (Comptes Rendus, Paris).
K. OKA: Note sur les familles de fonctions analytiques multiformes etc., 1934. (Journal of Science of the Hiroshima University).

[3]) For univalent domains in finite space, we have defined them in Memoir IV; see No. 10 of the present memoir.

[4]) As well as the converse. See
H. CARTAN-P. THULLEN: Regularitäts- und Konvergenzbereiche, 1932 (Mathematische Annalen). For the idea, see
H. CARTAN: Sur les domaines d'existence de fonctions de plusieurs variables complexes, 1931 (Bull. Société mathématique de France).

[5]) Previous memoirs in the present research: I. Rationally convex domains, 1936; II. Domains of holomorphy, 1937; III. The second COUSIN problem, 1939 (Journal of Science of the Hiroshima University); IV. Domains of holomorphy and rationally convex domains, 1941; V. The CAUCHY integral, 1941 (Japanese Journal of Mathematics).

[6]) It is also known, for example, that if in a univalent finite domain in the space of 2 complex variables, the first COUSIN problem is always solvable, the domain is necessarily a domain of holomorphy. See
H. CARTAN: Les problèmes de Poincaré et de Cousin pour les fonctions de plusieurs variables, 1934 (Comptes Rendus, Paris).

been shown to consist of domains of holomorphy, when they are in finite space, by H. CARTAN, P. THULLEN, H. BEHNKE and K. STEIN[7]. But outside this case, the problem remains more or less open up until now[8].

In the present memoir, we shall treat this problem. We limit ourselves to the space of 2 complex variables for simplicity, but the conclusion, I think, applies to spaces of any number of variables. We shall see that *finite univalent, pseudoconvex domains are domains of holomorphy*[9].

The problem we have just explained is the last of those which formed the subject of the present researches[10].

I. The Main Problem

1. The Problem. We shall continue to suppose that all domains in the present memoir are *univalent and lie in the finite part of space.*

In our researches, *the first Cousin problem* plays a fundamental role nearly everywhere. The essential part of this consists in the following problem.

Let us consider a bounded domain Δ in the space of two complex variables x, y. The upper and lower parts of this domain relative to the hyperplane $x_2=0$, where $x=x_1+ix_2$, i being the imaginary unit, will be denoted by Δ_1 and Δ_2 respectively. Given a holomorphic function $f(x,y)$ "in the neighbourhood" of the common boundary of Δ_1, Δ_2,[11] *find a pair of functions $F_1(x,y)$, $F_2(x,y)$, holomorphic in Δ_1 and Δ_2 respectively, which remain holomorphic at every point of Δ on $x_2=0$, in such a way as to have*

$$F_1(x,y)-F_2(x,y)=f(x,y)$$

identically.

In 1895, P. COUSIN solved this problem for cylindrical domains[12]. In 1934, H. CARTAN indicated that this problem is solvable for rationally convex do-

[7]) This is called the problem of JULIA. See H. CARTAN-P. THULLEN cited above.
H. BEHNKE-K. STEIN: Konvergente Folgen von Regularitätsbereichen und die Meromorphiekonvexität, 1938, (Math. Annalen); Approximation analytischer Funktionen in vorgegebenen Bereichen des Raumes von *n* komplexen Veränderlichen, 1939 (Göttinger Nachrichten); Die Konvexität in der Funktionentheorie mehrerer komplexer Veränderlichen, 1940 (Mitteilung der Mathematischen Gesellschaft in Hamburg).

[8]) For this problem, see: the book of H. BEHNKE-P. THULLEN: Theorie der Funktionen mehrerer komplexer Veränderlichen, 1934, pages 54–56; also H. BEHNKE-K. STEIN, 1940, cited above.

[9]) The author has published this result in the note: Sur les domaines pseudoconvexes, 1941 Proc. Imp. Acad. Tokyo).

[10]) The author undertook these researches thanks to the interesting work of H. BEHNKE-P. THULLEN cited above.

[11]) We say that a phenomenon occurs in the neighbourhood of a set *E* if it does on a certain open set containing *E* and its points of accumulation.
We say that a phenomenon occurs "in the interior" of an open set $\mathfrak{D}$ if it does on any bounded open set which is contained, together with its boundary, in $\mathfrak{D}$.

[12]) Acta Mathematica.

mains[13] by applying *the Weil integral*[14]. The same method can be used for domains of holomorphy because of what we have seen in Memoir V. And, more recently, we solved this problem by a different method for domains of holomorphy in Memoirs I, II.

We shall solve this problem for a pseudoconvex domain composed of two domains of holomorphy which overlap on each other, of which one contains Δ_1, the other Δ_2; the domain will be otherwise more or less arbitrary.

2. The Hypotheses. In the present section, we shall treat this problem when Δ satisfies certain conditions which we now explain; the significance of these conditions will be discussed in the section which follows.

Let us consider in the space (x, y) a *bounded* domain $\mathfrak{D}$ containing the origin; consider 3 parallel hyperplanes L, L_1, L_2 of the form

$$x_2=0, \quad x_2=a_1, \quad x_2=a_2, \quad a_1<0<a_2,$$

respectively, and suppose that each of these hyperplanes crosses $\mathfrak{D}$. We shall denote the part of $\mathfrak{D}$ above L_1 by $\mathfrak{D}_1$, that below L_2 by $\mathfrak{D}_2$ and that between L_1 and L_2 by $\mathfrak{D}_3$. *We suppose that the connected components of the open sets* $\mathfrak{D}_1$, $\mathfrak{D}_2$ *are domains of holomorphy.* Then, the same is necessarily true of the open set $\mathfrak{D}_3$.

Consider ν holomorphic functions

$$X_j(x, y) \quad (j=1,2,\ldots,\nu)$$

in $\mathfrak{D}_3$.

1°. *We suppose that the set of points of* $\mathfrak{D}_3$ *satisfying*

$$|X_j(x, y)| \leqq 1 \quad (j=1,2,\ldots,\nu)$$

contains no point in a neighbourhood of the intersection of the boundary of $\mathfrak{D}$ *with L, and that for some sufficiently small positive number* ε *the set of points of* $\mathfrak{D}_3$ *satisfying*

$$|X_j(x, y)| > 1-\varepsilon$$

contains no points in a neighbourhood of the hyperplanes L_1, L_2 *for any* j *in* $(1,2,\ldots,\nu)$.

The points of $\mathfrak{D}$ which either do not belong to $\mathfrak{D}_3$ or satisfy the inequalities

$$|X_j(x, y)| < 1, \quad (j=1,2,\ldots,\nu)$$

form an open set. *We suppose that this set has a connected component containing the origin and extending beyond* L_1 *and* L_2. It is this domain which we shall call Δ.

As we shall see later, this hypothesis rests on the theorem of H. CARTAN-P. THULLEN that *any (finite) domain of holomorphy is convex with respect to functions holomorphic on the domain*[15].

---

[13]) More or less arbitrary; the same remark applies in what follows.
[14]) H. CARTAN, cited above.
[15]) Math. Ann. 1932.

2°. Let x_0, y_0 be new variables. *We suppose that we have, identically,*

$$(X_j - X_j^0) R = (x - x_0) P_j + (y - y_0) Q_j \qquad (j = 1, 2, \ldots, \nu),$$

where X_j^0 stands for $X_j(x_0, y_0)$, P_j, Q_j and R are holomorphic functions of the variables x, y, x_0, y_0 when $(x, y) \in \mathfrak{D}_3$, $(x_0, y_0) \in \mathfrak{D}_3$, and the function R reduces to 1 when $x = x_0$ and $y = y_0$.

Concerning this hypothesis, we refer to *the lemma of Memoir V.*

3°. *We suppose that the derivative $\partial X_j / \partial y$ does not vanish identically for any j.* Let Σ_j be the analytic variety defined in $\mathfrak{D}_3$ by

$$x_2 = 0, \qquad |X_j| = 1;$$

by our hypothesis, Σ_j, if it is non-empty, is 2 dimensional.

We suppose that the intersections of the varieties Σ_j, Σ_k $(j \neq k; j, k = 1, 2, \ldots, \nu)$, if they are non-empty, are 1 dimensional.

We shall denote the part of Δ above L by Δ_1, that below L by Δ_2, and that between L_1 and L_2 by Δ_3. Δ_3 obviously *consists of domains of holomorphy*[16]. We shall see later that this is also true for Δ_1 and Δ_2.

Let us consider the portion of the hyperplane L in Δ. Let S be the set of points consisting of this portion and its points of accumulation. By hypothesis 1°, S is contained in $\mathfrak{D}_3$. Let σ be the boundary of S (considered as being a set on the hyperplane). σ clearly lies on the sum of the varieties Σ_j $(j = 1, 2, \ldots, \nu)$. We denote the part of σ on Σ_j by σ_j. The σ_j are of dimension at most 2. The intersections of the varieties σ_j, σ_k $(j \neq k;\ j, k = 1, 2, \ldots, \nu)$ are of dimension at most 1 by the preceding hypothesis. To simplify the reasoning later on, *we shall suppose, in addition, that for each j, the derivative $\partial X_j / \partial y$ is non-zero on σ_j, except possibly at a finite number of points.*

We shall see later that we can always find a domain Δ with these properties, arbitrarily close to the given domain $\mathfrak{D}$.

3. The Case of Domains of Holomorphy. We shall first solve the problem for Δ_3 by using *the method due to A. Weil and to H. Cartan.*

Let $\varphi(x, y)$ be a holomorphic function given in a neighbourhood of S. We shall define, precisely, the following double integral

$$(1) \qquad I(x_0, y_0) = \frac{-1}{4\pi^2} \sum_j \int_{\sigma_j} \psi_j(x, y; x_0, y_0)\, \varphi(x, y)\, dx\, dy$$

extended over the two dimensional part of σ; here we have set

$$\psi_j(x, y; x_0, y_0) = \frac{Q_j}{(x - x_0)(X_j - X_j^0)},$$

$(j = 1, 2, \ldots, \nu)$. For any j, $\partial X_j / \partial y$ is non-zero on the variety σ_j except, perhaps at a finite number of points; let us take a point P on σ_j different from the

[16]) More precisely, it consists of connected components which are domains of holomorphy.

exceptional points but otherwise arbitrary. Near P, we can represent the analytic variety Σ_j by means of real parameters u, v in the form

$$x=x(u,v), \qquad y=y(u,v),$$

where the right hand terms are power series in u, v, in such a way the correspondence between (x,y) and (u,v) is one-to-one. We can, for instance, take (x,θ_j) as parameters, θ_j being the argument of X_j. We regard (u,v) as being a point in the plane given by rectangular axes in cartesian geometry. To the boundary of σ_j correspond a finite number of analytic arcs in the (u,v) plane. We choose (u,v) such that

$$\frac{\partial(x,\theta_j)}{\partial(u,v)}>0;$$

then we have, by definition,

$$\int_{\sigma_j}\psi_j\varphi\,dx\,dy=\iint\psi_j\varphi\frac{\partial(x,y)}{\partial(u,v)}du\,dv$$

in a neighbourhood of P, the second term being a double integral extended over the portion of the (u,v) plane corresponding to the 2 dimensional part of σ.

Let (x_0,y_0) be a point such that, for an arbitrary j, $\psi_j(x,y;x_0,y_0)$ is holomorphic with respect to x,y in a neighbourhood of σ_j. The integral $I(x_0,y_0)$ is then well-defined and finite[17].

When $(x,y)\in\sigma_j$, $\psi_j(x,y;x_0,y_0)$ is holomorphic with respect to (x_0,y_0) in Δ_3 except on L because $|X_j|=1$. Consequently, *the function* $I(x_0,y_0)$ *is holomorphic on* Δ_3 *except on* L.[18]

We shall study the behaviour of $I(x_0,y_0)$ in the neighbourhood of L.[19] Let (ξ,η) be any point on the portion of L in Δ. We draw a small circle (γ) around ξ in the x-plane and denote by σ',σ'_j the parts of σ,σ_j respectively which are in (γ). We consider the integral

$$I_1(x_0,y_0)=\frac{-1}{4\pi^2}\sum_j\int_{\sigma'_j}\psi_j(x,y;x_0,y_0)\,\varphi(x,y)\,dx\,dy,$$

$(j=1,2,\ldots,\nu)$. The difference $I-I_1$ is obviously holomorphic on the open set $x_0\in(\gamma)$, $(x_0,y_0)\in\Delta_3$. It therefore suffices to study I_1 in place of I.

---

17) Let us show that the double integral $\int_{\sigma_j}f(x,y)\,dx\,dy$ is well-defined and finite, $f(x,y)$ being a function holomorphic in a neighbourhood of σ_j. Let P be a point of σ_j such that $\partial X_j/\partial y=0$; there are only finitely many points on σ_j with this property. We consider a hypersphere with centre P of sufficiently small radius ρ. Let σ' be the part of σ_j in this hypersphere. We find easily that $\int_{\sigma'}|dx\,dy|$ tends to 0 with ρ. The same is therefore true for $\int_{\sigma'}f\,dx\,dy$. The statement results immediately from this.

18) This is because, by very definition of the integral, $I(x_0,y_0)$ is the limit of a sequence of functions holomorphic in the open set $(x_0,y_0)\in\Delta_3$, $(x_0,y_0)\notin L$, converging uniformly near any point.

19) This will be done directly; one can also take $R\psi_j$ in place of ψ_j and apply to the corresponding integral the theorem of A. WEIL expounded in Memoir V.

We set

$$\Phi_j(x,y;x_0,y_0)=\frac{Q_j}{X_j-X_j^0}\varphi(x,y);$$

denote by Γ the projection on the y-plane of the intersection of σ with $x=\xi$. The function $Q_j/(X_j-X_j^0)$ is holomorphic with respect to (x_0,y_0) when $(x_0,y_0)\in\Delta_3$ and $(x,y)\in\sigma$ because $|X_j|=1$. $\varphi(x,y)$ is holomorphic in a neighbourhood of S; $\Phi_j(x,y;x_0,y_0)$ is therefore holomorphic at every point of the set

$$x=\xi, \quad y\in\Gamma; \quad x_0=\xi, \quad y_0=\eta.$$

In the y-plane, we consider a circle (γ') around η and an open set E containing Γ. In order that $\Phi(x,y;x_0,y_0)$ be holomorphic on the open set

$$\text{(I)} \qquad x\in(\gamma), \quad y\in E; \quad x_0\in(\gamma), \quad y_0\in(\gamma'),$$

and that σ' be contained in $[(\gamma),E]$, it is sufficient to first take (γ') and E sufficiently close to η and to Γ respectively, and then to choose (γ) sufficiently close to ξ.

Under these conditions, set

$$\Psi_j(y;x_0,y_0)=\Phi_j(x_0,y;x_0,y_0).$$

This function is also holomorphic on the set (I) and satisfies identically

$$\Phi_j-\Psi_j=(x-x_0)\,\chi(x,y;x_0,y_0)$$

where χ is a function holomorphic on the set (I). Consider the integral

$$I_2(x_0,y_0)=\frac{-1}{4\pi^2}\sum_j\int_{\sigma'_j}\frac{\Psi_j}{x-x_0}\,dx\,dy,$$

$(j=1,2,\dots,\nu)$. Because of the identity above, I_1-I_2 is holomorphic on the dicylinder $[(\gamma),(\gamma')]$. It is therefore sufficient to study I_2 instead of I_1.

Now, in the relation

$$(X_j-X_j^0)\,R=(x-x_0)\,P_j+(y-y_0)\,Q_j,$$

if we set $x=x_0$, we obtain

$$(X_j-X_j^0)\,R=(y-y_0)\,Q_j.$$

Hence

$$I_2=\frac{-1}{4\pi^2}\sum_j\int_{\sigma'_j}\frac{R(x_0,y;x_0,y_0)}{(x-x_0)(y-y_0)}\varphi(x_0,y)\,dx\,dy.$$

We shall now study the variety σ' of integration. Let l be the diameter of the circle (γ) on the real axis, and let x' be an arbitrary point of l. Let us consider, for any j, the projection on the y-plane of the intersection of σ with $x=x'$ and that of σ_j; we shall denote them by $\Gamma_{x'}$ and $\Gamma_{x'}^{(j)}$ respectively. The set $\Gamma_{x'}^{(j)}$, if it is non-empty, lies on

$$|X_j(x',y)|=1.$$

Two cases are possible. Either this relation holds identically, in which case $\Gamma_{x'}^{(j)}$ consists of a finite number of points since $\partial X_j(x', y)/\partial y$ vanishes for any point of $\Gamma_{x'}^{(j)}$; or this relation is an equation representing a finite number of analytic curves in the y-plane; in this generic case, *the figure* $\Gamma_{x'}^{(j)}$ *is composed of a finite number of analytic arcs* (and of points).

The intersection of the analytic varieties Σ_j, Σ_k $(j \neq k; j, k = 1, 2, \ldots, \nu)$, if it is non-empty, is of dimension 1; consequently, the analytic variety given by

$$|X_j(x', y)| = 1, \quad |X_k(x', y)| = 1$$

can contain only a finite number of points in the neighbourhood of S, with the possible exception of a finite number of points on the diameter l. We suppose, from now on, *that* x' *is situated outside the exceptional points* (if they exist). *The figures* $\Gamma_{x'}^{(j)}, \Gamma_{x'}^{(k)}$ *then do not have any arcs in common.* $\Gamma_{x'}$ is the sum of the ν figures $\Gamma_{x'}^{(j)}$ having the preceding properties.

Let s be an arbitrary arc of $\Gamma_{x'}$, and s' a suitable subarc of s. We suppose that s' is contained in only one of the figures $\Gamma_{x'}^{(j)}$, say $\Gamma_{x'}^{(1)}$, and that $\partial X_1(x', y)/\partial y$ does not vanish on s'. We can always achieve this by a suitable s'. We then have $|X_1(x', y)| < 1$ on one side of s'. As for the other X_j, we have $|X_j(x', y)| \leqq 1$ by very definition of σ, and equality can hold at only finitely many points in view of what we saw above. We can therefore suppose that $|X_j(x', y)| < 1$ $(j = 2, 3, \ldots, \nu)$. Thus, the arc s' is situated on the boundary of the open set

$$\text{(II)} \qquad |X_j(x', y)| < 1 \quad (j = 1, 2, \ldots, \nu).$$

Moreover, since $\partial X_1/\partial y \neq 0$ and $|X_j| < 1$ $(j = 2, 3, \ldots, \nu)$, the arc in (x, y)-space given by $x = x'$, $y \in s'$ lies on a regular piece of the hypersurface $|X_1| = 1$, and this piece consequently belongs to the boundary of a single connected component of the open set

$$|X_j(x, y)| < 1, \quad (x, y) \in \mathfrak{D}_3 \quad (j = 1, 2, \ldots, \nu).$$

Since the above arc belongs to the boundary of Δ, this connected component is necessarily contained in Δ. Consequently, the boundary of the connected component of (II) containing the arc s' lies entirely on $\Gamma_{x'}$; here s' was a part of an arbitrary arc of $\Gamma_{x'}$. Therefore, *the figure* $\Gamma_{x'}$ *consists entirely of closed curves* (and of points).

Let $(\Gamma_{x'})$ be the portion of the y-plane bounded by the closed curves $\Gamma_{x'}$. The configuration we have just seen also requires that *the projection on the y-plane of the intersection of* Δ *with* $x = x'$ *be given by* $(\Gamma_{x'})$; this is necessarily true for every point x' of l (exceptional or not).

Now, (ξ, η) being an interior point of S, we take $[(\gamma), (\gamma')]$ sufficiently close to (ξ, η), so that every $(\Gamma_{x'})$ contains (γ') and the functions $\varphi(x_0, y)$, $R(x_0, y; x_0, y_0)$ are holomorphic in a neighbourhood of the

$$y \in (\Gamma_{x'}); \quad x_0 \in (\gamma), \quad y_0 \in \Gamma_{x'}.$$

Let (x_0, y_0) be a point of $[(\gamma), (\gamma')]$ outside L, but otherwise arbitrary. Let x' be a point of l, different from the exceptional points (which are finite in number) but otherwise arbitrary. We then have, by CAUCHY's theorem

$$\frac{1}{2\pi i}\int_{\Gamma_{x'}}\frac{R(x_0,y;x_0,y_0)}{(x'-x_0)(y-y_0)}\varphi(x_0,y)\,dy=\frac{\varphi(x_0,y)}{x'-x_0},$$

the integral being taken in the positive sense, because $R=1$ when $x=x_0$, $y=y_0$. From this, it immediately results that

$$I_2(x_0,y_0)=\frac{1}{2\pi i}\int_l\frac{\varphi(x_0,y_0)}{x-x_0}\,dx\,^{20)},$$

where the integral is taken positively since $\partial(x_1,\theta_j)/\partial(u,v)>0$.

Therefore, *the integral $I(x_0,y_0)$ is a solution of the problem for the domain Δ_3.*

4. The Case of Pseudoconvex Domains. We shall modify the integral (1). Consider the variety of integration σ. Each σ_j lies on Σ'_j, the piece (connected or not) of the analytic variety defined by

$$x_2=0,\quad |X_j(x,y)|=1,\quad |X_p(x,y)|\leqq 1\quad (p=1,2,\ldots,\nu).$$

We see immediately that Σ'_j is the limit of a decreasing sequence of open sets consisting of domains of holomorphy. Let V_j be one of the sets in this sequence; V_j contains σ_j in its interior.

For any j in $(1,2,\ldots,\nu)$, we shall find *a meromorphic function $\Phi_j(x,y;x_0,y_0)$ on $(x,y)\in V_j$, $(x_0,y_0)\in\mathfrak{D}_1$ in such a way that Φ_j has the same poles as ψ_j when $(x_0,y_0)\in\mathfrak{D}_3$, and is holomorphic when $(x_0,y_0)\notin\mathfrak{D}_3$.* The open set $(V_j,\mathfrak{D}_1)$ (i.e. $(x,y)\in V_j$, $(x_0,y_0)\in\mathfrak{D}_1$) consists of domains of holomorphy. As for the poles of $\psi_j(x,y;x_0,y_0)$, if (x,y) remains in a V_j sufficiently close to Σ'_j, by hypothesis 1°, ψ_j has no poles in the neighbourhood of the sets $(x_0,y_0)\in L_1$, $(x_0,y_0)\in L_2$. Hence, taking V_j sufficiently close to Σ'_j, we obtain the required function (Theorem II, Memoir II).

The function $\Phi_j-\psi_j$ is holomorphic in $(V_j,\mathfrak{D}_3)$; now $\mathfrak{D}_3$ is convex with respect to functions holomorphic on $\mathfrak{D}_1$ in view of the theorem of CARTAN-THULLEN mentioned in No. 2 as we see by taking into account the form of $\mathfrak{D}_3$. Hence, given a positive number ε and open sets V'_j, $\mathfrak{D}'_3$ completely interior to V_j, $\mathfrak{D}_3$ respectively, we can find *a holomorphic function $\Psi_j(x,y;x_0,y_0)$ on $(V_j,\mathfrak{D}_1)$ such that*

$$|\Phi_j-\psi_j-\Psi_j|<\varepsilon$$

on $(V'_j,\mathfrak{D}'_3)$. (No. 5, Memoir II.) We take $(V'_j,\mathfrak{D}'_3)$ so as to contain the closed set (σ_j,S).

We have thus obtained functions Φ_j, Ψ_j on $\mathfrak{D}_1$; we set

$$A_j=\Phi_j-\Psi_j-\psi_j.$$

---

[20]) To check this, it is enough to take (x_1,θ_j) as parameters and to reduce the double integral I_2 to two simple integrals taken successively; the possibility of doing this is obvious.

We construct similarly functions B_j ($j=1,2,\ldots,\nu$) with respect to $\mathfrak{D}_2$, and consider the following integrals

(2)
$$J_1(x_0,y_0)=\frac{-1}{4\pi^2}\sum_j \int_{\sigma_j} (\psi_j+A_j)\,\varphi(x,y)\,dx\,dy$$
$$J_2(x_0,y_0)=\frac{-1}{4\pi^2}\sum_j \int_{\sigma_j} (\psi_j+B_j)\,\varphi(x,y)\,dx\,dy$$

($j=1,2,\ldots,\nu$), where $\varphi(x,y)$ is an arbitrary function holomorphic in a neighbourhood of S.

When $(x,y)\in\sigma_j$, the function ψ_j+A_j, being equal to $\Phi_j-\Psi_j$, is holomorphic relative to (x_0,y_0) in Δ_1; hence *$J_1(x_0,y_0)$ is holomorphic in Δ_1*. Similarly, *$J_2(x_0,y_0)$ is holomorphic in Δ_2*.

The functions $J_1(x_0,y_0)$, $J_2(x_0,y_0)$ remain holomorphic at every point of L in Δ, because this is the case for the functions given by the integral (1), and the functions A_j, B_j are holomorphic on $(V_j,\mathfrak{D}_3)$. From the properties of the integral (1), it also follows that we have

$$J_1(x_0,y_0)-J_2(x_0,y_0)=\varphi(x_0,y_0)-\frac{1}{4\pi^2}\sum_j \int_{\sigma_j} (A_j-B_j)\,\varphi(x,y)\,dx\,dy$$

identically ($j=1,2,\ldots,\nu$). We also find, from this same relation, that J_1-J_2 remains holomorphic in a neighbourhood of S. We set

$$f(x_0,y_0)=J_1(x_0,y_0)-J_2(x_0,y_0).$$

We now regard f as given and φ as unknown in this relation; we then have an integral equation of the form

(3)
$$\varphi(x_0,y_0)=\lambda\sum_j \int_{\sigma_j} K_j(x,y;x_0,y_0)\,\varphi(x,y)\,dx\,dy+f(x_0,y_0),$$

where we have set
$$K_j=\frac{A_j-B_j}{4\pi^2},$$

and $\lambda=1$, $j=1,2,\ldots,\nu$. We are looking for a holomorphic solution in a neighbourhood of S.[21] If we can find a function of this nature, the solution of our problem would be given by the integrals (2) as is obvious.

We apply the usual method of successive approximation, choosing ε to be sufficiently small. Let us set, formally

$$\varphi(x_0,y_0)=\varphi_0(x_0,y_0)+\lambda\varphi_1(x_0,y_0)+\ldots+\lambda^p\varphi_p(x_0,y_0)+\ldots$$

where we consider λ as a complex parameter. Substituting this series in Equation (3) and formally equating the coefficients of the same power of λ in the two terms, we have the relations

$$\varphi_0=f_0,\quad \varphi_1=K(\varphi_0),\ldots,\varphi_{p+1}=K(\varphi_p),\ldots,$$

[21] In reality, it would be enough to find a solution which is holomorphic on σ, because, then it would be holomorphic on S by virtue of a theorem of F. SEVERI. Lincei, 1932.

where

$$K(\varphi_p)(x_0, y_0) = \sum_j \int_{\sigma_j} K_j(x, y; x_0, y_0)\, \varphi_p(x, y)\, dx\, dy.$$

The functions $\varphi_p(x_0, y_0)$ thus determined successively are holomorphic in a neighbourhood of S since this is the case with $f(x_0, y_0)$ and with $K_j(x, y; x_0, y_0)$ when $(x, y) \in \sigma_j$.

Substituting these functions φ_p in the formal series, we obtain a series of functions holomorphic on a neighbourhood of $(x, y) \in S$. If this series converges uniformly on a neighbourhood of S for $|\lambda| < 1 + \varepsilon'$, ε' being some positive number, the sum $\varphi(x_0, y_0)$ satisfies the equation (3) for $|\lambda| \leqq 1$ as is evident. Thus, it remains only to examine the convergence.

Let U be an open set contained completely in the interior of $\mathfrak{D}'_3$, containing S and such that f is holomorphic in a neighbourhood of U; U certainly exists. Let M_p be the upper bound of $|\varphi_p|$ on U, and let

$$N = \sum_j \int_{\sigma_j} |dx\, dy|.$$

Then, on U,

$$|K(\varphi_p)(x_0, y_0)| \leqq \frac{\varepsilon}{2\pi^2} N \cdot M_p$$

since

$$|A_j| < \varepsilon, \quad |B_j| < \varepsilon \quad \text{on} \quad (V'_j, \mathfrak{D}_3)$$

for any j. Consequently, we have

$$M_p \leqq \left(\frac{\varepsilon N}{2\pi^2}\right)^p M_0.$$

Suppose therefore that

$$\varepsilon < \frac{2\pi^2}{N}.$$

The series in question then converges uniformly on U for $|\lambda| < 1 + \varepsilon'$, ε' being a sufficiently small positive number. We therefore see that:

The problem of No. 1 is solvable for a domain Δ as in No. 2.

II. An Intermediary Result

5. Preliminary Propositions. In the present section, we shall reformulate the result we have just obtained with the help of the well known work of H. Cartan-P. Thullen[22] and of a recent Theorem of H. Behnke-K. Stein. We shall begin by making some simple remarks on this work.

Theorem of H. Behnke-K. Stein. *Given an increasing sequence of domains of holomorphy in the space of several complex variables, the limit of the sequence is again a domain of holomorphy*[23].

22) Math. Ann. 1932.

23) For bounded (univalent) domains, see Behnke-Stein, Math. Ann., 1938.

I think that this theorem has been proved precisely only for bounded domains. We shall therefore examine the case where the limit domain $\mathfrak{D}$ is not bounded. Taking a point M of the domain $\mathfrak{D}$ as centre we draw a polycylinder (γ_p) of radius p, p being an arbitrary positive number. We denote by $\mathfrak{D}_p$ the connected component of the part of $\mathfrak{D}$ in (γ_p) which contains M. Now, each $\mathfrak{D}_p$, being bounded and of the same nature as $\mathfrak{D}$, is a domain of holomorphy. Consequently, taking into account the form of $\mathfrak{D}_p$, $\mathfrak{D}_p$ is convex with respect to functions holomorphic in $\mathfrak{D}_{p+q}$ because of the theorem of CARTAN-THULLEN; here q is an arbitrary positive number. Consequently, as we said above, given a holomorphic function on $\mathfrak{D}_p$, we can find a holomorphic function on $\mathfrak{D}_{p+q}$ arbitrarily close to the given function on a given domain completely interior to $\mathfrak{D}_p$ (No. 5, Memoir II). From this, it follows by the same reasoning as in the case of bounded domains[24] that we can make the approximation with functions holomorphic on $\mathfrak{D}$. The domain $\mathfrak{D}$ is consequently convex with respect to functions holomorphic on $\mathfrak{D}$ and so is a domain of holomorphy, thanks to H. CARTAN and P. THULLEN.

Given a bounded domain of holomorphy $\mathfrak{D}$ in (x,y)-space and a function χ holomorphic on the domain, let us consider, for any positive number r, the set of points P such that the "boundary distance"[25] of P with respect to $\mathfrak{D}$ is bigger than $r|\chi(P)|$; we denote this set by $\mathfrak{D}(\chi,r)$. Then:

1°. *$\mathfrak{D}(\chi,r)$ consists of domains of holomorphy.*

2°. *If $1/\chi$ is holomorphic and bounded in $\mathfrak{D}$, $\mathfrak{D}(\chi,r)$ is completely interior to $\mathfrak{D}$ and possesses the following property:*

(α) *we can find an open set containing $\mathfrak{D}(\chi,r)$ completely in its interior, arbitrarily close to $\mathfrak{D}(\chi,r)$, and convex with respect to functions holomorphic on $\mathfrak{D}$.*[26]

1°. This is an obvious consequence of a theorem of H. CARTAN-P. THULLEN. Let us start by proving the first part. $\mathfrak{D}$ being a domain of holomorphy, we can find a function $\Phi(x,y)$ having $\mathfrak{D}$ as its exact domain of existence. With this function, we form the family of functions

$$\Phi[x+\rho\chi(x,y)e^{i\theta},\ y+\rho'\chi(x,y)e^{i\theta'}],$$

where ρ, ρ', θ, θ' are real parameters such that $|\rho|\leqq r$, $|\rho'|\leqq r$. The set $\mathfrak{D}(\chi,r)$ is open (if it is not empty). Let P be an arbitrary point of this set; let us describe the dicylinder with centre P and radius $r|\chi(P)|$. The values taken by the family at the centre P are those assumed by $\Phi(x,y)$ in this dicylinder. It follows immediately from this that the above family of functions has $\mathfrak{D}(\chi,r)$ as its exact domain of existence (whether or not it is connected). The set consists therefore of domains of holomorphy thanks to CARTAN and THULLEN.

2°. Suppose that $1/\chi$ is holomorphic and bounded in $\mathfrak{D}$; $\mathfrak{D}(\chi,r)$ is then completely interior to $\mathfrak{D}$ since $\mathfrak{D}$ is bounded. Let F be any closed set in the

---

24) See: Hilfssatz 1 of the paper cited above.

25) That is to say, the upper bound of radii ρ such that the dicylinders with centre P and equal radii ρ are contained in $\mathfrak{D}$. (We shall use this terminology for any open set, connected or not.)

26) We can also take 2 holomorphic functions φ, ψ and consider the set $\mathfrak{D}(\varphi,\psi,r)$ defined by dicylinders of the form $|x-x_0|<r|\varphi(x_0,y_0)|$, $|y-y_0|<r|\psi(x_0,y_0)|$ without changing the conclusion.

interior of $\mathfrak{D}$ which contains $\mathfrak{D}(\chi, r)$ and has property (α). Every set $\mathfrak{D}(1, \rho)$, ρ being any positive number, has property (α) by CARTAN-THULLEN; a set F as above therefore certainly exists. Consider the intersection F_0 of all the F. F_0 is a closed set containing $\mathfrak{D}(\chi, r)$. Any intersection F_1 of finitely many of the sets F is a set of the same nature. We can find an F_1 arbitrarily close to F_0. F_0 therefore has property (α). Consequently, it is sufficient to prove that F_0 consists of $\mathfrak{D}(\chi, r)$ and its points of accumulation. Suppose that this were not so. Let r' be the largest positive number such that F_0 is contained in the closure of $\mathfrak{D}(\chi, r')$. We have $r' < r$; $\mathfrak{D}(\chi, r')$ contains $\mathfrak{D}(\chi, r)$ completely in its interior. Let M be a point of F_0 such that the "boundary distance" of M with respect to $\mathfrak{D}$ equals $r'|\chi(M)|$. Choose a positive number r'' less than r' and so close to r' that the "boundary distance" of M with respect to $\mathfrak{D}(\chi, r'')$ is less than half that of any point of $\mathfrak{D}(\chi, r)$. Since $\mathfrak{D}(\chi, r'')$ consists of domains of holomorphy, the theorem of CARTAN-THULLEN requires the existence of a holomorphic function $f(x, y)$ in $\mathfrak{D}(\chi, r'')$ such that

$$|f(M)| > 1, \quad |f(x, y)| < 1 \quad \text{on} \quad \mathfrak{D}(\chi, r).$$

But then, since F_0 has property (α), we can find a holomorphic function $\varphi(x, y)$ in $\mathfrak{D}$ such that

$$|\varphi(x, y)| < |\varphi(M)|$$

for any point (x, y) in a neighbourhood of $\mathfrak{D}(\chi, r)$, by a proposition we have used often. This contradicts the definition of F_0. Q.E.D.

We denote $\mathfrak{D}(1, r)$ *by* $\mathfrak{D}^{(r)}$; $\mathfrak{D}^{(r)}$ is then the set of points whose "boundary distances" with respect to $\mathfrak{D}$ are bigger than r.

6. Arranging the Hypotheses. We shall see how to arrange for the three hypotheses of No. 2 to be fulfilled.

Let us again consider a *bounded* domain $\mathfrak{D}$ in (x, y)-space containing the origin and consider 3 parallel hyperplanes L, L_1, L_2 of the form

$$x_2 = 0, \quad x_2 = a_1, \quad x_2 = a_2, \quad a_1 < 0 < a_2$$

respectively, such that each of them crosses $\mathfrak{D}$. We denote by $\mathfrak{D}_1$ the part of $\mathfrak{D}$ above L_1, by $\mathfrak{D}_2$ the part below L_2 and by $\mathfrak{D}_3$ the part between L_1 and L_2.

We shall suppose that the open sets $\mathfrak{D}_1$, $\mathfrak{D}_2$ *consist of domains of holomorphy.* We begin again with this single assumption.

We consider in $\mathfrak{D}$ the sets

$$A = \mathfrak{D}(\chi_1, \rho), \quad B = \mathfrak{D}(\chi_2, \rho), \quad C = \mathfrak{D}^{(\rho)},$$

where

$$\chi_1 = e^{-ix}, \quad \chi_2 = e^{ix},$$

and ρ is a positive number, small enough to satisfy several later conditions. We remark, above all, that χ_1 and χ_2 are never zero. We denote by (a), (b), (c) respectively the part of A, B, C on a hyperplane of the form $x_2 = x_2'$, where x_2' is a constant.

1°. If $x'_2=0$, we have

$$(a)=(b)=(c)$$

since $|\chi_1|=|\chi_2|=1$.

2°. If $x'_2>0$, since $|\chi_1|>1>|\chi_2|$, hence $\rho|\chi_1|>\rho>\rho|\chi_2|$, we have

$$(a)\leqq(c)\leqq(b).$$

More precisely, if (a) is non-empty, then so is (c) and contains (a) completely in its interior; similarly, if (c) is non-empty, then so is (b) and contains (c) completely in its interior.

3°. If $x'_2<0$, we have, similarly,

$$(b)\leqq(c)\leqq(a).$$

Let P, Q be two fixed points of $\mathfrak{D}$ such that one of them is below L_1 and the other is above L_2. We choose ρ so small that the part common to the open sets A, B contains P, Q and the origin in the same connected component. We call this connected component G. We take 2 real numbers b_1, b_2 such that

$$a_1<b_1<0<b_2<a_2,$$

and denote the hyperplanes $x_2=b_1$, $x_2=b_2$ by L'_1 and L'_2 respectively. The domain G and the hyperplanes L, L'_1, L'_2 will correspond to $\mathfrak{D}$, L, L_1 and L_2 of No. 2. We denote by G_1 the part of G above L'_1, by G_2 the part below L'_2 and by G_3 the part between L'_1 and L'_2. When ρ tends to 0, G tends to the given domain $\mathfrak{D}$.

We shall show that *the set* G_1, *as well as the set* G_2, *consists of domains of holomorphy*. Let A_1 be the part of A above L'_1, B_2 the part of B below L'_2. Since $\mathfrak{D}_1$ consists of domains of holomorphy, the same is true for A_1 in view of the proposition of the last section (ρ being assumed to be small enough). And G_1, the set in question, is contained in A_1. Let us look at the boundary of G_1; it contains boundary points of A_1 and boundary points of B_2; let M be a boundary point of G_1 of this second kind. M is necessarily below L, or on L, because of the relation between A and B. In view of the proposition mentioned above, B_2 possesses property (α) relative to functions holomorphic on $\mathfrak{D}_2$ since $\mathfrak{D}_2$ consists of domains of holomorphy. (ρ is assumed to be small enough.) Let F be an arbitrary closed set contained in G_1 and L''_2 a hyperplane parallel to L and lying between L and L'_2. Under these conditions, we can easily find a holomorphic function $f(x,y)$ in $\mathfrak{D}_2$ such that

$$f(M)=1, \qquad |f(x,y)|<1$$

on the part of F below L''_2 and in a neighbourhood of the part of L''_2 in A_1. We shall find a meromorphic function $\Phi(x,y)$ in A_1 having the poles

$$\frac{1}{f(x,y)-1}$$

in the portion below L''_2 and holomorphic elsewhere. Since A_1 consists of domains of holomorphy and the given poles do not lie near L''_2, $\Phi(x,y)$ certainly exists. This function Φ has M as a pole and is holomorphic at every point of

F; here M is an arbitrary point of the common boundary of G_1 and B_2. Thanks to CARTAN and THULLEN, we can therefore find an open set consisting of domains of holomorphy, which is contained in G_1 and contains F. Since F is an arbitrary closed set in G_1, G_1 itself consists of domains of holomorphy in view of the theorem of BEHNKE-STEIN. Similarly for G_2. We also remark that *the set G_1, as well as G_2, is the limit of a decreasing sequence of open sets consisting of domains of holomorphy*, the proof of which is immediate because of the form of G_1 and that of G_2.

We take a positive number δ greater than ρ and sufficiently close to it and consider

$$\mathfrak{D}' = \mathfrak{D}_3^{(\rho)}, \quad \mathfrak{D}'' = \mathfrak{D}_3^{(\delta)} \quad (\rho < \delta).$$

$\mathfrak{D}''$ is contained completely in the interior of $\mathfrak{D}'$; since $C = \mathfrak{D}^{(\rho)}$, C and $\mathfrak{D}'$ coincide between L_1' and L_2' and near these hyperplanes. (ρ is assumed to be sufficiently small.) Since $\mathfrak{D}_3$ consists of domains of holomorphy, it follows from the theorem of CARTAN-THULLEN that to each boundary point of $\mathfrak{D}'$, there correspond a small dicylinder (γ) and a holomorphic function $f(x, y)$ in $\mathfrak{D}_3$ such that

$$|f| > 1 \quad \text{on} \quad (\gamma), \qquad |f| < 1 \quad \text{on} \quad \mathfrak{D}''.$$

By the BOREL-LEBESGUE lemma, we can cover the boundary of $\mathfrak{D}'$ by a finite number of these (γ). Let

$$X_j(x, y) \qquad (j = 1, 2, \ldots, \nu)$$

be the functions $f(x, y)$ corresponding to these (γ). The set G_3 is contained completely in the interior of $\mathfrak{D}_3$; consequently, *the functions X_j are holomorphic in a neighbourhood of G_3*.

Let us examine *hypothesis 1°* (No. 2). 1°. Near the hyperplane L, G is composed of one or several connected components of $C = \mathfrak{D}'$. Hence, there is no point satisfying

$$|X_j(x, y)| \leqq 1 \qquad (j = 1, 2, \ldots, \nu)$$

in a neighbourhood of the intersection of the boundary of G with L.

2°. The part of G on L_2' is completely interior to the part of C on L_2', consequently to that of $\mathfrak{D}'$ on L_2'. Hence, if δ is chosen sufficiently close to ρ, there is no point in the neighbourhood of the portions of L_1', L_2' in G which satisfies one of the ν conditions

$$|X_j(x, y)| > 1 - \varepsilon,$$

ε being a sufficiently small positive number.

3°. If we take away from G those points of G_3 which do not satisfy the conditions

$$|X_j(x, y)| < 1 \qquad (j = 1, 2, \ldots, \nu),$$

we obtain an open set. This set tends to G when δ tends to ρ. We choose δ so close to ρ that this set contains the points P, Q and the origin in the same connected component, and call Δ this connected component. We have thus formed in G a domain Δ satisfying condition 1° and arbitrarily close to G.

7. Let us pass now to *hypothesis 2°*. We have seen the following in Memoir V:

"If V is a domain of holomorphy in the space (x, y), then, given a positive number ε and a domain V_0 contained completely in the interior of V, we can find a holomorphic function R of the complex variables x, y; x_0, y_0 in (V_0, V_0), which reduces to 1 for $x = x_0$, $y = y_0$ in such a way that to every holomorphic function $f(x, y)$ in V, there corresponds a holomorphic function $\varphi(x, y)$ on V_0 such that $|f - \varphi| < \varepsilon$ and

$$(1) \qquad (\varphi - \varphi_0)R = (x - x_0)P + (y - y_0)Q$$

identically; here φ_0 means $\varphi(x_0, y_0)$ and P, Q are holomorphic functions of the variables x, y; x_0, y_0 in (V_0, V_0)."

We shall apply this lemma to $\mathfrak{D}_3$ which consists of domains of holomorphy by taking $\mathfrak{D}_3$ for V and an open set containing $\mathfrak{D}'(=\mathfrak{D}_3^{(\rho)})$ completely in its interior for V_0. Then, if ρ is small enough, V_0 contains G_3 completely in its interior. Let us consider the family consisting of all functions $\varphi(x, y)$ holomorphic and having the property (1) on V_0. We start with this family of functions $\varphi(x, y)$ in place of the family of functions $f(x, y)$ of No. 6. We then arrive, by using the same method, at a new domain Δ realising hypotheses 1° and 2° simultaneously.

Let us look now at *hypothesis 3°*. 1°. We can, as above, start with a family of functions $\varphi(x, y)$ such that the derivative $\partial\varphi/\partial y$ is not identically zero; we will then have a domain Δ satisfying the two conditions 1, 2 and such that the ν derivatives $\partial X_j/\partial y$ are not identically zero. We shall make our discussions with domains of this nature.

2°. Let us consider in G_3 the ν analytic varieties Σ_j:

$$x_2 = 0, \qquad |X_j(x, y)| = 1.$$

Because of the property of Δ concerning the derivatives $\partial X_j/\partial y$, the Σ_j have dimension at most 2. We shall make Δ satisfy the condition that the intersection of the varieties Σ_j, Σ_k ($j \neq k$; $j, k = 1, 2, \ldots, \nu$) be of dimension at most 1. We take 2 positive numbers r, r' larger than 1/2 and consider the variety T_{jk} in G_3:

$$x_2 = 0, \qquad |X_j(x, y)| = r, \qquad |X_k(x, y)| = r'.$$

The analytic variety defined by $x_2 = 0$, $|X_j| = r$ in G_3 is obviously at most 2 dimensional; consequently the variety T_{jk} is at most 1 dimensional, except perhaps for a finite number of values of r', r being regarded as fixed. For our present goal, it is sufficient to make a suitable modification of the form

$$(2) \qquad Y_j = \alpha_j X_j \quad (j = 1, 2, \ldots, \nu),$$

α_j being positive numbers, since for the new condition to hold, it suffices to choose the numbers $\alpha_1, \alpha_2, \ldots, \alpha_\nu$ successively, each time avoiding a finite number of values; the three earlier properties of Δ are not changed by this modification, so long as one takes the α_j sufficiently close to 1.

3°. We have now to make Δ satisfy the condition that $\partial X_j/\partial y$ be non-zero on the variety Σ_j, except, perhaps, at a finite number of points, j being arbitrary. Now, the characteristic variety $\partial X_j/\partial y=0$ may contain characteristic planes of the form $x=\xi$ where ξ is a point on the real axis. To avoid this in G_3, it is sufficient to take the functions

$$X_j(x+\beta, y)$$

instead of the $X_j(x,y)$, β being a suitable sufficiently small purely imaginary number. As for the effect of this modification, we find easily that hypothesis 1° for Δ continues to be fulfilled, but not necessarily hypothesis 2°. Since the modification (2) does not affect this latter hypothesis, we can reverse the order of these two modifications. By making the second modification first and applying the earlier argument, we find that hypotheses 1° and 2° are satisfied, and, in addition, the analytic variety defined in G_3 by

$$x_2=0, \quad \frac{\partial X_j}{\partial y}=0$$

is, for any j, at most 1 dimensional. Consequently, the intersection of this variety with $|X_j|=r$, where r is a positive number bigger than 1/2, is made up of a finite number of points, except perhaps for a finite number of values of r. Consequently the last two hypotheses (2° and 3°) can now be fulfilled simultaneously by a suitable modification of the form (2), in such a way that Δ continues to satisfy the other conditions.

Thus:

We can find a domain Δ, arbitrarily close to the given domain $\mathfrak{D}$, having the character described in No. 2.

8. Amalgamation of Contiguous Domains of Holomorphy. Let us consider Δ_1. It is an open set in G_1; G_1 consists of domains of holomorphy. The boundary points of Δ_1 belong to the boundary of G_1 or to L or to one of the ν hypersurfaces $|X_j(x,y)|=1$. Let us consider the third possibility. Let M be a boundary point of Δ_1 of this kind, but otherwise arbitrary; we shall suppose, to fix our ideas, that $|X_1(M)|=1$. The function $X_1(x,y)$ is holomorphic in a neighbourhood of G_3. We shall find a meromorphic function $\Phi(x,y)$ in G_1 having the poles

$$g(x,y)=\frac{1}{X_1(x,y)-X_1(M)}$$

in G_3 and remaining holomorphic elsewhere. By condition 1° on Δ, $g(x,y)$ is holomorphic at every point on the part of L_2' in G; hence $\Phi(x,y)$ exists. Consequently Δ_1 *consists of domains of holomorphy*, thanks to Cartan and Thullen. Moreover, G_1 can be approximated from the exterior by open sets consisting of domains of holomorphy (No. 6). Hence, we find in the same way that *there exists an open set, arbitrarily close to Δ_1, containing Δ_1 completely in its interior, and consisting of domains of holomorphy*. The same is the case for Δ_2.

Let $(\mathfrak{p})$ be poles given in a neighbourhood of Δ; there exists a meromorphic function $\Phi_1(x,y)$ having the poles $(\mathfrak{p})$ in a neighbourhood of Δ_1. Similarly, let

$\Phi_2(x, y)$ be a function corresponding to Δ_2. Let $f = \Phi_1 - \Phi_2$; f is a holomorphic function in a neighbourhood of the common boundary S of Δ_1 and Δ_2. We can find the solution $F_1(x, y)$, $F_2(x, y)$ of the problem of Section I on Δ corresponding to the function f. Let us set

$$\Phi = \Phi_1 - F_1 \quad \text{or} \quad = \Phi_2 - F_2$$

according as $(x, y) \in \Delta_1$ or $\in \Delta_2$. $\Phi(x, y)$ is meromorphic in Δ and has the poles $(\mathfrak{p})$. We have thus seen that *if we are given poles in the neighbourhood of Δ, we can find a meromorphic function having these poles on Δ.*

We shall prove from this that Δ is a domain of holomorphy[27]. The boundary of Δ consists of 3 parts:

1°. Let M be a boundary point of Δ such that one at least of the v functions $X_j(x, y)$ has absolute value 1. Suppose that $|X_1(M)| = 1$ to fix our ideas. Let Δ_0 be a given domain in the interior of Δ. We choose a point N of Δ so close to M that the characteristic surface $X_1(x, y) = X_1(N)$, defined in a neighbourhood of G_3, neither passes through Δ_0, nor approaches arbitrarily close to the hyperplanes L_1', L_2'. N certainly exists because of condition 1° on Δ. We can then find a meromorphic function on Δ which has the poles

$$\frac{1}{X_1(x, y) - X_1(N)}$$

on Δ_3 and is holomorphic elsewhere. $\Phi(x, y)$ is holomorphic on Δ_0 and has N as a pole, N being a point in a certain neighbourhood of M, but otherwise arbitrary.

2°. Let M' be an arbitrary point of the boundary common to Δ, G_1. Since G_1 can be approximated from the exterior by open sets consisting of domains of holomorphy, if we take into account condition 1° on Δ, it follows from the theorem of CARTAN-THULLEN that if we take a point N' in Δ, sufficiently close to M', we can find a holomorphic function $f(x, y)$ in a neighbourhood of G_1 such that

$$f(N') = 1 \quad \text{and} \quad |f(x, y)| < 1$$

on the part Δ_0 above L and in a neighbourhood of the part of L in Δ. There therefore exists a meromorphic function $\Psi(x, y)$ on Δ having the poles

$$\frac{1}{f(x, y) - 1}$$

on the portion above L, and holomorphic elsewhere. $\Psi(x, y)$ is holomorphic on Δ_0 and has a pole at N', N' being an arbitrary point of Δ in a certain neighbourhood of M'.

3°. The same result holds for the common boundary of Δ, G_2.

There therefore exists, by CARTAN-THULLEN, an open set in Δ containing Δ_0 and consisting of domains of holomorphy, where Δ_0 is an arbitrary domain in the interior of Δ. Consequently, by the BEHNKE-STEIN theorem, Δ is a domain of holomorphy; and the same is true of $\mathfrak{D}$.

[27]) The corresponding result due to H. CARTAN (Note 6, Introduction) cannot be applied to the present case.

We thus obtain:

Given a bounded domain $\mathfrak{D}$ and 2 hyperplanes of the form $x_2=a_1$, $x_2=a_2$, $a_1<a_2$, x_2 being the imaginary part of x, if the part of $\mathfrak{D}$ above $x_2=a_1$ and that below $x_2=a_2$ consist of domains of holomorphy, then $\mathfrak{D}$ is a domain of holomorphy.

9. Domains Pseudoconvex in the Sense of H. Cartan. As we have often mentioned, it was H. CARTAN who introduced the notion of global convexity in our theory. Now, in the same memoir[28], he considered the following type of domain. If $\mathfrak{D}$ is the domain in question, *the part of $\mathfrak{D}$ in a sufficiently small hypersphere around any (finite) boundary point of $\mathfrak{D}$ consists of domains of holomorphy.* He called such domains (everywhere) *pseudoconvex.* We shall apply what we have just seen above to this kind of domain.

Let $\mathfrak{D}$ be an arbitrary domain of this type. Let us first suppose that $\mathfrak{D}$ is bounded. We divide the x-plane into equal rectangles ω by two systems of straight lines parallel to the real and imaginary axis respectively; we divide the y-plane similarly into rectangles ω'. If we divide the space (x, y) into sufficiently small (ω, ω'), it follows from the definition given above that for any domain of holomorphy Δ sufficiently close to an arbitrary one of the (ω, ω'), the part of $\mathfrak{D}$ in Δ consists of domains of holomorphy. Hence, from the preceding proposition, it is easy to see that $\mathfrak{D}$ is a domain of holomorphy.

If $\mathfrak{D}$ is not bounded, we draw a hypersphere (γ_p) of radius p about a point M of $\mathfrak{D}$, p being an arbitrary positive integer. Let $\mathfrak{D}_p$ be the connected component of the part of $\mathfrak{D}$ in (γ_p) which contains M. By the preceding result, $\mathfrak{D}_p$ is a domain of holomorphy. The sequence $\mathfrak{D}_p$ $(p=1,2,\ldots)$ of domains converges to $\mathfrak{D}$. By virtue of the theorem of BEHNKE-STEIN, $\mathfrak{D}$ is therefore also a domain of holomorphy. Thus:

Any domain pseudoconvex in the sense of H. Cartan is a domain of holomorphy.

The author believes that this theorem, as well as the essential part of the proof, applies to an arbitrary number of complex variables.

III. A Complementary Problem

10. Preliminaries. *Let $\mathfrak{D}$ be a domain in the space (x, y) and let E be the complement of $\mathfrak{D}$. We shall call $\mathfrak{D}$ pseudoconvex if E satisfies the continuity theorem due to F. Hartogs in the neighbourhood of an arbitrary point (ξ, η) of the boundary of E, and if this property remains invariant under one-to-one pseudoconformal transformations in the neighbourhood of (ξ, η).*

The first condition means that if there is no point of E on the characteristic plane $x=\xi$ in the neighbourhood of (ξ, η), excepting (ξ, η) itself, then, to any given positive number ρ, there exists a sufficiently small positive number r such that for

[28]) Bull. Soc. Math. France, 1931.

any x' in $|x-\xi|<r$, there exists at least one y' in $|y-\eta|<\rho$ for which (x',y') belongs to E.[29]

The principal properties of pseudoconvex domains were studied by F. HARTOGS and E.E. LEVI, and will be found treated very systematically in the memoir of G. JULIA cited earlier[30].

Let us consider in the space (x,y) a bounded domain Δ such that its boundary S is given by $\varphi(x_1,x_2,y_1,y_2)=0$, where $x=x_1+ix_2$ and $y=y_1+iy_2$, in such a way that one has $\varphi<0$ at points of Δ and $\varphi>0$ at exterior points; here φ is a real function, *well-defined*[31] and continuous in a neighbourhood of S, and having continuous partial derivatives up to second order. Under these conditions, E.E. LEVI has proved that for Δ to be pseudoconvex, it is *necessary* that we have $L(\varphi)\geqq 0$ on S, where

$$\begin{aligned} L(\varphi)= & \left(\frac{\partial^2\varphi}{\partial x_1^2}+\frac{\partial^2\varphi}{\partial x_2^2}\right)\left[\left(\frac{\partial\varphi}{\partial y_1}\right)^2+\left(\frac{\partial\varphi}{\partial y_2}\right)^2\right]+\left(\frac{\partial^2\varphi}{\partial y_1^2}+\frac{\partial^2\varphi}{\partial y_2^2}\right)\left[\left(\frac{\partial\varphi}{\partial x_1}\right)^2+\left(\frac{\partial\varphi}{\partial x_2}\right)^2\right] \\ & -2\left(\frac{\partial^2\varphi}{\partial x_1\,\partial y_1}+\frac{\partial^2\varphi}{\partial x_2\,\partial y_2}\right)\left(\frac{\partial\varphi}{\partial x_1}\frac{\partial\varphi}{\partial y_1}+\frac{\partial\varphi}{\partial x_2}\frac{\partial\varphi}{\partial y_2}\right) \\ & -2\left(\frac{\partial^2\varphi}{\partial x_1\,\partial y_2}-\frac{\partial^2\varphi}{\partial x_2\,\partial y_1}\right)\left(\frac{\partial\varphi}{\partial x_1}\frac{\partial\varphi}{\partial y_2}-\frac{\partial\varphi}{\partial x_2}\frac{\partial\varphi}{\partial y_1}\right). \end{aligned}$$

This is a consequence of the following proposition: *If $L(\varphi)<0$ at a point P of S, we can find a characteristic surface passing through P, regular at P, and remaining in Δ in a neighbourhood of P except at P.* Let us remark that $L(-\varphi) = -L(\varphi)$.

From which it follows, because of the result that we have proved, that *if $L(\varphi)>0$ everywhere on S, Δ is a domain of holomorphy.*

Hence, *taking into account the theorem of H. Behnke-K. Stein, the question is if an arbitrary pseudoconvex domain can be approximated from the interior by a sequence of domains of this character* ($L(\varphi)>0$). This is the problem we shall study in this section.

Now, we do not necessarily have $L(\varphi_1+\varphi_2)>0$ even if $L(\varphi_1)>0$ and $L(\varphi_2)>0$. To treat our problem, we shall first search for another condition playing the same role which remains invariant with respect to addition of functions.

Let us consider a pseudoconvex domain $\mathfrak{D}$ in the space (x,y). We shall denote by $\mathfrak{D}(\xi)$ the projection onto the y-plane of the intersection of $\mathfrak{D}$ with the characteristic plane $x=\xi$, ξ being a constant. $\mathfrak{D}(\xi)$ is an open set of the y-plane. We denote by $R_\eta(\xi)$ the distance (in the usual sense) of a point η of $\mathfrak{D}(\xi)$ to the boundary of $\mathfrak{D}(\xi)$. This function $R_y(x)$ defined on $\mathfrak{D}$ corresponds to the

---

29) This definition is a little different from the one we gave in Memoir IV but they are equivalent. See the memoir of G. JULIA cited below.

30) Acta Math. 1926. In what follows, the author will often content himself with referring the reader to this memoir.

31) That is to say, to each point (x,y), there corresponds a value of φ, finite or infinite, but only one.

radius of holomorphy of F. Hartogs. Hartogs has proved that

$$-\log R_y(x)$$

is a subharmonic function with respect to x, (the determination of the logarithm being real).

Let us make a remark about the definition of subharmonic functions. Let A be a domain in the x-plane. We call a function $\varphi(x)$ *subharmonic* with respect to x on A if it is real, well-defined on A and satisfies the following conditions: 1°. $e^{\varphi(x)}$ is finite and upper semicontinuous on A. 2°. Let (δ) be an arbitrary domain completely interior to A, let (γ) be the boundary of (δ), and let $u(x)$ be a function harmonic with respect to x (i.e. with respect to x_1, x_2) in (δ) which remains continuous up to the boundary (γ); then, if

$$\varphi(x) \leqq u(x)$$

on (γ), the same is true on (δ). Subharmonic functions defined thus can take the value $-\infty$. *We shall understand that the constant* $-\infty$ *is included*[32].

Concerning the *Hartogs radius*, we shall find that *this function is subharmonic on every characteristic plane, with respect to x or to y.*[33] By studying functions having this property, we shall find that the corresponding differential condition has the required nature. And we shall see, for the problem formulated above, that every pseudoconvex domain can be approximated from the interior by a sequence of domains of this new character. This is sufficient for our purposes.

11. A New Class of Real Functions. Let us consider a domain $\mathfrak{D}$ in (x, y)-space, pseudoconvex or not.

We shall call a real, well-defined function $\varphi(x, y)$ *on* $\mathfrak{D}$ *pseudoconvex with respect to* (x, y) *if it satisfies the following conditions:* 1°. *The function* $e^{\varphi(x,y)}$ *is finite and upper semicontinuous with respect to* x, y *in* $\mathfrak{D}$. 2°. *On any characteristic plane* L *passing through a point of* $\mathfrak{D}$, $\varphi(x, y)$ *is a subharmonic function of* x *or of* y *on the portion of* L *in* $\mathfrak{D}$.

There are properties of *pseudoconvex functions* corresponding immediately to properties of *subharmonic functions*; of these we shall only mention those which will be used in what follows.

1°. *If* $\varphi_1(x, y)$, $\varphi_2(x, y)$ *are two pseudoconvex functions, so is the sum* $\varphi_1(x, y) + \varphi_2(x, y)$.

2°. *The same is true of the supremum* $\psi(x, y)$ *of the quantities* $\varphi_1(x, y)$, $\varphi_2(x, y)$.

3°. *Given a sequence* $\varphi_n(x, y)$ $(n = 1, 2, \ldots)$ *of pseudoconvex functions in a domain* $\mathfrak{D}$, *if the sequence* e^{φ_n} *either converges uniformly, or decreases, to* e^{φ} *in the interior of* $\mathfrak{D}$, $\varphi(x, y)$ *is pseudoconvex in* $\mathfrak{D}$.

---

[32]) For subharmonic functions see e.g.: F. Riesz: Sur les fonctions subharmoniques et leur rapport avec la théorie du potentiel, I, 1926 (Acta Math.).

[33]) See No. 11 below.

Let $\mathfrak{D}$ be a pseudoconvex domain in the space (x, y), *and let* $R_y(x)$ *be the Hartogs radius with respect to* $\mathfrak{D}$. *The function*

$$\varphi(x, y) = -\log R_y(x)$$

is pseudoconvex in $\mathfrak{D}$ *(the determination of the logarithm being real).*

First, $1/R_y(x)$ is a well-defined finite function on $\mathfrak{D}$. To examine the semicontinuity, let us first suppose that $\mathfrak{D}$ is bounded. Let (ξ, η) be an arbitrary point of $\mathfrak{D}$, and (ξ_p, η_p) $(p=1,2,\ldots)$ a sequence of points of $\mathfrak{D}$ tending to (ξ, η), but otherwise arbitrary. Let α be any one of the limits of the sequence $R_{\eta_p}(\xi_p)$ $(p=1,2,\ldots)$; we shall prove that

$$\alpha \geqq R_\eta(\xi). \tag{1}$$

Let Y_p be a boundary point of the domain $\mathfrak{D}(\xi_p)$ such that

$$|Y_p - \eta_p| = R_{\eta_p}(\xi_p);$$

Y_p certainly exists because $R_{\eta_p}(\xi_p)$ is the distance of the point η_p to the boundary of the bounded domain $\mathfrak{D}(\xi_p)$. The sequence Y_p necessarily has a limit Y_0 such that

$$|Y_0 - \eta| = \alpha,$$

since α is one of the limits of the sequence $|Y_p - \eta_p|$. Since every point of the sequence (ξ_p, Y_p) is on the boundary of $\mathfrak{D}$, so is the limit point (ξ, Y_0). Y_0 is therefore on the boundary, or in the exterior, of $\mathfrak{D}(\xi)$; from which relation (1) follows. $R_y(x)$ is thus lower semicontinuous in $\mathfrak{D}$.

If $\mathfrak{D}$ is not bounded, we can find a sequence $\mathfrak{D}_p$ $(p=1,2,\ldots)$ of bounded pseudoconvex domains which increases and tends to $\mathfrak{D}$. Let R_p be the HARTOGS radius with respect to $\mathfrak{D}_p$. The function $1/R_p$ is upper semicontinuous by the above, and the sequence $1/R_p$ is decreasing. The limit $1/R_y(x)$ is therefore upper semicontinuous in $\mathfrak{D}$.

We shall now examine the second condition. Let us take an arbitrary point P of $\mathfrak{D}$, which we shall suppose to be the origin to simplify the notation. Thanks to HARTOGS, the function $\varphi(x, 0)$ is subharmonic with respect to x in a neighbourhood of the origin.

We transform $\mathfrak{D}$ into $\mathfrak{D}'$ in the space (X, Y) by

$$X = x, \qquad Y = y - ax,$$

a being an arbitrary constant, and consider the HARTOGS radius $R'_Y(X)$ with respect to $\mathfrak{D}'$. For any fixed x, $Y = y - ax$ is just a translation in the y-plane; we therefore have

$$R'_Y(x) = R_y(x).$$

The function $-\log R'_0(X)$ is subharmonic in a neighbourhood of $X=0$ because $\mathfrak{D}'$ is a pseudoconvex domain containing the origin $X=0$, $Y=0$. The function $\varphi(x, ax)$ is therefore subharmonic near $x=0$.

It remains only to examine $\varphi(0, y)$ near $y=0$. $R_y(0)$ represents the distance of the point y to the boundary of $\mathfrak{D}(0)$ and $\mathfrak{D}(0)$ contains the origin $y=0$. Hence $1/R_y(0)$ is continuous. This function reduces to the constant 0 if $\mathfrak{D}(0)$

coincides with the y-plane. Let us therefore suppose that $\mathfrak{D}(0)$ has (finite) boundary points; let η be an arbitrary one of these points. Then, we obviously have

$$\varphi(0, y) = \max[-\log|y-\eta|],$$

where the symbol "max" represents the upper bound of the quantities inside the brackets, and the determination of the logarithm is real. The functions $-\log|y-\eta|$ are harmonic with respect to y, from which it follows easily that $\varphi(0, y)$ is a subharmonic function, in view of the properties of subharmonic functions corresponding to properties 2, 3 of pseudoconvex functions. Q.E.D.

Let $\mathfrak{D}$ be a pseudoconvex domain and let $d(x, y)$ be the distance (in the usual sense) of the point (x, y) in $\mathfrak{D}$ to the boundary of $\mathfrak{D}$. The function $-\log d(x, y)$ is pseudoconvex in $\mathfrak{D}$ (the determination of the logarithm being real).

This proposition will be true if it is true for bounded domains, because, then, $-\log d(x, y)$ can be expressed as the limit of a decreasing sequence of pseudoconvex functions. We therefore suppose that $\mathfrak{D}$ is bounded.

We consider the transformation T:

$$X = ax - by, \qquad Y = \bar{b}x + \bar{a}y$$

with 2 numbers a, b such that $|a|^2 + |b|^2 = 1$; here $\bar{a}, \bar{b}$ are the conjugates of a, b respectively. This transformation evidently leaves the distance invariant. Let $\mathfrak{D}'$ be the transform of $\mathfrak{D}$ by T, and let $R'_y(X)$ be the *Hartogs radius* with respect to $\mathfrak{D}'$. The function $-\log R'_Y(X)$ is pseudoconvex with respect to X, Y in $\mathfrak{D}'$ (the logarithm being real); consequently, by very definition,

$$\psi(x, y) = -\log R'_Y(X)$$

is pseudoconvex in x, y in $\mathfrak{D}$.

Now let (x', y') be a point of $\mathfrak{D}$, (x'', y'') a boundary point, and let (X', Y'), (X'', Y'') be their transforms by T respectively. If we choose (a, b) such that

$$a(x' - x'') = b(y' - y''),$$

we have

$$|Y' - Y''| \geqq R'_{Y'}(X')$$

since $X' = X''$; hence

$$\frac{-1}{2}\log[|x' - x''|^2 + |y' - y''|^2] \leqq \psi(x', y').$$

It follows immediately from this that the function $-\log d(x, y)$ can be expressed as the limit of a sequence of pseudoconvex functions (property 2) converging uniformly in the interior of $\mathfrak{D}$. Hence, by property 3, $-\log d(x, y)$ is pseudoconvex in $\mathfrak{D}$.

This function $-\log d(x, y)$ has the following special properties: 1°. It is *continuous* in $\mathfrak{D}$, so long as $\mathfrak{D}$ does not coincide with (finite) (x, y)-space. 2°. If the point (x, y) of $\mathfrak{D}$ approaches the boundary of $\mathfrak{D}$ in an arbitrary way, *this function always tends to* $+\infty$.

12. Differential Condition. Suppose that we are given a real continuous function $\varphi(x_1, x_2, y_1, y_2)$ in an arbitrary domain $\mathfrak{D}$ in the space (x, y), which has continuous partial derivatives up to second order; we shall look for a condition that φ be pseudoconvex in x, y.

We take an arbitrary point P in $\mathfrak{D}$, which we shall suppose to be the origin to simplify the notation. Let us consider a characteristic plane of the form

$$y = a x, \qquad a = \alpha + i\beta$$

on which we have

$$\Phi(x_1, x_2) = \varphi(x_1, x_2, y_1, y_2),$$

where

$$y_1 = \alpha x_1 - \beta x_2, \qquad y_2 = \beta x_1 + \alpha x_2;$$

$\Phi(x_1, x_2)$ is defined in a neighbourhood of the origin. We shall calculate

$$\Delta\Phi = \frac{\partial^2 \Phi}{\partial x_1^2} + \frac{\partial^2 \Phi}{\partial x_2^2}.$$

If we set

$$A = \frac{\partial^2 \varphi}{\partial x_1^2} + \frac{\partial^2 \varphi}{\partial x_2^2}, \qquad D = \frac{\partial^2 \varphi}{\partial y_1^2} + \frac{\partial^2 \varphi}{\partial y_2^2},$$

$$B = \frac{\partial^2 \varphi}{\partial x_1 \partial y_1} + \frac{\partial^2 \varphi}{\partial x_2 \partial y_2}, \qquad C = \frac{\partial^2 \varphi}{\partial x_1 \partial y_2} - \frac{\partial^2 \varphi}{\partial x_2 \partial y_1},$$

we obtain without difficulty

(1) $$\Delta\Phi = A + 2B\alpha + 2C\beta + D(\alpha^2 + \beta^2).$$

We consider A, B, C, D as real constants and α, β as real parameters, and study the form

(2) $$U(\alpha, \beta) = A + 2B\alpha + 2C\beta + D(\alpha^2 + \beta^2).$$

We first calculate *the lower bound m of U.*

1°. If $D \leqq 0$, we have $m = A$ if $B = C = D = 0$; otherwise we have $m = -\infty$.

2°. If $D > 0$, there certainly exists at least one (α, β) for which $U = m$; this (α, β) is given by

$$\frac{1}{2}\frac{\partial U}{\partial \alpha} = B + D\alpha = 0; \qquad \frac{1}{2}\frac{\partial U}{\partial \beta} = C + D\beta = 0.$$

Hence

(3) $$mD = AD - (B^2 + C^2).$$

We then look for the condition that the form $U(\alpha, \beta)$ be *non-negative.* For this, it is first necessary that $D \geqq 0$, since otherwise $m = -\infty$. Moreover:

1° if $D = 0$, we must have $B = C = 0$ and $A \geqq 0$;

2° if $D > 0$, we must have $mD \geqq 0$. Thus,

(4) $$A \geqq 0, \qquad D \geqq 0, \qquad AD - (B^2 + C^2) \geqq 0$$

gives the *necessary* condition.

Suppose, conversely, that condition (4) is satisfied. Then, there are only two possibilities; either $D=0$, which implies $m=A\geqq 0$, since then, necessarily $B=C=0$; or $D>0$, and we then have $mD\geqq 0$ by (3), and consequently $m\geqq 0$. The condition (4) is thus *sufficient.*

Let us come back to our original problem, and let us suppose that the function φ is pseudoconvex in $\mathfrak{D}$. Φ being then subharmonic with respect to x at the origin, we necessarily have $\Delta\Phi\geqq 0$ at $x=0$, and this for any characteristic plane of the form $y=ax$. The condition (4) is therefore necessary at P by what we have just seen, P being an arbitrary point of $\mathfrak{D}$.

Suppose, conversely, that condition (4) is satisfied in $\mathfrak{D}$. For an arbitrary point P of $\mathfrak{D}$, φ is then subharmonic near P on any characteristic plane of the form $y=ax+b$ passing through P, because we necessarily have $\Delta\Phi\geqq 0$. It is then necessarily so for any characteristic plane of the form $x=ay+b$, because condition (4) is symmetric in x, y. The function φ is thus pseudoconvex in $\mathfrak{D}$. We have thus proved:

Let $\varphi(x_1,x_2,y_1,y_2)$ be a real continuous function having partial derivatives up to second order in a domain $\mathfrak{D}$. For φ to be pseudoconvex with respect to (x,y) in $\mathfrak{D}$, it is necessary and sufficient that we have

$$\frac{\partial^2\varphi}{\partial x_1^2}+\frac{\partial^2\varphi}{\partial x_2^2}\geqq 0,\qquad \frac{\partial^2\varphi}{\partial y_1^2}+\frac{\partial^2\varphi}{\partial y_2^2}\geqq 0,$$

$$V(\varphi)=\left(\frac{\partial^2\varphi}{\partial x_1^2}+\frac{\partial^2\varphi}{\partial x_2^2}\right)\left(\frac{\partial^2\varphi}{\partial y_1^2}+\frac{\partial^2\varphi}{\partial y_2^2}\right)-\left(\frac{\partial^2\varphi}{\partial x_1\,\partial y_1}+\frac{\partial^2\varphi}{\partial x_2\,\partial y_2}\right)^2$$
$$-\left(\frac{\partial^2\varphi}{\partial x_1\,\partial y_2}-\frac{\partial^2\varphi}{\partial x_2\,\partial y_1}\right)^2\geqq 0.$$ [34)]

13. The Main Property. We continue to consider a real continuous function $\varphi(x_1, x_2, y_1, y_2)$ having partial derivatives up to second order in a domain $\mathfrak{D}$. *We suppose that*

$$A>0,\quad V=AD-(B^2+C^2)>0$$

in $\mathfrak{D}$, from which it follows that $D>0$.

Let P be an arbitrary point of $\mathfrak{D}$. *Let us first suppose that*

$$\left(\frac{\partial\varphi}{\partial x_1}\right)^2+\left(\frac{\partial\varphi}{\partial x_2}\right)^2+\left(\frac{\partial\varphi}{\partial y_1}\right)^2+\left(\frac{\partial\varphi}{\partial y_2}\right)^2\neq 0$$

at the point P. We shall examine the sign of the expression $L(\varphi)$ of E.E. Levi. If we set

$$a=\left(\frac{\partial\varphi}{\partial x_1}\right)^2+\left(\frac{\partial\varphi}{\partial x_2}\right)^2,\qquad d=\left(\frac{\partial\varphi}{\partial y_1}\right)^2+\left(\frac{\partial\varphi}{\partial y_2}\right)^2,$$

$$b=\frac{\partial\varphi}{\partial x_1}\frac{\partial\varphi}{\partial y_1}+\frac{\partial\varphi}{\partial x_2}\frac{\partial\varphi}{\partial y_2},\qquad c=\frac{\partial\varphi}{\partial x_1}\frac{\partial\varphi}{\partial y_2}-\frac{\partial\varphi}{\partial x_2}\frac{\partial\varphi}{\partial y_1},$$

34) This condition is also necessary and sufficient for φ to be subharmonic on every regular characteristic surface.

we have

$$L(\varphi)=Ad+Da-2Bb-2Cc.$$

In what follows, we shall understand by a, b, c, d and A, B, C, D, their values at the point P. By hypothesis, at least one of a, d is non-zero. Suppose, to fix our ideas, that

$$d>0.$$

We have

$$ad-(b^2+c^2)=0.$$

If we set

$$\alpha=\frac{-b}{d}, \quad \beta=\frac{-c}{d},$$

we therefore have

$$\frac{L(\varphi)}{d}=A+2B\alpha+2C\beta+D(\alpha^2+\beta^2).$$

Let us look at the form $U(\alpha,\beta)$ of No. 12. If $D>0$, the lower bound m is given by $mD=V$. Consequently, again if $V>0$, we always have $U(\alpha,\beta)>0$. Hence we have $L(\varphi)>0$.

Thus $d>0$ and $L(\varphi)>0$ at P. Thanks to E.E. Levi, we can therefore find a characteristic surface* of the form $y=f(x)$ which passes through P and such that we have $\varphi>\varphi(P)$ on the surface near P, excepting P itself; here $f(x)$ is a function holomorphic in a neighbourhood of the projection of P on the x-plane; we can even choose $f(x)$ among polynomials of second degree in x. In the case $a>0$, the situation is entirely similar.

We shall now study the *exceptional case*. Suppose that we have

$$\frac{\partial\varphi}{\partial x_1}=\frac{\partial\varphi}{\partial x_2}=\frac{\partial\varphi}{\partial y_1}=\frac{\partial\varphi}{\partial y_2}=0$$

at the point P. Suppose, to simplify notation that P is the origin and that $\varphi(P)=0$. We take a characteristic plane of the form

$$y=ax, \quad a=\alpha+i\beta$$

and consider

$$\Phi(x_1,x_2)=\varphi(x_1,x_2,y_1,y_2),$$

where

$$y_1=\alpha x_1-\beta x_2, \quad y_2=\beta x_1+\alpha x_2.$$

Since φ has partial derivatives up to second order, we have

$$\Phi=\tfrac{1}{2}(\mu_1 x_1^2+2\mu_2 x_1 x_2+\mu_3 x_2^2)+(\varepsilon_1 x_1^2+2\varepsilon_2 x_1 x_2+\varepsilon_3 x_2^2)$$

near $x=0$, where μ_1,μ_2,μ_3 are real constants depending on a, and $\varepsilon_1,\varepsilon_2,\varepsilon_3$ are well-defined real functions of x_1,x_2 tending to 0 with x. In order that the characteristic plane $y=ax$ does not pass through any point of $\varphi\leqq 0$ near the origin, excepting the origin, it is sufficient that $\Phi>0$ in a sufficiently small circle of the x-plane around the origin, except at the centre. And for this, it is sufficient that

$$(1)\qquad \mu_1>0, \quad \mu_1\mu_3-\mu_2^2>0.$$

By a simple calculation we have

$$(2)\qquad \mu_1=\frac{\partial^2\varphi}{\partial y_1^2}\alpha^2+\frac{\partial^2\varphi}{\partial y_2^2}\beta^2+2\frac{\partial^2\varphi}{\partial y_1\,\partial y_2}\alpha\beta+2\frac{\partial^2\varphi}{\partial x_1\,\partial y_1}\alpha+2\frac{\partial^2\varphi}{\partial x_1\,\partial y_2}\beta+\frac{\partial^2\varphi}{\partial x_1^2},$$

$$\mu_3=\frac{\partial^2\varphi}{\partial y_2^2}\alpha^2+\frac{\partial^2\varphi}{\partial y_1^2}\beta^2-2\frac{\partial^2\varphi}{\partial y_1\,\partial y_2}\alpha\beta+2\frac{\partial^2\varphi}{\partial x_2\,\partial y_2}\alpha-2\frac{\partial^2\varphi}{\partial x_2\,\partial y_1}\beta+\frac{\partial^2\varphi}{\partial x_2^2},$$

$$\mu_2=\frac{\partial^2\varphi}{\partial y_1\,\partial y_2}(\alpha^2-\beta^2)-\left(\frac{\partial^2\varphi}{\partial y_1^2}-\frac{\partial^2\varphi}{\partial y_2^2}\right)\alpha\beta+\left(\frac{\partial^2\varphi}{\partial x_1\,\partial y_2}+\frac{\partial^2\varphi}{\partial x_2\,\partial y_1}\right)\alpha$$

$$-\left(\frac{\partial^2\varphi}{\partial x_1\,\partial y_1}-\frac{\partial^2\varphi}{\partial x_2\,\partial y_2}\right)\beta+\frac{\partial^2\varphi}{\partial x_1\,\partial x_2},$$

where it is understood that the derivatives represent their values at the origin.

Let us consider (α,β) as being a point in the plane by means of rectangular coordinates in cartesian geometry. We shall distinguish 2 cases according to the form of the equation $\mu_2=0$. If

$$\frac{\partial^2\varphi}{\partial y_1\,\partial y_2}=0, \quad \frac{\partial^2\varphi}{\partial y_1^2}=\frac{\partial^2\varphi}{\partial y_2^2},$$

then, we have

$$\frac{\partial^2\varphi}{\partial y_1^2}>0, \quad \frac{\partial^2\varphi}{\partial y_2^2}>0,$$

because $D>0$; consequently, condition (1) is satisfied by the point $(\alpha,0)$ if we take α sufficiently large. *Let us therefore suppose that*

$$\left(\frac{\partial^2\varphi}{\partial y_1^2}-\frac{\partial^2\varphi}{\partial y_2^2}\right)^2+4\left(\frac{\partial^2\varphi}{\partial y_1\,\partial y_2}\right)^2>0.$$

The equation $\mu_2=0$ then represents a hyperbola, or, as a limiting case, a pair of (real) lines crossing each other. As for the equations $\mu_1=0$, $\mu_3=0$, they are effectively of second degree since $\partial^2\varphi/\partial y_1^2$ and $\partial^2\varphi/\partial y_2^2$ cannot both vanish. We first find:

$$\mu_1+\mu_2=D(\alpha^2+\beta^2)+2B\alpha+2C\beta+A.$$

$$(3)\qquad \therefore\ \mu_1+\mu_2>0.$$

Suppose first that $\mu_2=0$ is a hyperbola. We let (α,β) trace out one of the branches of the hyperbola and look at the sign of μ_1. If μ_1 changes sign at a point M, then condition (1) is certainly satisfied at a suitable point close to M because $\mu_2=0$ and $\mu_3>0$ at M by (3). The same applies to the sign of μ_3, even in the case when $\mu_2=0$ represents a pair of lines. We suppose, therefore, that if $\mu_2=0$ represents a hyperbola, μ_1 *and* μ_3 *do not change sign* on each of the branches of the hyperbola, and if $\mu_2=0$ represents a pair of lines, this is so on the whole curve. Under these conditions, we shall prove the existence of a point (α, β) satisfying condition (1). This is sufficient.

We shall denote by

$$\lambda_1(\alpha, \beta), \quad \lambda_2(\alpha, \beta), \quad \lambda_3(\alpha, \beta)$$

the terms of second degree in α, β of μ_1, μ_2, μ_3 respectively.

Consider the transformation

$$\alpha' = -\beta, \quad \beta' = \alpha,$$

which is a rotation of the plane about the origin by $\pi/2$. We have

$$\lambda_2(\alpha', \beta') = -\lambda_2(\alpha, \beta).$$

Hence, $\mu_2 = 0$ represents either a hyperbola or 2 *perpendicular* lines. We then have

$$\lambda_1(\alpha', \beta') = \lambda_3(\alpha, \beta), \quad \lambda_3(\alpha', \beta') = \lambda_1(\alpha, \beta),$$

and once again

$$\lambda_1 + \lambda_3 = D(\alpha^2 + \beta^2).$$

Hence, the portion $\lambda_1 > 0$ has *a bigger angle* at the origin than the portion $\lambda_1 < 0$; the same holds for λ_3. Let us consider the case when $\mu_2 = 0$ represents a rectangular hyperbola. If the point (α, β) moves out indefinitely in a certain sense along any one of the branches of the hyperbola, this point cannot remain in the portion $\lambda_1 \leqq 0$ in view of what we have seen above; consequently, there is at least one point on this branch for which $\mu_1 > 0$. We therefore have $\mu_1 \geqq 0$ everywhere on $\mu_2 = 0$ by our hypothesis above. The same is true of μ_3, and applies also to the case when $\mu_2 = 0$ represents 2 perpendicular lines. We thus have $\mu_1 \geqq 0$, $\mu_3 \geqq 0$ everywhere on $\mu_2 = 0$. The curve $\mu_2 = 0$ cannot coincide with either $\mu_1 = 0$ or $\mu_3 = 0$, since the asymptotes are different. We can therefore find a point (α, β) satisfying the conditions $\mu_1 > 0$, $\mu_3 > 0$, $\mu_2 = 0$, and consequently the required condition $\mu_1 \mu_3 - \mu_2^2 > 0$.

Let $\varphi(x_1, x_2, y_1, y_2)$ be a continuous real function having continuous partial derivatives up to second order, and such that

$$\frac{\partial^2 \varphi}{\partial x_1^2} + \frac{\partial^2 \varphi}{\partial x_2^2} > 0, \quad V(\varphi) > 0$$

in a domain $\mathfrak{D}$. For any point P of $\mathfrak{D}$, we can find a characteristic surface passing through P, regular at P, and remaining in the portion $\varphi > \varphi(P)$ in a neighbourhood of P, P itself being excluded.

14. We shall first make some remarks on the above proposition. Let us consider a hypersphere S around a point P in the interior of the domain $\mathfrak{D}$, and the portion $\varphi < \varphi(P)$ in S. If it is non-empty, we shall denote one of its connected components by Δ. We shall prove that Δ is a domain of holomorphy.

Consider the (univalent) domains of holomorphy containing Δ. Thanks to H. CARTAN-P. THULLEN[35], there exists a smallest such domain, which we shall call G. G is necessarily contained in S. *Suppose that* $\Delta \neq G$. Because of the form of Δ, G contains at least one point at which φ takes the value $\varphi(P)$. Let α be the upper bound of the values taken by φ on G; φ cannot attain α in the domain G by virtue of the preceding proposition. We therefore have $\varphi(P) < \alpha$.

Let Q be a boundary point of G such that $\varphi(Q) = \alpha$. By the preceding proposition, there exists a characteristic surface $F(x, y) = 0$ passing through Q and remaining in the portion $\varphi > \alpha$ near Q, except for Q itself. Let us draw a hypersphere (δ) around Q so small that F is holomorphic in a neighbourhood of (δ), that the intersection of $F = 0$ with the boundary of (δ) lies in the exterior of G, and that (δ) contains no point of Δ; (δ) certainly exists. Let us take a positive number ε so small that the intersection of the hypersurface $|F| = \varepsilon$ with the boundary of (δ) is also situated in the exterior of G; we consider the set E composed of the points of G which either do not belong to (δ) or satisfy the condition $|F| > \varepsilon$. E is an open set containing Δ; every connected component of E is pseudoconvex in the sense of H. CARTAN and is consequently a domain of holomorphy. This contradicts the definition of G. We therefore have $\Delta = G$. Thus, we have:

Under the conditions of the preceding proposition, if S is a hypersphere about the point P contained in the interior of the domain $\mathfrak{D}$, the portion of the set $\varphi < \varphi(P)$ in S, if it is not empty, consists of domains of holomorphy.

Let $\varphi(x, y)$ be a real *continuous* function in a domain $\mathfrak{D}$. Let (x', y') be a point of $\mathfrak{D}$ such that the dicylinder (γ):

$$|x - x'| < r, \qquad |y - y'| < r$$

is contained in the interior of $\mathfrak{D}$, but otherwise arbitrary; here r is a radius given à priori. We shall denote by $\varphi_1(x', y')$ the average of the values taken by φ on (γ). The function $\varphi_1(x, y)$ is defined on $\mathfrak{D}_1 = \mathfrak{D}^{(r)}$ (the set of points of $\mathfrak{D}$ such that the "boundary distances" with respect to $\mathfrak{D}$ are larger than r). We denote this operation by

$$\varphi_1(x, y) = A_r[\varphi(x, y)];$$

this is analogous to a well-known operation in the plane[36]. From this, the following properties of A_r will be obvious:

$A_r(\varphi)$ has continuous partial derivatives of the first order in x_1, x_2, y_1, y_2.

If φ has continuous partial derivatives of the first order, then $A_r(\varphi)$ has continuous partial derivatives of the second order.

Consider

$$A_r[A_r(\varphi)] = A_r^2(\varphi).$$

The function $A_r^2(\varphi)$ is defined and continuous on $\mathfrak{D}_2 = \mathfrak{D}_1^{(r)}$ and has continuous partial derivatives with respect to x_1, x_2, y_1, y_2 up to second order, by the above remarks.

[35]) See: H. CARTAN-P. THULLEN cited above.

[36]) See e.g.: F. RIESZ, cited earlier.

When r tends to 0, $\mathfrak{D}_1$ and $\mathfrak{D}_2$ tend to $\mathfrak{D}$, and $A_r(\varphi)$ and $A_r^2(\varphi)$ tend to φ uniformly in the interior of $\mathfrak{D}$.

If the function $\varphi(x, y)$ is *pseudoconvex*, then $A_r(\varphi)$ is also pseudoconvex by properties 1,3 and consequently, so is $A_r^2(\varphi)$.

Let $\psi(x, y)$ be a continuous pseudoconvex function having continuous partial derivatives with respect to x_1, x_2, y_1, y_2, up to second order. By what we have seen in No. 12, we then have

$$\frac{\partial^2 \psi}{\partial x_1^2} + \frac{\partial^2 \psi}{\partial x_2^2} \geqq 0, \qquad \frac{\partial^2 \psi}{\partial y_1^2} + \frac{\partial^2 \psi}{\partial y_2^2} \geqq 0, \qquad V(\psi) \geqq 0$$

on $\mathfrak{D}$. Let us consider

$$\chi(x, y) = \psi(x, y) + \varepsilon(x_1^2 + y_1^2),$$

ε being a positive number, however small. We find that

$$\frac{\partial^2 \chi}{\partial x_1^2} + \frac{\partial^2 \chi}{\partial x_2^2} > 0, \qquad V(\chi) > 0.$$

Recapturing, we have thus proved:

Let $\varphi(x, y)$ be a continuous pseudoconvex function in a domain $\mathfrak{D}$. Given a positive number ε and a domain $\mathfrak{D}_0$ completely interior to $\mathfrak{D}$, there exists a continuous pseudoconvex function $\chi(x_1, x_2, y_1, y_2)$ in $\mathfrak{D}_0$, having continuous partial derivatives up to second order, and such that

$$|\varphi - \chi| < \varepsilon, \qquad \frac{\partial^2 \chi}{\partial x_1^2} + \frac{\partial^2 \chi}{\partial x_2^2} > 0, \qquad V(\chi) > 0.$$

Let us now consider a pseudoconvex domain $\mathfrak{D}$ in the space (x, y). Let us first suppose that $\mathfrak{D}$ is bounded. Consider the function

$$\varphi(x, y) = -\log d(x, y)$$

of No. 11, where $d(x, y)$ is the distance of a point (x, y) in $\mathfrak{D}$ to the boundary of $\mathfrak{D}$. Since $\varphi(x, y)$ is continuous and pseudoconvex, we obtain a function χ having the properties indicated in the above proposition. To determine $\mathfrak{D}_0$ and ε, we take 3 real numbers such that

$$\alpha < \beta < \gamma,$$

and we really have points satisfying $\varphi < \alpha$. We choose $\mathfrak{D}_0$ sufficiently large so as to contain the set $\varphi < \gamma$ completely in its interior; $\mathfrak{D}_0$ certainly exists since, $\mathfrak{D}$ being bounded, the set $\varphi < \gamma$ is completely interior to $\mathfrak{D}$. We choose ε so small that the set $\chi < \beta$ contains the set $\varphi < \alpha$ and is contained in the set $\varphi < \gamma$. By the first of the preceding propositions, the set $\chi < \beta$ consists of domains pseudoconvex in the sense of H. Cartan, and consequently of domains of holomorphy. When α tends to $+\infty$, the set $\varphi < \alpha$ tends to $\mathfrak{D}$. By virtue of the *Theorem of H. Behnke-K. Stein*, $\mathfrak{D}$ is therefore a domain of holomorphy.

When $\mathfrak{D}$ is not bounded, we can find, as we have seen often, an increasing sequence of bounded pseudoconvex domains tending to $\mathfrak{D}$. By the result above, $\mathfrak{D}$ is therefore a domain of holomorphy.

Theorem. *In the space of 2 complex variables, every univalent, finite, pseudoconvex domain is a domain of holomorphy.*

The author believes that this conclusion too is true independently of the number of complex variables.

Commentaire de H. Cartan

Soit D un domaine de $\mathbb{C}^n$. Hartogs et Levi ont donné des conditions *nécessaires* pour que D soit domaine d'holomorphie; ces conditions ont un caractère *local* au voisinage de chaque point-frontière de D. Le problème était resté ouvert de savoir si réciproquement ces conditions entraînent que D est un domaine d'holomorphie. OKA se propose de résoudre ce problème; pour simplifier il se borne au cas $n=2$. Le cas général a été ensuite résolu par BREMERMANN et par NORGUET (indépendamment l'un de l'autre).

OKA prouve d'abord le théorème suivant (n° 9): si un domaine $D \subset \mathbb{C}^2$ est "pseudo-convexe au sens de H. CARTAN" (i.e.: si tout point-frontière de D possède un voisinage ouvert V tel que $V \cap D$ soit un domaine d'holomorphie), alors D est un domaine d'holomorphie. Cela résulte de l'énoncé suivant (que nous appelons "lemme de recollement"): "Soit D un domaine borné de $\mathbb{C}^2$ (coordonnées complexes x et y) tel que l'intersection de D avec l'ouvert $\operatorname{Im} y<1$ et l'intersection de D avec l'ouvert $\operatorname{Im} y>-1$ soient réunions disjointes de domaines d'holomorphie; alors D est un domaine d'holomorphie."

La preuve du lemme de recollement met en jeu des techniques assez compliquées (y compris l'intégrale d'ANDRÉ WEIL), mais les outils techniques seront repris dans le Mémoire IX d'OKA (n°s 26 à 29). Au cours de la démonstration, OKA utilise plusieurs fois le théorème de BEHNKE-STEIN qui dit que la réunion d'une suite croissante de domaines d'holomorphie est un domaine d'holomorphie.

Une fois prouvé le théorème du n° 9, il reste à en déduire le résultat qui fait l'objet essentiel du Mémoire VI, à savoir: "tout domaine $D \subset \mathbb{C}^2$, pseudo-convexe au sens de Hartogs, est un domaine d'holomorphie". C'est à cette occasion que OKA introduit la notion de "fonction pseudo-convexe", introduite indépendamment par P. LELONG sous la dénomination de "fonction plurisousharmonique", devenue classique, et que nous adoptons dans le présent commentaire.

Rappelons d'abord la définition: une fonction φ dans un ouvert de $\mathbb{C}^n$ est *plurisous-harmonique* si elle est à valeurs réelles (y compris la valeur $-\infty$), semi-continue supérieurement, et si sa restriction à tout morceau de droite complexe est sous-harmonique (alors sa restriction à tout morceau de sous-variété analytique de dimension un est sous-harmonique). Si φ est de classe C^2, la condition de plurisous-harmonicité est la suivante: en chaque point $z \in \mathbb{C}^n$ de l'ouvert où φ est définie, la forme hermitienne (en les variables complexes ξ_i)

$$\sum_{i,j} \frac{\partial^2 \varphi}{\partial z_i \partial \bar{z}_j} \xi_i \bar{\xi}_j$$

est $\geqq 0$. Si cette forme est partout définie positive, on dit que φ est *strictement* plurisous-harmonique. On doit à E.E. LEVI le résultat suivant: si D est un domaine qui, au voisinage de chacun des points P de la frontière ∂D, est défini par une inégalité $\varphi(z) < \varphi(P)$ (avec φ strictement plurisous-harmonique au voisinage de P), alors D est "pseudo-convexe au sens de H. CARTAN"; il en résulte que D est un domaine d'holomorphie, grâce au théorème prouve par OKA au n°9. La méthode utilisée par OKA va consister à ramener le cas général à ce cas particulier.

Il s'agit de prouver que tout domaine D, pseudo-convexe au sens de HARTOGS, est un domaine d'holomorphie. Bien sûr il faut d'abord préciser ce qu'on entend par "pseudo-convexe au sens de HARTOGS". Or la définition que donne OKA (n° 10 en note de bas de page) est un peu bizarre et diffère de la définition usuelle; Oka affirme qu'elle lui est équivalente.

Pour clarifier la situation, nous allons rappeler la définition de HARTOGS lui-même. Pour cela, introduisons quelques notations: K désignera le disque-unité compact $|z| \leqq 1$ dans $\mathbb{C}$, ∂K son bord $|z|=1$, et $0 \in K$ son centre $z=0$. On notera a_0 un point de $\mathbb{C}^{n-1}$, et λ désignera n'importe quel isomorphisme analytique d'un voisinage ouvert (non précisé) de $K \times \{a_0\}$ dans $\mathbb{C}^n$ sur un ouvert de $\mathbb{C}^n$; B désignera une boule ouverte de $\mathbb{C}^{n-1}$, de centre a_0, telle que $K \times B$ soit contenu dans l'ouvert de définition de λ. On dit que D est *pseudo-convexe au sens de Hartogs* si D satisfait à la condition

($H(D)$) *Si* $\lambda(\partial K \times B) \subset D$ *et* $\lambda(K \times \{a_0\}) \subset D$, *alors* $\lambda(K \times B) \subset D$.

Il est clair que $H(D)$ et $H(D')$ entraînent $H(D \cap D')$. Comme toute boule ouverte de $\mathbb{C}^n$ est pseudo-convexe au sens de Hartogs, on voit que tout D pseudo-convexe au sens de Hartogs est réunion d'une suite croissante d'ouverts pseudo-convexes au sens de Hartogs et *bornés*.

HARTOGS a introduit la condition $H(D)$ en prouvant que tout domaine d'holomorphie satisfait à cette condition.

En fait, la propriété $H(D)$ ne dépend que des propriétés *locales* de D au voisinage des points de la frontière ∂D. En effet, disons que D est *pseudo-convexe en un point-frontière* $P \in \partial D$ s'il existe un voisinage ouvert V de P satisfaisant à la condition suivante:

($H(D, P, V)$) *Si* $\lambda(K \times B) \subset V$, $\lambda(0, a_0) = P$, *et s'il existe des* $a \in B$ *arbitrairement voisins de* a_0 *tels que* $\lambda(K \times \{a\}) \subset D$, *alors* $\lambda(\partial K \times \{a_0\})$ *rencontre la frontière* ∂D.

On peut montrer: pour que D soit pseudo-convexe au sens de Hartogs, il faut et il suffit que D soit pseudo-convexe en chaque point P de ∂D. Nous dirons désormais simplement: D est pseudo-convexe.

Ces définitions étant posées, revenons, avec OKA, au cas $n=2$. On veut montrer que tout domaine D pseudo-convexe est un domaine d'holomorphie. En vertu du théorème de BEHNKE-STEIN, il suffit de le prouver lorsque D est pseudo-convexe et *borné*.

Soit D un domaine de $\mathbb{C}^2$, et soit $(x_0, y) \in D$. Pour chaque x assez voisin de x_0, on a $(x, y) \in D$, et on peut considérer la borne supérieure des $r>0$ tels que D contienne les (x, y') satisfaisant à $|y'-y| \leqq r$. Soit $R_y(x)$ ce nombre (qui peut être infini si D n'est pas borné). HARTOGS a montré que si D est pseudo-convexe, $-\log R_y(x)$ est une fonction sous-harmonique de x.

De là OKA déduit (n° 11): si D est pseudo-convexe, la distance $d_D(x, y)$ d'un point $(x, y) \in D$ à la frontière ∂D est une fonction continue, et $\varphi_D(x, y) = -\log d_D(x, y)$ est plurisous-harmonique dans D. Supposons désormais D *pseudo-convexe et borné*. L'ensemble des $(x, y) \in D$ tels que $\varphi_D(x, y) \leqq a$ (a réel quelconque) est un compact K_a. Soit D_a l'ouvert $\varphi_D(x, y) < a$; ayant choisi une fois pour toutes un point $(x_0, y_0) \in D$, à tout $a > \varphi_D(x_0, y_0)$ associons la composante connexe D'_a de D_a qui contient (x_0, y_0). D est réunion des D'_a. On va montrer que chaque D'_a est contenu dans un domaine d'holomorphie Δ lui-même contenu dans un D_b, et par suite contenu dans D'_b. Il s'ensuivra que D est réunion d'une suite croissante de domaines d'holomorphie, et est donc un domaine d'holomorphie (d'après BEHNKE-STEIN).

Donnons-nous un $b>a$, et soit $\varepsilon = (b-a)/2$. Un procédé classique de régularisation permet d'approcher uniformément sur K_b, à ε près, la fonction φ_D par une fonction plurisous-harmonique ψ de classe C^2, qu'on peut même supposer *strictement* plurisous-harmonique dans un voisinage de K_b. Soit $c = (a+b)/2$. L'ouvert défini par $\psi(x, y) < c$ contient D_a et est contenu dans D_b; d'après LEVI, cet ouvert est "pseudo-convexe au sens de H. CARTAN", donc ses composantes connexes sont des domaines d'holomorphie d'après OKA (n° 9 du présent mémoire). La composante connexe Δ qui contient D'_a est un domaine d'holomorphie contenu dans D'_b; ceci achève la démonstration.

Une dernière remarque: le raisonnement d'OKA (n° 14) est incorrect, car l'enveloppe d'holomorphie d'un domaine univalent n'est pas nécessairement univalente.

VII. On Some Arithmetical Notions

Sur quelques notions arithmétiques

Bulletin de la Société Mathématique de France **78** (1950), p. 1–27

Introduction. We are now embarked on thinking hard about the subject; among other things, understanding the nature of the difficulties we have met on the path that we have followed[1], and studying the form of the difficulties we shall meet in continuing. We present here one of the results of this reflection.

It will be noticed that there are already certain arithmetical notions in Theorem II of Memoir I (fundamental lemma), in Theorem I of Memoir II, and in condition (β) (WEIL's condition) in Memoir V. We shall meet another such notion when we allow points of ramification in our domains without which we could not even treat algebraic functions. This makes us begin by studying these notions.

Let us suppose that certain arithmetical notions, those of congruence and ideal for example, are transplanted from the field of polynomials to that of analytic functions. Since functions can no longer always be continued to all of (finite) space, we meet new problems. H. CARTAN discovered a phenomenon of this nature[2], and in the present memoir, we shall find, at the end, several theorems and a rather complex problem of the same nature (see No. 7). These theorems are indispensable to me to be able to treat the problems we have been interested in since Memoir I on domains containing ramification points; they are also useful for less complicated domains.

Having found ourselves face to face with the beautiful problems introduced by F. HARTOGS and his successors, we should like, in turn, to bequeath new problems to those who will follow us. The field of analytic functions of several variables happily extends into diverse branches of mathematics, and we might be permitted to dream of the many types of new problems in store for us[3].

"In the present memoir, we shall only treat *univalent domains without points at infinity*, and this condition will, in general, not be repeated."

1. Congruences and Equivalence. Theorem of H. Cartan. Among the problems generated by transplanting arithmetic notions to the field of analytic functions, there is *one type* of problem in which one asks to obtain *globally*

---

[1]) The preceding memoirs are: I. Rationally convex domains, 1936; II. Domains of holomorphy, 1937; III. The second COUSIN problem, 1939 (Journal of Science of Hiroshima University); IV. Domains of holomorphy and rationally convex domains, 1941; V. The CAUCHY integral, 1941 (Japanese Journal of Mathematics); VI. Pseudoconvex domains, 1942 (Tohoku Math. Journal).

[2]) H. CARTAN: Sur les matrices holomorphes de n variables complexes, 1940, p. 1–26 (Journal de mathématiques, Vol. 19); we owe much to this memoir.

[3]) The author would like to express here his sincere thanks to HUJU-KAI for his help since around the time of Memoir VI.

something we are given *locally* (as in the COUSIN problems); we shall collect together some problems of this type.

Let us consider two holomorphic functions f, φ in a domain D in the space of n complex variables $x_1, x_2, \ldots, x_n$; let $(F_1, F_2, \ldots, F_n)$ be a finite collection of holomorphic functions on D. The space of the variables will be denoted briefly by (x); the function f will also be denoted by $f(x)$.

Suppose that we have a relation of the form

$$f-\varphi=\alpha_1 F_1+\alpha_2 F_2+\ldots+\alpha_p F_p,$$

the α_i $(i=1,2,\ldots,p)$ being holomorphic functions on D; we shall then call the functions f, φ *congruent* with respect to (F) in D, and this relation will be denoted by

$$f \equiv \varphi \bmod (F).$$

We shall say that the functions f, φ are congruent *at a point* P of D, if they are congruent in a neighbourhood of P. Now, even if they are congruent at every point of D, they are not necessarily congruent globally on D; and one of the problems mentioned is then as follows:

Problem (C$_1$). *Suppose given a finite collection $(F_1, F_2, \ldots, F_p)$ of holomorphic functions and a holomorphic function $\Phi(x)$ in a neighbourhood of a closed (bounded) set E which satisfies the relation $\Phi(x) \equiv 0 \bmod (F)$ at every point of E; find p holomorphic functions $A_i(x)$ in a neighbourhood of E in such a way that $\Phi = \sum A_i F_i$ $(i=1,2,\ldots,p)$ identically.*

This problem is concerned with expressing the function in terms of others; there is also the problem of existence.

Problem (C$_2$). *Let us consider a finite collection of holomorphic functions $(F_1, F_2, \ldots, F_p)$ in the neighbourhood of a closed set E in the space (x). Suppose given, for any point P of E, a polycylinder (γ) around P and a holomorphic function $\varphi(x)$ on (γ). Suppose that for each pair $[(\gamma), (\gamma')]$ of contiguous polycylinders with intersection (δ), we have the relation $\varphi(x) \equiv \varphi'(x) \bmod (F)$ between the corresponding functions at every point of (δ). We ask to find a holomorphic function $\Phi(x)$ in a neighbourhood of E such that $\Phi(x) \equiv \varphi(x) \bmod (F)$ at every point P of E.*

We shall explain some terms related to the word "polycylinder". If E_i is a set in the x_i-plane, any set of the form $x_i \in E_i$ $(i=1,2,\ldots,n)$ will be called a *cylindrical set*, and the sets E_i themselves will be called the *components* of this set. A closed cylindrical set E will be called *simply connected* if, for each i, the complement of the component E_i is connected if we consider it on the Riemann sphere. A cylindrical set is called a polycylinder if all its components are circles.

Let us now consider two finite collections of holomorphic functions on a domain D in the space (x); we denote these by $(f_1, f_2, \ldots, f_p)$, $(\varphi_1, \varphi_2, \ldots, \varphi_q)$.

Suppose that we have the two relations

$$\varphi_i \equiv 0 \bmod (f), \quad f_j \equiv 0 \bmod (\varphi), \quad (i=1,2,\ldots,p; j=1,2,\ldots,q)$$

globally on D. We shall then say that the systems of functions (f), (φ) are equivalent in D; we denote these simultaneous relations by

$$(f) \sim (\varphi).$$

If we have only one of the two relations, for instance the first, we shall write this

$$(\varphi) \subset (f).$$

We shall call the systems (f), (φ) equivalent *at a point P* of D, if they are equivalent in some neighbourhood of P; similarly with the relation $(\varphi) \subset (f)$.

Regarding equivalence, we have the following

Problem (E). *Suppose that we have the geometric configuration of problem* (C_2), *and that for each* (γ), *we are given a finite system* (f) *of holomorphic functions, such that for each non-empty intersection* (δ) *we have the relation* $(f) \sim (f')$ *between the systems corresponding to the polycylinders* (γ), (γ') *at every point of* (δ). *We ask to find a finite system* (F) *of holomorphic functions in a neighbourhood of the closed set E such that* $(F) \sim (f)$ *at every point P of E.*

These are the problems we wished to collect together.

H. Cartan has proved the following theorem, relating to problem (E), in the memoir cited earlier:

H. Cartan's Theorem. *Consider in the space* (x), *two closed cylindrical sets* Δ', Δ'' *which have the same components in the planes of all the variables but one, and have a non-empty intersection* $\Delta' \cap \Delta'' = \Delta_0$ *which is simply connected. We suppose given a finite system* (f) *of holomorphic functions in a neighbourhood of* Δ' *a finite system* (f'') *of holomorphic functions in a neighbourhood of* Δ'' *and a global relation* $(f') \sim (f'')$ *in a neighbourhood of* Δ_0. *Under these circumstances, we can find a finite system* (f) *of holomorphic functions in a neighbourhood of the union* $\Delta = \Delta' \cup \Delta''$ *so that* $(f) \sim (f^{(i)})$ *in a neighbourhood of* $\Delta^{(i)}$ $(i=1,2)$.

In this paper, H. Cartan proved this theorem by carrying over the notion of *matrices* to analytic functions. If we denote by p, q, r the number of functions in the systems $(f'), (f''), (f)$ respectively, one always has

$$p < r, \quad q < r$$

by his method; he also paid much attention to this phenomenon.

We shall apply *Cartan's theorem* to problem (E). Let us consider a circle (α_i) in the x_i-plane $(i=1,2,\ldots,n-1)$ and three circles (β_j) $(j=1,2,3)$ in the x_n-plane which overlap two by two. We then consider the closed polycylinder Δ_j $(j=1,2,3)$ defined by having as components the closed circles $(\bar{\alpha}_i)$, $(\bar{\beta}_j)$ and suppose given a finite system (f_j) of holomorphic functions in the neighbourhood of Δ_j in such a way that any two of the systems (f_j) are globally equivalent in the neighbourhood of the corresponding intersection.

We shall try to apply CARTAN's theorem under these conditions. This is certainly possible for $\Delta=\Delta_1\cup\Delta_2$, and we find a finite system (f) of functions holomorphic in a neighbourhood of Δ such that $(f)\sim(f_j)$ globally in the neighbourhood of Δ_j $(j=1,2)$. But we cannot always apply the theorem to $\Delta\cup\Delta_3$ because, depending on the geometric configuration, it is not obvious whether $(f),(f_3)$ are globally equivalent in the neighbourhood of the corresponding intersection. However, if problem (C_1) is solvable for any closed polycylinder (i.e. on a neighbourhood), this second application of Cartan's theorem is clearly possible. From this, it follows that:

If the problem (C_1) *is solvable for any closed polycylinder, so is problem* (E).

The proof, being easy and standard, will be omitted. We shall explain the restriction to polycylinders. In the present memoir, *we shall treat all problems exclusively for closed polycylinders*; intrinsic properties of the problems themselves make it possible to extend them to less restricted closed domains; in addition, of course, the notion of polycylinder is simple.

As for the relation between problems (C_1) and (C_2), we have the following:

Consider the same geometric configuration as in Cartan's theorem, except that the intersection Δ_0 *is no longer required to be simply connected. Suppose given a finite system* $(F_1, F_2, \dots, F_p)$ *of functions holomorphic in a neighbourhood of* Δ *such that the problem* (C_1) *with respect to* (F) *is solvable in the neighbourhood of* Δ_0. *Under these conditions, if we are given holomorphic functions* f' *in the neighbourhood of* Δ', f'' *in the neighbourhood of* Δ'', *such that* $f'-f''\equiv 0 \bmod (F)$ *at every point of a neighbourhood of* Δ_0, *we can find a holomorphic function* f *in a neighbourhood of* Δ *for which* $f\equiv f' \bmod (F)$ *at each point in a neighbourhood of* Δ', $f\equiv f'' \bmod (F)$ *at each point near* Δ''.

In fact, the function

$$\varphi(x)=f'(x)-f''(x)$$

is holomorphic in the neighbourhood of Δ_0 and we have

$$\varphi(x)\equiv 0 \bmod (F)$$

at every point of Δ_0; solving problem (C_1), φ can be represented as

$$\varphi=\alpha_1 F_1+\alpha_2 F_2+\dots+\alpha_p F_p,$$

where the α_i $(i=1,2,\dots,p)$ are holomorphic functions in a neighbourhood of Δ_0.

Since Δ_0 is cylindrical, we can use *the Cousin integral* to find holomorphic functions, $a_i(x)$ in a neighbourhood of Δ', $b_i(x)$ in one of Δ'' in such a way that

$$a_i(x)-b_i(x)=\alpha_i(x) \qquad (i=1,2,\dots,p)$$

identically; let us form the functions

$$\psi' = f' - (a_1 F_1 + \dots + a_p F_p), \qquad \psi'' = f'' - (b_1 F_1 + \dots + b_p F_p);$$

ψ' is then holomorphic in a neighbourhood of Δ', ψ'' in a neighbourhood of Δ''. Now, for Δ_0, we have

$$\psi' - \psi'' = \varphi - (\alpha_1 F_1 + \dots + \alpha_p F_p) = 0;$$

ψ' and ψ'' are therefore parts of the same function $f(x)$ which is holomorphic in a neighbourhood of Δ; we have $f = \psi' \equiv f' \bmod (F)$ in a neighbourhood of Δ', and similarly for Δ''. Q.E.D.

From this we immediately deduce:

If the problem (C_1) *is solvable for any closed polycylinder, so is problem* (C_2).

Problems (C_1), (C_2), (E) *are therefore reduced to problem* (C_1).

One will recognize in Theorem II of Memoir I, Theorem I of Memoir II and condition (β) of Memoir V problems (C_2), (E) and (C_1) respectively.

2. Ideals with Indeterminate Domains. We shall now carry over the notation of "ideal" to analytic functions, as did also H. CARTAN.

Let us consider a domain D in the space (x) and a set (I) of functions holomorphic on D. Suppose (I) has the following two properties:

1°. If $f(x) \in (I)$ and $\alpha(x)$ is a holomorphic function on D, then $\alpha f \in (I)$;

2°. If $f'(x) \in (I)$, $f''(x) \in (I)$, then $f' + f'' \in (I)$.

We shall call (I) a *holomorphic ideal with domain D*; but we shall investigate further.

Consider domains (δ) (connected or not) and holomorphic functions $f(x)$ on (δ); we consider a set (I) of pairs (f, δ). Instead of saying that $(f, \delta) \in (I)$, we shall sometimes say that $f \in (I)$ on δ. Suppose that this set has the following two properties:

1°. If $(f, \delta) \in (I)$ and $\alpha(x)$ is a holomorphic function on the domain δ' (connected or not), then, we have, $\alpha f \in (I)$ on $\delta \cap \delta'$;

2°. If $(f, \delta) \in (I)$, $(f', \delta') \in (I)$, then, we have, $f + f' \in (I)$ on $\delta \cap \delta'$.

We shall call (I) a *holomorphic ideal with indeterminate domains.* We shall sometimes abbreviate this to ideal with indeterminate domains, or, simply, ideal.

It follows from the definition that:

If $(f, \delta) \in (I)$ *for the ideal* (I) *and* $\delta \supset \delta'$, *we necessarily have* $(f, \delta') \in (I)$.

We shall, therefore, say that a function belongs to the ideal (I) with indeterminate domains *at a point P* if it belongs to (I) on a neighbourhood of P.

Let us consider the following properties of an ideal (I) with indeterminate domains:

Property (T_1). If $(f,\delta)\in(I)$, $(f,\delta')\in(I)$, we necessarily have $f\in(I)$ on $\delta\cup\delta'$.

Property (T_2). If $\delta_1\subset\delta_2\subset\ldots$ and $(f,\delta_1)\in(I)$, $(f,\delta_2)\in(I),\ldots$ and δ_0 is the limit of the sequence $\delta_1,\delta_2,\ldots$, then, we necessarily have $(f,\delta_0)\in(I)$.

Let us examine these notions.

Example 1. Let us consider the circle (C) around the origin, of radius 1, in the plane of a complex variable x; we consider in (C) an arbitrary domain δ of diameter d (which is the upper bound of the distances between two points of δ). Let (I) be the set of (f,δ) for which $d\leqq\frac{1}{2}$. (I) is an ideal *having property* (T_2) *but not property* (T_1).

Example 2. Consider, in the circle (C), a sequence of circles (C_i) $(i=1,2,\ldots)$ around the origin with radius r_i increasing and tending to 1. Let (I) be the set of (f,δ) where δ is contained in at least one circle (C_i) of the sequence. Then (I) is an ideal *with property* (T_1) *but not property* (T_2).

We shall now carry over the concept of "finite basis of an ideal" to an ideal with indeterminate domains and analyse this notion without referring to properties (T_1)(T_2) and to problem (C_1).

In the space (x), let us consider an ideal (I) with indeterminate domains, a domain D and a finite number of holomorphic functions $F_1(x), F_2(x),\ldots,F_p(x)$ on D. Suppose that they possess the following properties:

1°. Each of the functions F_i $(i=1,2,\ldots,p)$ belongs to (I) at every point of D.

2°. For any point P of D, if a function $f(x)$ belongs to (I) at P, we necessarily have $(f)\subset(F_1,F_2,\ldots,F_p)$ at P.

We call (F) a *finite pseudobasis of* (I) *on* D.

We now pose

Problem (I). *Given an ideal* (I) *with indeterminate domains in the space* (x) *and a closed set E, find a finite pseudobasis of* (I) *in a neighbourhood of E.*

We consider an ideal (I) with indeterminate domains and two domains D, D' in the space (x) such that $D'\subset D$. If a finite system (F) of holomorphic functions gives a finite pseudobasis of (I) on D, we find at once that it gives a finite pseudobasis also on D'.

We shall therefore say that a finite system (F) of holomorphic functions gives a pseudobasis of an ideal (I) with indeterminate domains *at a point P* if this is the case in a neighbourhood of the point P. We shall sometimes call (F) a *local pseudobasis.*

We can then pose also the following

Problem (J). *Given an ideal* (I) *with indeterminate domains in the space* (x) *and a point P in* (x), *find a finite pseudobasis of* (I) *at P.*

Problem (J) is a special case of problem (I); there is however the following relation between these two problems.

Consider, in the space (x), *a problem* (I) *relative to an ideal* (I) *and a closed set E which is a closed polycylinder. Suppose that, for any point P of E, the problem* (J) *for the ideal* (I) *at the point P is solvable. Suppose also that problem* (C_1) *is always solvable for a closed polycylinder. One then finds that problem* (I) *is solvable for the ideal* (I) *and the polycylinder E.*

In fact, consider any point P of E; the problem (J) being solvable at P for the ideal (I), one can find a polycylinder (γ) around P and a finite system (f) of holomorphic functions on (γ) such that (f) gives a pseudobasis of (I) on (γ). Let us examine a pair $[(\gamma),(\gamma')]$ of contiguous polycylinders. The corresponding systems $(f),(f')$ of functions are equivalent at each point of the intersection $(\gamma)\cap(\gamma')$ since they both give pseudobases. We want to find a finite system (F) of holomorphic functions in a neighbourhood of E such that $(F)\sim(f)$ at every point of E. Now, this is a problem (E) for the closed polycylinder E; because of our hypothesis that problem (C_1) is always solvable for a closed polycylinder, the system (F) exists as we have seen. Since $(F)\sim(f)$ at any point of E, (F) gives a pseudobasis at any point of E, thus is a pseudobasis for E. Q.E.D.

Problem (J) is thus the part of problem (I) not covered by problem (C_1). However, it is not always solvable, as seen by the following example.

Example 3. Let us consider two hyperspheres (C), (γ) about the origin in the space of two complex variables x, y, such that $(C)\supset(\gamma)$. We denote by Σ_0 the part of the characteristic surface $x=y$ which is either between the two hypersurfaces C, γ or an γ. [C, γ denote the boundaries of (C), (γ) respectively.] Consider the set (I) of pairs (f, δ), where $\delta\subset(C)$ and such that $f/(x-y)$ is holomorphic at every point of the intersection $\Sigma_0\cap\delta$. (I) is an ideal (having properties (T_1) and (T_2)), but since the set of zeros of $(f,\delta)\in(I)$ contains $\Sigma_0\cap\delta$, *this ideal cannot have finite local pseudobases at any point of* γ.

One cannot therefore always solve problem (J) without further conditions. We shall see further diverse kinds of counterexamples later.

We shall see that problem (J) on closed polycylinders can be solved for two kinds of ideals with indeterminate domains. One of them will be dealt with in the present memoir.

The other kind consists of *geometric ideals with indeterminate domains* (which correspond to the ideal of a variety in the space of polynomials). These are indispensable when we have to deal with domains admitting points of ramification. The proof that problem (J) is solvable on closed polycylinders for this kind of ideal requires, besides the results of the present memoir, some notions concerning domains with ramification. We shall therefore treat this question in a later memoir.

3. Homogeneous Linear Functional Equations and Formulae for Their Solutions. We shall scrutinize the circle of questions around problem (C_1) using the notions introduced above.

1°. *Homogeneous linear functional equations.* Let $F_i(x)$ $(i=1,2,\ldots,p)$ be p holomorphic functions, not identically zero, in a domain D in the space (x). Consider the equation

$$(1)\qquad A_1 F_1 + A_2 F_2 + \ldots + A_p F_p = 0,$$

where the A_i are unknown functions. A system of holomorphic functions $[A_1(x), A_2(x), \ldots, A_p(x)]$ on a domain δ (connected or not) contained in D which satisfies equation (1) identically in (δ) will be called a *(holomorphic) solution of the equation on* δ. We call any equation of this nature a *homogeneous linear functional equation.*

Given equation (1) consider for every solution $(A_1, A_2, \ldots, A_p)$ on δ the pair (A_1, δ); let (I_1) be the set of pairs so obtained. I say that (I_1) *is an ideal.*

For, if $(A_1,\delta)\in(I_1)$ and $\alpha(x)$ is holomorphic on a domain δ' (connected or not), we have $\alpha A_1 \in (I_1)$ on $\delta\cap\delta'$; if $(A_1,\delta)\in(I_1)$, $(A_1',\delta')\in(I_1)$, we have $A_1 + A_1' \in (I_1)$ on $\delta\cap\delta'$.

We do not see immediately whether (I_1) satisfies properties $(T_1), (T_2)$.

This was the kind of ideal we had in mind at the end of the preceding section. Problem (J) for these ideals is as follows:

Problem (K). *Given a homogeneous linear functional equation in a domain D in the space (x) and a point P of D, find a finite pseudobasis of the corresponding ideal (I_1) at P.*

2°. *Formulae for solutions.* We propose now to find a representation of the solutions of equation (1). We considered the ideal (I_1) relative to the equation (1). Let us now consider the equation

$$(2)\qquad A_2 F_2 + A_3 F_3 + \ldots + A_p F_p = 0,$$

and the set (I_2) of the corresponding pairs (A_2, δ). (I_2) is an ideal by what we saw above. We continue this process; we have finally to consider the equation $A_p F_p = 0$. The corresponding ideal $(I_p) = (0)$ [consists of the single element 0] since the function F_p is not identically zero.

We now pose the following composite problem:

Problem (λ). *Find finite pseudobases of the ideals $(I_1), (I_2), \ldots, (I_p)$ on any closed set E contained in the domain D consisting of functions which belong to the corresponding ideal globally in a neighbourhood of E.*

Let us suppose that problem (λ) has been solved. We then have pseudobases

$$(\Phi_1, \Phi_2, \ldots, \Phi_q), (\Phi_1', \Phi_2', \ldots, \Phi_r'), \ldots, (0)$$

of the ideals $(I_1), (I_2), \ldots, (I_p)$ respectively, of the kind indicated. Let P be an arbitrary point in the neighbourhood of E, and $[A_1(x), A_2(x), \ldots, A_p(x)]$ any solution of the Equation (1) at P (i.e. in a neighbourhood of P). (Φ) being a pseudobasis of (I_1), the function $A_1(x)$ can be represented, near P, in the form

$$A_1 = C_1 \Phi_1 + C_2 \Phi_2 + \dots + C_q \Phi_q,$$

the C_i $(i=1,2,\dots,q)$ being holomorphic functions.

Let us next consider $A_2(x)$. Since each Φ_i belongs globally to the ideal (I_1) in a neighbourhood of E, the Equation (1) has q solutions of the form $(\Phi_i, \Psi_{i2}, \dots, \Psi_{ip})$ $(i=1,2,\dots,q)$ in a neighbourhood of E. Let us set

$$B_j = C_1 \Psi_{1j} + C_2 \Psi_{2j} + \dots + C_q \Psi_{qj} \qquad (j=2,3,\dots,p),$$

and one obtains a solution of equation (1) of the form $(A_1, B_2, \dots, B_p)$ near P. If we now set $A'_j = A_j - B_j$, $(A'_2, A'_3, \dots, A'_p)$ is a solution of equation (2) near P; consequently, in a neighbourhood of P, the function $A_2(x)$ will be represented in the form

$$A_2 = B_2 + C'_1 \Phi'_1 + C'_2 \Phi'_2 + \dots + C'_r \Phi'_r,$$

where the C'_i $(i=1,2,\dots,r)$ are holomorphic functions, and B_2 can be represented as above. And we can continue thus.

We see now that the solution $[A_1(x), A_2(x), \dots, A_p(x)]$ of equation (1) can be represented, near P, as follows:

$$(3) \qquad A_i = C_{i1} \pi_{i1} + C_{i2} \pi_{i2} + \dots + C_{iN} \pi_{iN} \qquad (i=1,2,\dots,p)$$

with certain identities of the form $C_{ij} = C_{kl}$ $(k=1,\dots,p; l=1,\dots,N)$; here π_{ij} $(i=1,2,\dots,p; j=1,2,\dots,N)$ are certain well determined functions holomorphic in a neighbourhood of E, and the C_{ij} are holomorphic functions in a neighbourhood of P which depend on the solution $(A_1, A_2, \dots, A_p)$. For example, we have for A_1:

$$\pi_{1i} = \Phi_i, \qquad \pi_{1j} = 0 \qquad (i=1,2,\dots,q; j=q+1, q+2, \dots, N);$$

for A_2 we have

$$\pi_{2i} = \Psi_{i2}, \qquad \pi_{2j} = \Phi'_{j-q}, \qquad \pi_{2k} = 0$$
$$(i=1,2,\dots,q; j=q+1,\dots,q+r; k=q+r,\dots,N),$$

and

$$C_{1i} = C_{2i} \qquad (i=1,2,\dots,q).$$

Let us consider, conversely, any point P in a neighbourhood of E and an arbitrary collection (C) of functions holomorphic at P as in (3) (respecting, naturally, the identities indicated). To this collection (C), we make correspond a system (A) of holomorphic functions at P by (3); this system obviously satisfies the functional equation (1).

Suppose, in general, that to the equation (1), there corresponds a collection of functions π_{ij} $(i=1,2,\dots,p; j=1,2,\dots,N)$ holomorphic on a domain D' (connected or not) given in D, and identities of the form $C_{ij} = C_{kl}$ as in (3), which have the following properties:

1°. Any solution $[A_1(x), A_2(x), \dots, A_p(x)]$ of the equation (1) at an arbitrary point P of D' can be represented in the form (3), where the C_{ij} are suitable holomorphic functions at P (respecting, of course, the identities indicated).

2°. If C_{ij} are functions holomorphic at a point P of D' and satisfying the indicated identies, then the expression (3) gives a solution (A) of equation (1) at P. Under these conditions we shall say that *the expression* (3) *is a (linear) formula for the solutions of the functional equation* (1) *on the domain* D'. We shall call the collection (π) the *kernel* of this formula.

Thus, the problem we have just been discussing can be formulated as follows:

Problem (L). *Given a homogeneous linear functional equation in a domain and a closed set E in the domain, find a formula for the solutions of the equation in a neighbourhood of E.*

We have proved the following:

Lemma 1. *If problem* (λ) *is solvable for a certain homogeneous linear functional equation, so is the corresponding problem* (L).

3°. *Application to problem* (C_1). Problem (C_1) is related to finding a formula for the solutions of a homogeneous linear functional equation by the following proposition.

Lemma 2. *Consider two closed cylindrical sets Δ', Δ'' in the space* (x). *Suppose that they intersect, and that all their components but one are the same, and set $\Delta' \cup \Delta'' = \Delta$, $\Delta' \cap \Delta'' = \Delta_0$. Consider a holomorphic function Φ and a finite collection $(F_1, F_2, \ldots, F_p)$ of holomorphic functions on a neighbourhood of Δ. Suppose that we have relations*

$$\Phi = A_1' F_1 + A_2' F_2 + \ldots + A_p' F_p$$

and

$$\Phi = A_1'' F_1 + A_2'' F_2 + \ldots + A_p'' F_p$$

where A_i', A_i'' $(i=1,2,\ldots,p)$ are functions holomorphic in a neighbourhood of Δ', Δ'' respectively. Suppose also that we have a formula for the solutions of the functional equation

$$A_1 F_1 + A_2 F_2 + \ldots + A_p F_p = 0 \tag{4}$$

in the neighbourhood of Δ_0 which has the following properties: 1°. *The kernel of this formula consists of functions holomorphic on Δ;* 2°. *Any solution of the equation in a neighbourhood of Δ_0 can be represented globally in a neighbourhood of Δ_0 in the form* (3). *Under these conditions, Φ can be represented in the form*

$$\Phi = B_1 F_1 + B_2 F_2 + \ldots + B_p F_p,$$

where the B_i $(i=1,2,\ldots,p)$ are functions holomorphic in a neighbourhood of Δ.

In fact, let

$$A_i = \sum_j C_{ij} \pi_{ij} \quad (i=1,2,\ldots,p; j=1,2,\ldots,N) \tag{5}$$

with some identities among (C)

be the formula in question, having the properties indicated. Let us consider the functions

$$\gamma_i = A'_i - A''_i \quad (i=1,2,\ldots,p)$$

which are holomorphic in a neighbourhood of Δ_0 and satisfy there the equation (4). They can therefore be represented, globally in the neighbourhood of Δ_0, in the form

$$\gamma_i = \sum d_{ij}\pi_{ij},$$

the d_{ij} being holomorphic in a neighbourhood of Δ_0.

Since Δ_0 is cylindrical, we can find, using *the Cousin integral*, functions $a_{ij}(x)$, $b_{ij}(x)$ holomorphic in a neighbourhood of Δ', Δ'' respectively, such that

$$a_{ij} - b_{ij} = d_{ij}$$

identically. Further, since the functions (d) satisfy the identities in the formula (5) for solutions, we can choose (a) and (b) so as to satisfy them also. Let us then set

$$\alpha_i = \sum a_{ij}\pi_{ij}, \qquad \beta_i = \sum b_{ij}\pi_{ij};$$

we have

$$\alpha_i - \beta_i = \gamma_i;$$

moreover, we have

$$\sum \alpha_i F_i = 0, \quad \sum \beta_i F_i = 0$$

identically in a neighbourhood of Δ_0 by our requirements on the formula (5) for solutions of (4), and, since the functions (π) are holomorphic in a neighbourhood of Δ, the functions (α) are holomorphic in a neighbourhood of Δ', and satisfy there the equation (4); the functions (β) have similar properties relative to Δ''.

We set

$$B'_i = A'_i - \alpha_i, \qquad B''_i = A''_i - \beta_i \quad (i=1,2,\ldots,p).$$

The functions (B'), (B'') are holomorphic, respectively, in a neighbourhood of Δ', Δ'' and we have

$$\Phi = B'_1 F_1 + B'_2 F_2 + \ldots + B'_p F_p$$

in a neighbourhood of Δ',

$$\Phi = B''_1 F_1 + B''_2 F_2 + \ldots + B''_p F_p,$$

in a neighbourhood of Δ''. Now, since

$$B'_i - B''_i = (A'_i - A''_i) - (\alpha_i - \beta_i) = \gamma_i - \gamma_i = 0$$

for each i, the functions B'_i and B''_i are but parts of the same function B_i holomorphic in a neighbourhood of Δ. Q.E.D.

4°. *Application of the theorem of H. Cartan.*

Lemma 2. *Let us suppose that, in the geometric configuration of Lemma 2, the intersection Δ_0 is simply connected and has the property that the problem* (C_1) *is always solvable in a neighbourhood of Δ_0. We consider p holomorphic functions*

$F_i(x)$ $(i=1,2,\dots,p)$ in a neighbourhood of Δ, not identically zero, and the corresponding problem (λ). Suppose that problem (λ) is solved in neighbourhoods of Δ' and of Δ''. Then, problem (λ) is solvable in a neighbourhood of Δ.

If the conclusion in this lemma is established, we can use Lemma 1 to find a formula for the solution of equation (1) with the properties required in Lemma 2.

We shall prove this proposition by induction on p. For $p=1$, there is only the single ideal $(I_1)=0$ in problem (λ), and it is obviously solved, so that the proposition is true in this case. Let us therefore consider the case p, assuming the proposition true for $1,2,\dots,p-1$.

Let $(\Phi'_1, \Phi'_2, \dots, \Phi'_q)$ be a pseudobasis for (I_1) in a neighbourhood of Δ', such that each Φ'_i belongs globally to (I_1) on this neighbourhood; and let $(\Phi''_1, \Phi''_2, \dots, \Phi''_q)$ be a similar pseudobasis for Δ''. The systems (Φ'), (Φ'') of functions are equivalent at every point of Δ_0. Since, by hypothesis, any problem (C_1) is solvable for Δ_0, the equivalence is global. Since Δ_0 is simply connected, one finds, *by Cartan's theorem*, a system $(\Phi_1, \Phi_2, \dots, \Phi_s)$ of holomorphic functions in a neighbourhood of Δ such that $(\Phi)\sim(\Phi')$ globally for Δ', $(\Phi)\sim(\Phi'')$ globally for Δ''.

Since Φ'_i $(i=1,2,\dots,q)$ belongs globally to (I_1) in a neighbourhood of Δ', we have, globally on this neighbourhood,

$$\Phi'_i F_1 \equiv 0 \bmod (F_2, F_3, \dots, F_p) \qquad (i=1,2,\dots,q);$$

the same is therefore true for $\Phi_i F_1$ $(i=1,2,\dots,s)$, and an analogous result holds for Δ''. Now, since the functional equation

$$A_2 F_2 + A_3 F_3 + \dots + A_p F_p = 0$$

is in the case $(p-1)$, we can find a formula for its solutions, having the properties stated in Lemma 2 on a neighbourhood of Δ_0. By Lemma 2, we therefore have

$$\Phi_i F_1 \equiv 0 \bmod (F_2, F_3, \dots, F_s), \qquad (i=1,2,\dots,s)$$

globally on a neighbourhood of Δ, that is to say, the functions Φ_i belong globally to (I_1) on this neighbourhood.

The system (Φ) of functions, which has the above properties, is a pseudobasis of (I_1) in a neighbourhood of Δ' since $(\Phi)\sim(\Phi')$; the same holds for Δ''. (Φ) is thus a pseudobasis for (I_1) in a neighbourhood of Δ. The part of problem (λ) relating to the ideal (I_1) is thereby solved; the rest of problem (λ) is solvable by induction hypothesis, and the proposition is proved.

4. Reduction of the Problems to the Local Problem (K). Let us take a closed circle $(\bar{C}_i)$ in the x_i plane $(i=1,2,\dots,n)$, and consider the closed polycylinder having the $(\bar{C}_i)$ as components. We shall denote by E_0 an arbitrary point belonging to the closed polycylinder $(\bar{C})$.

We separate the variable x_n into its real and imaginary parts, $x_n = X + iY$, and consider a line of the form

$$x_i = x_i^0 \quad (i=1,2,\dots,n-1), \qquad Y = Y^0,$$

x_i^0 being complex numbers, Y^0, real. Let E_1 be the part of this line lying in the closed polycylinder $(\bar{C})$; in general, E_1 is a segment; sometimes it reduces to a point.

We introduce $E_2, E_3, \ldots$ similarly by successively raising the real dimension, and terminating with $E_{2n} \cdot E_2$, for example is a closed cylindrical set with a point of $(\bar{C}_i)$ as component in the x_i-plane for $i=1,2,\ldots,n-1$, and $(\bar{C}_n)$ as component in the x_n-plane; E_{2n} stands for the polycylinder $(\bar{C})$.

We shall apply the lemmas of No. 3 to this geometric configuration and find that problem (C_1) and (λ) can be solved successively for $E_0, E_1, \ldots, E_{2n}$, *so long as problem* (K) *is always solvable.*

Case of E_0. Problem (C_1) is trivially solvable in the neighbourhood of a point; by our hypothesis about problem (K), problem (λ) is also solvable.

Case of E_1. We consider any one of the sets E_1 and denote it by the same symbol E_1; we can suppose that it is a real segment.

1°. Consider holomorphic functions $F_1(x), F_2(x), \ldots, F_p(x)$, not identically zero, in a neighbourhood of E_1 and for which problem (λ) is solvable at every point of E_1.

The set E_1 is a cylindrical set whose component in $(x_1, x_2, \ldots, x_{n-1})$-space is a point which we denote by Q. The component in x_n-space is a segment which we denote by l; l is horizontal, and its left extremity will be denoted m_0, its right extremity by m_q; we introduce $(q-1)$ points $m_1, m_2, \ldots, m_{q-1}$ on this segment, from left to right, and denote the closed segment (m_{i-1}, m_i) by l_i $(i=1,2,\ldots,q)$. Let M_i be the cylindrical set (Q, m_i) in the space (x), L_i the set (Q, l_i). We set

$$L_1 \cup L_3 \cup L_5 \cup \ldots = \Delta_1,$$
$$L_2 \cup L_4 \cup L_6 \cup \ldots = \Delta_2;$$

we then have

$$\Delta = \Delta_1 \cup \Delta_2 = E_1,$$
$$\Delta_0 = \Delta_1 \cap \Delta_2 = M_1 \cup M_2 \cup \ldots \cup M_{q-1}.$$

Now, problem (λ) is solvable for the system (F) at every point of E_1; hence, if the segments L_i are sufficiently small, the problem is solvable in a neighbourhood of each of them *(as is obvious by the Borel-Lebesgue lemma).* It is therefore solvable for Δ_1 and for Δ_2. Since problem (C_1) is solvable for Δ_0, it follows from Lemma 3 that problem (λ) is always then solvable for E_1. (F) and (E_1) being arbitrary, *problem* (λ) *is always solvable in a neighbourhood of any segment* E_1.

2°. Let us now consider problem (C_1) for a given set E_1. Let $(F_1, F_2, \ldots, F_p)$ be a system of holomorphic functions, $f(x)$ a holomorphic function in a neighbourhood of E_1, and consider problem (C_1) for (F), f and E_1. Since this problem is solvable near any point of E_1, we can find, as in problem (λ) above, sets $\Delta_1, \Delta_2, \Delta$ and Δ_0 so that problem (C_1) is solvable for Δ_1 and for Δ_2. Now, as seen above, problem (λ) for the functions (F) is solvable in a neighbourhood of E_1 and any problem (C_1) is solvable in a neighbourhood of the intersection

Δ_0. Hence, by Lemma 1, we can find a formula for the solutions of the linear equation corresponding to the functions (F) having the properties indicated in Lemma 2. By Lemma 2, problem (C_1) is solvable for the set E_1. Since (F) and E_1 are arbitrary, *any problem* (C_1) *is solvable in the neighbourhood of any segment* E_1.

Case of E_2. We repeat the above argument.

1°. *Problem* (λ). Consider a set E_2 and denote by Q its component in $(x_1, x_2, \ldots, x_{n-1})$-space; its component in the x_n-plane will be $(\bar{C}_n)$. Let us also consider holomorphic functions $F_1(x), F_2(x), \ldots F_p(x)$, not identically zero, and the corresponding problem (λ) in the neighbourhood of E_2. This problem has already been solved for any E_1 contained in E_2.

Between the lower and the upper extremities of the closed circle $(\bar{C}_n)$, we consider $(q-1)$ horizontal lines which partition $(\bar{C}_n)$ into q closed sets; we denote these sets by $\alpha_1, \alpha_2, \ldots, \alpha_q$ (from bottom to top). Denote the closed polycylinder (Q, α_i) $(i = 1, 2, \ldots, q)$ in the space (x) by A_i and set

$$A_1 \cup A_3 \cup A_5 \cup \ldots = \Delta_1,$$
$$A_2 \cup A_4 \cup A_6 \cup \ldots = \Delta_2;$$

we then find that

$$\Delta = \Delta_1 \cup \Delta_2 = E_2$$

and that the intersection $\Delta_0 = \Delta_1 \cap \Delta_2$ consists of a finite number of sets E_1.

Since problem (λ) is solvable in a neighbourhood of an arbitrary E_1-subset of E_2, the problem will be solvable in the neighbourhood of each A_i if the partition of E_2 is sufficiently fine (once again by the BOREL-LEBESGUE lemma), and consequently in neighbourhoods of Δ_1 and Δ_2. Since any problem (C_1) is solvable for any set E_1, and therefore for Δ_0, problem (λ) is solvable for the set E_2 by Lemma 3; again, since (F) and E_2 are arbitrary, *problem* (λ) *is always solvable in the neighbourhood of any set* E_2.

2°. *Problem* (C_1). We fix a set E_2 and a problem (C_1) for this set E_2. We can then realise the geometric configuration above in such a way that problem (C_1) is already solved in the neighbourhood of Δ_1 and of Δ_2. Since the corresponding problem (λ) is solvable for Δ and any problem (C_1) is solvable for Δ_0, problem (C_1) is solvable for the set E_2 by Lemmas 1 and 2. The problem (C_1) and the set E_2 being arbitrary, *problem* (C_1) *is always solvable in the neighbourhood of any set* E_2.

By continuing in this way, we find that: Problems (C_1) and (λ) are always solvable in the neighbourhood of a closed polycylinder, if problem (K) is always solvable.

From this, one obtains immediately the following intermediary result:

Problems (C_1), (C_2), (E) *and* (L) *are always solvable in the neighbourhood of a closed polycylinder if problem* (K) *is always solvable.*

5. The Remainder Theorem. The following theorem will furnish us with a useful tool in what follows.

The Remainder Theorem. *Consider a domain of the form* $[D,(C)]$ *in the space* $(x_1, x_2, \ldots, x_n, y)$ *where* D *is a (univalent, finite) domain in the space* (x) *and* (C) *is a circle in the* y*-plane. Let us consider a holomorphic function* $F(x,y)$ *in* $[D,(C)]$ *such that, for any point* (x^0) *of* D*, the equation* $F(x^0,y)=0$ *has* λ *roots in* (C)*, where* λ *is a finite number, independent of* (x^0)*. Then, any holomorphic function* $f(x,y)$ *in* $[D,(C)]$ *can be represented in the form*

$$f(x,y)=f_0(x,y)+\varphi(x,y)\,F(x,y),$$

where f_0 *and* φ *are holomorphic functions on* $[D,(C)]$; f_0 *is a polynomial with respect to* y *which is identically zero if* $\lambda=0$*, and is of degree at most* $\lambda-1$ *in* y *if* $\lambda>0$*. Furthermore, the representation is unique.*

We shall start by proving *the uniqueness*. If $\lambda=0$, we have $f_0=0$; suppose therefore that $\lambda>0$. In this case, if we had two representations with (f_0,φ) and (f_0',φ'), we would have

$$(f_0-f_0')+(\varphi-\varphi')F=0$$

identically. Now, for any point (x^0) of D, the equation $F(x^0,y)=0$ has λ roots in (C); hence, if we set $f_0-f_0'=\psi$, the equation $\psi(x^0,y)=0$ would have at least λ roots in (C); but, since ψ is a polynomial in y of degree at most $(\lambda-1)$, it must necessarily be identically zero, that is $f_0=f_0'$.

Let us look now at the form of $F(x,y)$. Let us fix a point (ξ) in D arbitrarily. Thanks to the theorem of WEIERSTRASS, when (x) is near (ξ) and y is in (C), we have

$$F(x,y)=\omega(x,y)\,F_0(x,y),$$

where ω and F_0 are holomorphic functions, ω is never zero and

$$F_0(x,y)=y^\lambda+\alpha_1(x)\,y^{\lambda-1}+\ldots+\alpha_\lambda(x) \qquad (\lambda\geqq 0),$$

the $\alpha_i(x)$ $(i=1,2,\ldots,\lambda)$ being holomorphic functions in a neighbourhood. Now, since such an expression for F is obviously unique, it holds globally on $[D,(C)]$.

Let us examine the special condition we imposed, that the equation $F(x^0,y)$ have always the same number λ of roots in (C). In the space of two complex variables x,y, let $F=y-x$, and $f(x,y)=f(y)$ where $f(y)$ is a holomorphic function of the single variable y having $|y|<1$ for its exact domain of existence. If we take the unit circle for (C), the condition is only valid for $|x|<1$. If now D is a (connected) domain in the x-plane which extends both outside and inside the circle, suppose we had a relation

$$f(y)=f_0+\varphi\cdot(y-x)$$

on $[D,(C)]$, f_0 being a function of x alone. For $x=y$, we would have

$$f(y)=f_0(y),$$

contradicting the hypothesis on the domain of existence of f.

The condition imposed on F is thus inevitable. This is one of the curious phenomena that one meets in the dealing with analytic functions when one leaves the field of functions of a single variable.

Let us consider the case of a single variable; this is when F and f are functions of a single variable y and the condition on F above is always satisfied. The theorem remains true in this case (if it is true for several variables).

We shall now prove the proposition. Suppose that

$$F(x, y)=y^\lambda+\alpha_1(x)\,y^{\lambda-1}+\ldots+\alpha_\lambda(x), \qquad \lambda>0,$$

where the α_i are holomorphic functions on D; suppose also that (C) is given by $|y|<1$. We then have

$$f=\beta_0+\beta_1 y+\beta_2 y^2+\ldots+\beta_m y^m+R_m$$

with

$$R_m=\beta_{m+1}\,y^{m+1}+\beta_{m+2}\,y^{m+2}+\ldots,$$

where the β_i $(i=0,1,\ldots)$ are holomorphic functions of (x) in D. We therefore have

$$(1) \qquad f(x,y)=A_m(x,y)\,F(x,y)+B_m(x,y)+R_m(x,y),$$

where $A_m(x,y)$ and $B_m(x,y)$ are holomorphic functions, and $B_m(x,y)$ is a polynomial in y of degree at most $\lambda-1$; we denote this polynomial by

$$B_m(x,y)=\gamma_1^{(m)}\,y^{\lambda-1}+\gamma_2^{(m)}\,y^{\lambda-2}+\ldots+\gamma_\lambda^{(m)},$$

the $\gamma_i^{(m)}$ $(i=1,2,\ldots,\lambda)$ being functions of (x), holomorphic in D. It is the behaviour of the sequence of functions $B_m(x,y)$ $(m=0,1,2,\ldots)$ which is in question, when $m\to\infty$; we shall omit the index m when it is fixed.

For a given point (x) of the domain D, the equation $F(x,y)=0$ has λ roots in the circle (C); we denote these roots by $y_1, y_2, \ldots, y_\lambda$. Substituting the root y_i $(i=1,2,\ldots,\lambda)$ in the equation (1), we have

$$f(x,y_i)-R(x,y_i)=B(x,y_i);$$

we abbreviate $B(x,y_i)$ to B_i. We then have

$$\gamma_1 y_i^{\lambda-1}+\gamma_2 y_i^{\lambda-2}+\ldots+\gamma_\lambda=B_i \qquad (i=1,2,\ldots,\lambda).$$

Thus, we have λ equations in the λ unknows $(\gamma_1,\gamma_2,\ldots,\gamma_\lambda)$.

Let us consider the determinant

$$\Delta=\begin{vmatrix} y_1^{\lambda-1} & y_1^{\lambda-2} & \ldots & 1\\ y_2^{\lambda-1} & y_2^{\lambda-2} & \ldots & 1\\ \cdots & \cdots & \cdots & \cdots\\ y_\lambda^{\lambda-1} & y_\lambda^{\lambda-2} & \ldots & 1 \end{vmatrix},$$

and the determinant Δ_p obtained by replacing the pth column of Δ by $(B_1, B_2, \ldots, B_\lambda)$ $(p=1,2,\ldots,\lambda)$; for example

$$\Delta_\lambda=\begin{vmatrix} y_1^{\lambda-1} & y_1^{\lambda-2} & \ldots & B_1\\ y_2^{\lambda-1} & y_2^{\lambda-2} & \ldots & B_2\\ \cdots & \cdots & \cdots & \cdots\\ y_\lambda^{\lambda-1} & y_\lambda^{\lambda-2} & \ldots & B_\lambda \end{vmatrix}.$$

Suppose now that $F(x,y)$ has no multiple factors. Then, the analytic variety

$$F(x,y)=0, \qquad \frac{\partial}{\partial y}F(x,y)=0$$

has complex dimension $(n-1)$; moreover, because of the form of F, the *projection* S of the variety on the space (x) [i.e. the set of (x') for which (x',y') is on the variety for a suitable value y'] is a characteristic surface. For any point (x) of D outside S we have $\Delta \neq 0$ because the roots y_i $(i=1,2,\ldots,\lambda)$ of the equation $F(x,y)=0$ are all distinct; consequently, we have

$$\gamma_i=\Delta_i/\Delta \qquad (i=1,2,\ldots,\lambda). \tag{2}$$

We shall estimate the absolute value of $\gamma_i^{(m)}(x)$ when m tends to infinity. Let us first consider a closed (bounded) set E contained in D and not containing any point of S, but otherwise arbitrary. The determinant Δ is independent of m and its absolute value admits a positive lower bound on E. As for $\Delta_i^{(m)}$, the function $B_m(x,y_i)$ $(i=1,2,\ldots,\lambda)$ obviously has an upper bound independent of m on E; so, therefore, does $\Delta_i^{(m)}$, and, because of (2), so does $\gamma_i^{(m)}$. Next consider any closed (bounded) set E' in D; since S is a characteristic surface and the $\gamma_i^{(m)}$ are holomorphic functions, the upper bound on the set E implies one on E'.

The sequence of functions $B_m(x,y)$ $(m=0,1,2,\ldots)$ forms therefore a normal family in D; let B_{p_i} $(i=1,2,\ldots)$ be a subsequence converging to a limit B_0; the limit is a holomorphic function of the variables (x,y) when (x) belongs to D and is a polynomial with respect to y of degree at most $\lambda-1$. Now, by relation (1), we have

$$f=A_{p_i}F+B_{p_i}+R_{p_i} \qquad (i=1,2,\ldots),$$

and, when i tends to infinity, B_{p_i} tends to B_0 and R_{p_i} tends to 0. A_{p_i} tends therefore to a limit $A_0(x,y)$ in $[D,(C)]$; this limit is meromorphic and certainly regular outside $F=0$. $A_0(x,y)$ is therefore holomorphic on $[D,(C)]$ (by the same reasoning as above, or by the theory of normal families), and one has the desired relation

$$f(x,y)=B_0(x,y)+A_0(x,y)\,F(x,y),$$

when $F(x,y)$ does not have multiple factors.

Let us now consider *the contrary case*. We introduce the function

$$\Phi(x,y,t)=F(x,y)+t, \qquad (|t|<\rho),$$

where t is a complex variable. The function Φ obviously has no multiple factors and is of the form

$$\Phi(x,y,t)=y^\lambda+a_1(x,t)\,y^{\lambda-1}+\ldots+a_\lambda(x,t) \qquad (\lambda>0).$$

As for the number of roots of the equation $\Phi(x',y,t')=0$, let D' be a bounded domain with $D'\Subset D$ but otherwise arbitrary; we can obviously choose ρ so small that for any point (x',t') in $(x')\in D'$, $|t'|<\rho$, the equation has λ roots in (C). The function $\Phi(x,y,t)$ thus satisfies the conditions of the proposition for the domain $[(x)\in D', |t|<\rho, |y|<1]$ and, in addition, has no multiple factors. By what we have just seen, we have

$$f(x,y)=G(x,y,t)+H(x,y,b)\,\Phi(x,y,t)$$

identically in this domain; here G and H are holomorphic, and G is a polynomial with respect to y of degree at most $\lambda-1$. Setting $t=0$, we obtain

$$f(x,y)=G(x,y,0)+H(x,y,0)\,F(x,y);$$

this is the desired relation for $[D',(C)]$; since D' is arbitrary and such a representation is unique, the relation holds for $[D,(C)]$. Q.E.D.

6. Solution of the Local Problem (K). We now devote ourselves to problem (K).

1°. Let E be the space of a finite number of complex variables (without points at infinity). Consider a domain D in E, p holomorphic functions F_1, $F_2, \ldots, F_p$ in D which are not identically zero in D, and an arbitrary point P of D. We shall study problem (K) for (F) at the point P.

Suppose that there is a formula for the solutions of the functional equation

(1) $$A_1F_1+A_2F_2+\ldots+A_pF_p=0$$

at P (i.e. in a neighbourhood of P), given by

(2) $$A_i=C_{i1}\pi_{i1}+C_{i2}\pi_{i2}+\ldots+C_{iN}\pi_{iN}\qquad(i=1,2,\ldots,p)$$

with certain identities among the C_{ij}.

We can also assume that the C_{1i} $(i=1,2,\ldots,N)$ are all different in this relation; if not, we could easily deform the expression for this to be the case. Under these circumstances, if we set $C_{11}=1$ and all the other C_{ij} equal to zero, we have $A_1=\pi_{11}$. The function π_{11} belongs therefore to the ideal (I_1) at the point P. Similarly, the same holds for π_{1i} $(i=2,3,\ldots,N)$; moreover, for any function A_1 belonging to the ideal (I_1) at P, we have $(A_1)\subset(\pi_{11},\pi_{12},\ldots,\pi_{1N})$. The system of functions (π_{1i}) is therefore a pseudobasis of the ideal (I_1) in a neighbourhood of P, that is to say, problem (K) is solved. We pose therefore the following:

Problem (μ). *Find a formula for the solutions of the functional equation* (1) *in a neighbourhood of the point P of the domain D.*

We have just seen that *it is sufficient to solve problem* (μ). In what follows, we shall treat the problem exclusively in this form.

2°. Let us look at a space (E) *of one complex dimension*; let x be the complex variable in E and suppose that P is the origin. If, among the functions $F_1(x)$, $F_2(x)$, ..., $F_p(x)$, there is one which is not zero at the origin, for example $F_p(0)\neq 0$, we obtain a formula for the solutions of (1) at the origin by setting

$$A_i=C_i\qquad(i=1,2,\ldots,p-1),$$

$$A_p=\frac{-1}{F_p}(C_1F_1+C_2F_2+\ldots+C_{p-1}F_{p-1}).$$

Let us therefore suppose that $F_1(0)=F_2(0)=\ldots=F_p(0)=0$. In this case, it is clear that there exists a number m such that all the functions $F_i(x)$ are divisible by x^m at the origin, and among these functions one at least, say $F_p(x)$, is not divisible by x^{m+1}. The above formula for the solutions is then still valid.

Problem (μ) is therefore solved for a space (E) of one complex dimension. It is therefore sufficient to solve it for a space (E) of dimension $(n+1)$, assuming it to be solvable for spaces of smaller dimension.

3°. Let us represent the space (E) by $(x_1, x_2, \ldots, x_n, y)$ and take the point P to the origin; we can do this in such a way that

$$F_i(0, y) \not\equiv 0 \quad (i=1,2,\ldots,p).$$

The equation $F_i(0, y)=0$ has, say, λ_i roots at $y=0$ ($\lambda_i \geqq 0$). If one among these λ_i $(i=1,2,\ldots,p)$ is zero, for example $\lambda_p=0$, this means that $F_p(0,0)\neq 0$ and we would have a formula for the solutions of the functional equation (1) of the same form as in the case $n=1$. We therefore suppose that $\lambda_i>0$ $(i=1,2,\ldots,p)$ and we shall abbreviate λ_p to λ, with the assumption that $\lambda_i \leqq \lambda$ $(i=1,2,\ldots, p-1)$.

We consider a small circle (C') around the origin in the y-plane and a polycylinder (C) around the origin in (x)-space which is sufficiently small in comparison with (C'); the functions $F_i(x, y)$ will then be holomorphic in $[(C),(C')]$ and, thanks to Weierstrass, can be represented in the form

$$F_i(x, y)=\Omega_i(x, y)\, \Phi_i(x, y) \quad (i=1,2,\ldots,p)$$

with

$$\Phi_i(x, y)=y^{\lambda_i}+a_{i1}(x)\, y^{\lambda_i-1}+\ldots+a_{i\lambda_i}(x),$$

where $a_{ij}(x)$ $(i=1,2,\ldots,p;\ j=1,2,\ldots,\lambda_i)$ are holomorphic, the $\Omega_i(x, y)$ are holomorphic and without zeros; in addition, the equation $\Phi_i(x', y)=0$ has, for any point (x') in (C), λ_i roots in (C'), and consequently has no roots either exterior to (C') or on the circumference.

Consider the functional equation

$$\text{(A)} \qquad B_1\Phi_1+B_2\Phi_2+\ldots+B_p\Phi_p=0,$$

the (B) being the unknown functions. If (2) is a formula for the solutions of equation (1), then we have a formula for the above equation (A) by setting

$$\text{(B)} \qquad B_i=\Omega_i(C_{i1}\pi_{i1}+C_{i2}\pi_{i2}+\ldots+C_{iN}\pi_{iN}) \quad (i=1,2,\ldots,p)$$

with the same system of identities for (C) *as in* (2).

In fact, let $(B(x, y))$ be a solution of (A) at a point Q near the origin; since $F_i=\Omega_i\Phi_i$ and consequently $A_iF_i=(A_i\Omega_i)\Phi_i$, the system $(B_1/\Omega_1,\ B_2/\Omega_2,\ \ldots,\ B_p/\Omega_p)$ satisfies equation (1) at Q. Consequently the functions (B) are represented in the form (B) above. Conversely, let (C) be a system of holomorphic functions at a point Q near 0, satisfying the identities indicated but otherwise arbitrary, and consider the system (B) of functions given by the above formula. Then the system $(B_1/\Omega_1,\ B_2/\Omega_2,\ \ldots,\ B_p/\Omega_p)$ satisfies equation (1) at Q; since $(B_i/\Omega_i)F_i=B_i\Phi_i$, the system (B) satisfies equation (A). Hence the expression (B) gives a formula for the solution of equation (A). In this reasoning, the equations (1) and (A) are obviously interchangeable. Thus, to solve problem (μ) at the origin for equation (1), it suffices to do this for equation (A).

4°. We therefore suppose that, in the functional equation

$$A_1F_1+A_2F_2+\ldots+A_pF_p=0, \tag{1}$$

the functions F_1, F_2, ..., F_p have the same properties relative to the polycylinder $[(C),(C')]$ as the functions $\Phi_1,\Phi_2,\ldots,\Phi_p$. Our aim is to find a formula

$$A_i=C_{i1}\pi_{i1}+C_{i2}\pi_{i2}+\ldots+C_{iN}\pi_{iN} \quad (i=1,2,\ldots,p) \tag{2}$$

with some identities among (C)

for the solutions of this equation (1) in a neighbourhood of the origin.

Let us consider a solution (A) of equation (1) at a point (x^0,y^0) of the polycylinder $[(C),(C')]$. Consider a small circle (γ') around y_0 in (C') and a polycylinder (γ) around (x^0) in (C), sufficiently small compared to (γ'). The functions $A_i(x,y)$ $(i=1,2,\ldots,p)$ are then holomorphic on the polycylinder $[(\gamma),(\gamma')]$, and the equation $F_p(x',y)=0$ has the same number μ of roots in (γ') for any point (x') in (γ); here $0\leqq\mu\leqq\lambda$. If we apply the *remainder theorem*, we have, on $[(\gamma),(\gamma')]$,

$$\text{(A)} \quad \begin{cases} A_i=A_i^0+\alpha_iF_p & (i=1,2,\ldots,p-1)\\ A_p=A_p^0-(\alpha_1F_1+\alpha_2F_2+\ldots+\alpha_{p-1}F_{p-1}), \end{cases}$$

where A_p^0 is defined by this equation, A_i^0, α_i $(i=1,2,\ldots,p-1)$ are holomorphic functions of (x,y), and A_i^0 are identically 0 if $\mu=0$, and are polynomials in y of degree at most $\mu-1$ if $\mu>0$; the system of functions (A^0) thus obtained is a solution of equation (1) in $[(\gamma),(\gamma')]$ because (A) is a solution at (x^0,y^0). The A_i^0 $(i=1,2,\ldots,p-1)$ are polynomials in y, and, for A_p^0, we have

$$A_p^0F_p=\Phi, \qquad \Phi=-(A_1^0F_1+A_2^0F_2+\ldots+A_{p-1}^0F_{p-1}),$$

so that A_p^0 is a rational function of y.

Now, the function $F_p(x,y)$ is not zero when (x) lies in (γ) and y is on the circumference γ' of (γ'). If $\lambda>\mu$, it can be decomposed into two functions

$$F_p=F'\cdot F'',$$

where F' and F'' are polynomials in y whose coefficients are holomorphic functions of the variables (x) in (γ), and the coefficients of the highest powers of y are equal to 1; moreover, if (x') is an arbitrary point of (γ), the equation $F'(x',y)=0$ has all its zeros in the circle (γ), while the equation $F''(x',y)=0$ has all its zeros in the exterior of (γ). If $\lambda=\mu$, we define $F'=F_p$, $F''=1$. The function $\Phi=A_p^0F_p$ is then divisible by $F'(x,y)$, since $A_p^0(x,y)$ is holomorphic on $[(\gamma),(\gamma')]$. Let us set

$$B_i=F''A_i^0 \qquad (i=1,2,\ldots,p); \tag{B}$$

B_p is then a polynomial with respect to y, holomorphic when $(x)\in(\gamma)$. The same is obviously also true for the other B_i; the system of functions (B) satisfies equation (1) on $[(\gamma),(\gamma')]$, hence whenever $(x)\in(\gamma)$.

Let us now set

$$\text{(C)} \qquad \begin{cases} B_i = B_i^0 + \beta_i F_p \quad (i=1,2,\ldots,p-1) \\ B_p = B_p^0 - (\beta_1 F_1 + \beta_2 F_2 + \ldots + \beta_{p-1} F_{p-1}), \end{cases}$$

where the B_i^0 and β_i $(i=1,2,\ldots,p-1)$ are polynomials in y, holomorphic when $(x)\in(\gamma)$ and the B_i^0 have degree at most $\lambda-1$ in y. The function B_p^0 defined by (C) is therefore a polynomial in y, holomorphic for $(x)\in(\gamma)$ and the system of functions (B^0) satisfies equation (1) for $(x)\in(\gamma)$. B_p^0 therefore satisfies

$$B_p^0 F_p = \Psi, \qquad \Psi = -(B_1^0 F_1 + B_2^0 F_2 + \ldots + B_{p-1}^0 F_{p-1});$$

here Ψ has degree at most $2\lambda-1$ in y since $\lambda_i \leqq \lambda$ $(i=1,2,\ldots,p-1)$. Thus B_p^0 is of degree at most $\lambda-1$.

We have thus seen that if we are given equation (1) with the F_i satisfying the condition indicated, then to any solution (A) at an arbitrary point (x^0, y^0) of $[(C),(C')]$, we can make correspond a solution (B^0) at (x^0, y^0) consisting of polynomials with respect to y of degree at most $(\lambda-1)$, and the correspondence is given by the three relations (A), (B) and (C).

These three relations together can be put in the form

$$\text{(D)} \qquad \begin{cases} A_i = \delta_p B_i^0 + \delta_i F_p \quad (i=1,2,\ldots,p-1) \\ A_p = \delta_p B_p^0 - (\delta_1 F_1 + \delta_2 F_2 + \ldots + \delta_{p-1} F_{p-1}), \end{cases}$$

where

$$\delta_i = \alpha_i + \frac{\beta_i}{F''}, \qquad \delta_p = \frac{1}{F''};$$

the (δ) are therefore holomorphic functions on $[(\gamma),(\gamma')]$.

Let us consider once again the functional equation (1). Let $[(x)\in(C_0), y\in(C_0')]$ be a polycylinder around the origin contained in the polycylinder $[(C),(C')]$. Suppose that we are given an expression

$$\text{(3)} \qquad B_i^0 = C_{i1}\theta_{i1} + C_{i2}\theta_{i2} + \ldots + C_{iN'}\theta_{iN'} \quad (i=1,2,\ldots,p)$$

with certain identities among the (C),

where the θ_{ij} $(i=1,2,\ldots,p;\ j=1,2,\ldots,N')$ are holomorphic functions of the variables (x, y) in $[(C_0),(C_0')]$, in such a way as to have the following properties:

1°. Let (B^0) be a solution of the equation (1) consisting of polynomials in y of degree at most $\lambda-1$, holomorphic at a point (x^0) of (C_0), but otherwise arbitrary. Then (B^0) can be represented in the form (3) at any point $(x)=(x^0)$, $y\in(C_0')$, where the (C) are suitable holomorphic functions, respecting, of course, the identities indicated.

2°. Consider holomorphic functions (C) at a point (x^0, y^0) of $[(C_0),(C_0')]$, respecting the indicated identities, but otherwise arbitrary. Then, the corresponding system (B^0) of functions satisfies equation (1) at (x^0, y^0).

Such an expression will, for the time being, be called *a conditional formula for the solutions*, and (B^0) will be called a *special solution*.

Suppose that we have a conditional formula (3) as above. I assert that (3), together with the relation (D) gives a (real) formula for the solutions of the equation (1) in the polycylinder $[(C_0), (C'_0)]$, if we regard the δ_i $(i=1,2,\ldots, p-1)$ and $\varepsilon_{ij}=(\delta_p\, C_{ij})$ $(i=1,2,\ldots,p;\ j=1,2,\ldots,N')$ as the independent unknown functions.

In fact, the union of these two relations takes the same form as (2), whose kernel consists of well determined functions (F) and (θ) holomorphic on $[(C_0), (C'_0)]$. Let us consider a solution (A) of equation (1) at an arbitrary point (x^0, y^0) of $[(C_0), (C'_0)]$; then, by what we have seen above, the solution is represented by the union of (3) and (D). Conversely, let us consider functions (δ, ε) holomorphic at a point (x^0, y^0) of $[(C_0), (C'_0)]$; by substituting the (ε) for (C) in formula (3), we form a system of functions (B^0) which satisfies the equation (1) at (x^0, y^0). Now, between this system (B^0) and the system (A) corresponding by the union to the functions (δ, ε), there is a relation given in (D) in which $\delta_p=1$; since (B^0) satisfies equation (1), so, then, does (A). Q.E.D.

It therefore suffices to show that if the functional equation has the property indicated, then we can find a conditional formula for its solutions in a neighbourhood of the origin.

5°. If (B^0) is a special solution of equation (1) at a point (x^0) of (C), we have

$$\text{(A)} \qquad B_1^0 F_1 + B_2^0 F_2 + \ldots + B_p^0 F_p = 0.$$

Suppose that

$$B_i^0 = \alpha_{i0}\, y^{\lambda-1} + \alpha_{i1}\, y^{\lambda-1} + \ldots + \alpha_{i,\lambda-1}$$
$$F_i = f_{i0}\, y^{\lambda} + f_{i1}\, y^{\lambda-1} + \ldots + f_{i\lambda} \qquad (i=1,2,\ldots,p),$$

where (α) and (f) are holomorphic functions of the variables (x) at (x^0); we then have

$$\sum B_i^0 F_i = \Phi_0\, y^{2\lambda-1} + \Phi_1\, y^{2\lambda-2} + \Phi_2\, y^{2\lambda-3} + \ldots + \Phi_{2\lambda-1} \qquad (i=1,2,\ldots,p),$$

where

$$\Phi_0 = \sum \alpha_{i0} f_{i0}$$
$$\Phi_1 = \sum \alpha_{i0} f_{i1} + \sum \alpha_{i1} f_{i0}$$
$$\Phi_2 = \sum \alpha_{i0} f_{i2} + \sum \alpha_{i1} f_{i1} + \sum \alpha_{i2} f_{i0}$$
$$\cdots\cdots\cdots\cdots\cdots\cdots\cdots\cdots\cdots\cdots$$
$$\Phi_{2\lambda-1} = \sum \alpha_{i,\lambda-1} f_{i\lambda};$$

the relation (A) is therefore equivalent to the following:

$$\text{(B)} \qquad \Phi_0=0, \quad \Phi_1=0, \quad \Phi_2=0, \ldots, \Phi_{2\lambda-1}=0.$$

In the relation (B), the (f) are well determined functions holomorphic on (C). If we regard the (α) as unknown functions, (B) *is a system of simultaneous (homogeneous linear) functional equations*, and the functions (α) above gives us a *solution* at (x^0). Now, as we shall see later, under the hypothesis that problem (μ) is always solvable in a space (E) of lower dimension, we can find a *formula for the solutions* of this system of simultaneous equations considered in the space (x).

Suppose that we can make correspond to the system (B) of simultaneous equations an expression

$$\text{(C)} \qquad \alpha_{ij} = \sum \beta_{ijk} \rho_{ijk} \qquad (i=1,\dots,p; j=0,\dots,\lambda-1; k=1,\dots,v)$$

with some identities among (β),

the ρ_{ijk} being well determined functions on a polycylinder (C_0) contained in (C), which has the following properties:

1°. Any solution (α) of equation (B) at any point (x^0) of (C_0) can be represented in the form (C) at (x^0).

2°. For any collection (β) of holomorphic functions at an arbitrary point (x^0) of (C_0), the corresponding system of functions (α) satisfies equation (B) at (x^0).

Such an expression is what we call a formula for the solutions of the simultaneous equations (B).

If we substitute the expression (C) for α_{ij} in B_i^0, we obtain

$$\text{(D)} \qquad B_i^0 = \sum \beta_{i0k}(\rho_{i0k} y^{\lambda-1}) + \sum \beta_{i1k}(\rho_{i1k} y^{\lambda-2}) + \dots + \sum \beta_{i,\lambda-1,k}(\rho_{i,\lambda-1,k})$$

$$(i=1,2,\dots,p;\ k=1,2,\dots,v),$$

with the same identities between the (β) *as in* (C);

here the $\rho_{i0k} y^{\lambda-1}, \dots$ are well determined holomorphic functions on (C_0), which we consider as constituting the kernel, and we regard the (β) as unknown functions.

Consider a special solution of equation (1), in which the functions are holomorphic at a point (x^0) in (C_0), but otherwise arbitrary. This means that the corresponding system (α) satisfies the system of simultaneous equations (B) at (x^0); (α) can therefore be represented in the form (C) at (x^0), and consequently, (B^0) can be represented in the form (D) at any point of the form (x^0, y).

Conversely, consider holomorphic functions (β) at a point (x^0, y^0) such that $(x^0) \in (C_0)$, but otherwise arbitrary. Let us form the system (B^0) of holomorphic functions of (x, y) at (x^0, y^0) by means of (D), and ask if this system (B^0) satisfies equation (A) at (x^0, y^0).

Now, this is obviously equivalent to forming the system (α) of holomorphic functions of the variables (x, y) by means of (C), using the same collection of functions (β), and asking if this (α) satisfies the system (B) of simultaneous equations. Let us, therefore, consider this latter problem.

If we substitute, in (C), the expansions of the (β) in a series of powers of $(y-y_0)$, we obtain the expansions of the (α). If we substitute these expansions in the functions Φ_i $(i=0,1,\dots,2\lambda-1)$ of the simultaneous equations (B), we obtain the expansions of the Φ_i; now each coefficient of Φ_i is obviously 0, and hence, so are the Φ_i.

Thus, the expression (D) gives a conditional formula for the solutions of (1) on $[(C_0), (C')]$. Thus, it remains only to prove the existence of the formula (C) of solutions for the system of simultaneous equations (B).

6°. Let us consider a domain D in the space (x) and holomorphic functions $A_{ij}(x)$ $(i=1,2,\ldots,q;\ j=1,2,\ldots,p)$ on this domain; (of these functions, some can be identically 0). Consider now the *system of simultaneous (homogeneous linear) functional equations*

$$(1)\qquad \begin{cases} A_{11}F_1 + A_{12}F_2 + \ldots + A_{1p}F_p = 0 \\ A_{21}F_1 + A_{22}F_2 + \ldots + A_{2p}F_p = 0 \\ \ldots\ldots\ldots\ldots\ldots\ldots\ldots\ldots\ldots\ldots \\ A_{q1}F_1 + A_{q2}F_2 + \ldots + A_{qp}F_p = 0 \end{cases}$$

with certain identities among the (A).

The (A) are the unknown function; between them, there are in general, certain identities (without this, the system would reduce to q separate equations; and would lead to the question of connections between the equations (1); however, we are not interested in this). Any system of holomorphic functions (A), respecting the identities indicated and satisfying equation (1), will be called a *solution.*

Let us consider an expression

$$(2)\qquad A_{ij} = \sum C_{ijk}\pi_{ijk} \qquad (k=1,\ldots,\lambda)$$

with certain identities among the (C),

the π_{ijk} being well determined holomorphic functions on D, and the C_{ijk} are variable functions. Suppose this expression has the following properties:

1°. Any solution (A) of equation (1) at an arbitrary point P of D can be represented in the form (2) at P (by choosing suitably the holomorphic functions (C) at P, respecting the identities).

2°. To every collection of holomorphic functions (C) at P (respecting the identities), the expression (2) makes correspond a solution (A) at P.

We then call the expression (2) a *formula for solutions* of the equation (1).

We now pose

Problem (M). *Find a formula for the solutions of the system of functional equations* (1) *in a neighbourhood of a closed domain* Δ *contained in* D.

We assert that *problem* (M) *in the space* (x) *is always solvable if problem* (λ) *in the same space is always solvable.*

Let us take the first r equations $(r \leqq q)$ of the system (1) with the identities which relate to these equations. This gives us a system of simultaneous equations which we denote by (E_r).

(E_1) is just one functional equation and the corresponding problem (M) is solvable by hypothesis. It is therefore sufficient to solve problem (M) for (E_{r+1}) $(r+1 \leqq q)$ assuming it to be solved for $(E_1), (E_2), \ldots, (E_r)$.

We replace the letters $A_{r+1,i}$ and F_i $(i=1,2,\dots,p)$ by new letters B_i, Φ_i in the last equation of (E_{r+1}); we then have

$$(3)\qquad \begin{cases} A_{11}F_1 + A_{12}F_2 + \dots + A_{1p}F_p = 0 \\ \dots\dots\dots\dots\dots\dots\dots\dots\dots\dots \\ A_{r1}F_1 + A_{r2}F_r + \dots + A_{rp}F_p = 0 \\ B_1\Phi_1 + B_2\Phi_2 + \dots + B_p\Phi_p = 0 \end{cases}$$

with certain identities among (A, B).

By hypothesis, we can find a formula for solutions of (E_r)

$$(4)\qquad A_{ij} = \sum C_{ijk}\Psi_{ijk} \quad (k=1,2,\dots,\mu)$$

with certain identities among the (C),

(this, in a neighbourhood of Δ). We are going to substitute this in the equation

$$(5)\qquad B_1\Phi_1 + B_2\Phi_2 + \dots + B_p\Phi_p = 0;$$

let us explain exactly this process.

We regard the B_i $(i=1,2,\dots,p)$ as being all different; let us look at B_1; if, for example, B_1 is not related to the (A) in (E_r) by the identities in (1), we keep the term $B_1\Phi_1$. Let us pass to B_2; if, for example, $B_2 = A_{11}$ according to the identities in (1), we replace $B_2\Phi_2$ by

$$\sum C_{11k}(\Psi_{11k}\Phi_2);$$

and so on. Thus, in place of (5), we will obtain an equation

$$(6)\qquad B_1\Phi_1 + C_{111}(\Psi_{111}\Phi_2) + C_{112}(\Psi_{112}\Phi_2) + \dots = 0;$$

we shall write this equation as

$$(7)\qquad D_1X_1 + D_2X_2 + \dots + D_sX_s = 0,$$

and, in general, there will be some identities among the (D); the (X) are well determined holomorphic functions on a neighbourhood of Δ.

This being a single equation, we can find a formula for solutions of equation (7) in a neighbourhood of Δ, say

$$(8)\qquad D_i = \sum \gamma_{ij}\theta_{ij} \quad (j=1,2,\dots,\nu)$$

with certain identities between the (γ).

In (8), each of the (D) stands either for one of the (B) or one of the (C). Let us take the A_{ij} in order: if it is not the case that this A is one of the (B), we keep for it the expression coming from (4).

In the contrary case, we substitute, in the expression for this A in (4), the expressions for the C_{ijk} involved in it which are taken from (8). We next take the B_i in order; if we find one of these (B) in (8), we write down this expression; if we do not find it in (8), then this B is an A, and we rewrite for this B the new expression for the corresponding A we have just obtained. The new expressions so obtained can be written

$$(9)\qquad \begin{cases} A_{ij}=\sum \delta_{ijk}\rho_{ijk} \\ B_i=\sum \delta'_{ik}\rho'_{ik} \qquad (i=1,\dots,p;\, j=1,\dots,r;\, k=1,\dots,t), \end{cases}$$

with certain identities among (δ, δ');

the (ρ, ρ') stand for the kernel, and (δ, δ') are the unknown functions. Among the identities indicated, some come from (4), some from (8), and others are new. We shall study the nature of this expression.

Let us consider an arbitrary solution (A, B) of the equation (3) at an arbitrary point P of Δ. Since (A) satisfies (E_r), it can be represented in the form (4) at P. Since (B) satisfies (5), if we form then the (D) according to the rule explained above, (D) will satisfy (7), and consequently can be represented in the form (8). Therefore (A, B) can be represented in the form (9) at P.

Conversely, consider an arbitrary collection of functions (δ, δ') holomorphic at a point P of Δ, and form the system of functions (A, B) according to (9). Let us first consider (A). The part of (9) which represents (A) was obtained by substituting (8) in (4), and is thus just a special case of (4). Hence (A) satisfies (E_r) at P. We have therefore only to verify that (B) satisfies (5) at P. Now, equation (7) is a special case of (5); hence if (7) is satisfied, the equation (5) will à fortiori be satisfied. Consider (9): the part representing (B) corresponds to (D) in (8) by a relation of the form

$$B=B=D \quad \text{or} \quad B=A=\sum C\Psi=\sum D\Psi;$$

the corresponding (D) clearly satisfy (7), and the (B) therefore satisfy (5) at P.

The expression (9) gives therefore a formula for the solutions of equation (3) in the neighbourhood of Δ. Thus, problem (M) in the space (x) is solvable for (E_{r+1}) under the hypothesis that it is solvable for (E_1), (E_2), ..., (E_r); the proposition is thus proved.

This finally completes the proof that *problem* (K) *is always solvable*; consequently, problems (C_1), (C_2), (E), (L) and (M) are solvable in the neighbourhood of a closed polycylinder.

7. Conclusions. Let us collect together the results obtained.

Theorem I. *Given a finite collection of holomorphic functions* $[F_1(x), F_2(x), \dots, F_p(x)]$, *and a holomorphic function* $\Phi(x)$ *in a neighbourhood of a closed polycylinder* Δ *in the space* (x) *such that* $\Phi(x)=0 \bmod (F)$ *at every point of* Δ, *we can find functions* $A_i(x)$ $(i=1,2,\dots,p)$ *holomorphic in a neighbourhood of* Δ *such that we have* $\Phi(x)=A_1(x)F_1(x)+A_2(x)F_2(x)+\dots+A_p(x)F_p(x)$ *identically. [A polycylinder is a cylindrical set whose components are circles.]*

Theorem II. *Let* $(F_1, F_2, \dots, F_p)$ *be a finite collection of functions holomorphic in a neighbourhood of a closed polycylinder* Δ. *Suppose that to each point* P *of* Δ, *there correspond a polycylinder* (γ) *around* P *and a holomorphic function* $\varphi(x)$ *on* (γ) *such that for each pair of polycylinders* (γ'), (γ'') *with intersection* (δ), *the corresponding functions* $\varphi'(x)$, $\varphi''(x)$ *satisfy* $\varphi'(x)\equiv\varphi''(x) \bmod (F)$ *at each point of*

(δ). *Then we can find a holomorphic function* $\Phi(x)$ *in a neighbourhood of* Δ *such that* $\Phi(x) \equiv \varphi(x) \bmod (F)$ *at any point* P *of* Δ.

Theorem III. *With the same geometric configuration as in Theorem* II, *suppose given on each* (γ) *a finite system* (f) *of holomorphic functions, in such a way that for each pair* (γ'), (γ''), *the corresponding systems* (f'), (f'') *are equivalent at each point of the intersection* (δ). *We can then find a finite system* (F) *of functions holomorphic in a neighbourhood of* Δ *such that* $(F) \sim (f)$ *at any point* P *of* Δ.

Theorem IV. *Given holomorphic functions* $F_i(x)$ $(i = 1, 2, \ldots, p)$ *in a neighbourhood of a closed polycylinder* Δ, *we can find a formula for the solutions of the functional equation* $A_1 F_1 + A_2 F_2 + \ldots + A_p F_p = 0$ *in a neighbourhood of* Δ. *The same is true also for systems of simultaneous homogeneous linear functional equations.*

We have restricted ourselves to closed polycylinders; these problems then become solvable for less restrictive closed sets by virtue of intrinsic properties. Besides the theorems stated above, we obtained the *remainder theorem* in No. 5. On the subject of these theorems, we shall be obliged to study them *quantitatively* if we hope to be able to apply them widely.

We have thus explained the results obtained. On the other hand, we shall speak of the problem we have been led to, viz:

Problem (J). *Given an ideal with indeterminate domain, find a finite local pseudobasis.*

As for this problem, I know almost nothing about it, not even an idea of what might be the most favourable attitude in its study. We only know that this problem cannot always be solved, without further conditions as we have seen a counter example in No. 2.

Problem (K) which we solved above is just a special case of this problem. This was essential in establishing the theorems stated above. We shall return to the general problem in another case and prove that the problem can be solved without further conditions for *a geometric ideal with indeterminate domains.* This will be indispensable in treating the problems we have been studying since Memoir I when we allow points of ramification to appear. These two examples will already show the importance of this problem.

Commentaire de H. Cartan

Ce Mémoire a été écrit en 1948 et publié en 1950 (Bull. Soc. Math. de France). Il est le résultat des réflexions auxquelles s'est livré OKA après la lecture du travail de H. CARTAN (J. de Math. 19, 1940, p. 1–26) dont il a dû avoir connaissance seulement après la guerre de 1939–1945. Il n'avait probablement pas connaissance à cette époque du travail de CARTAN sur les idéaux de fonctions analytiques (Ann. Ecole Normale Sup. 61, 1944, p. 149–197) où étaient notamment étudiés les "systèmes cohérents d'idéaux ponctuels". Les problèmes envisagés par CARTAN sont aussi considérés par OKA (quoique dans un langage différent), mais OKA va plus loin dans les résultats.

OKA introduit systématiquement la notion d'"idéaux de domaines indéterminés", notion qui est en substance équivalente à celle de faisceau d'idéaux introduite par CARTAN en 1949 (Bull. Soc. Math. de France 78, 1950, p. 29–64), laquelle a prévalu depuis.

OKA pose ici une série de problèmes fondamentaux. Le problème (J), en termes de faisceaux, est le suivant: "un faisceau analytique d'idéaux est-il cohérent?". OKA donne lui-même un contre-exemple. Il semble qu'à cette époque il savait que le faisceau d'idéaux défini par un sous-ensemble analytique est cohérent (cf. les 5 dernières lignes du Mémoire); mais il n'a pas publié de démonstration, ce résultat ayant été entre temps publié par CARTAN dans son article de 1950.

Le problème (K) est résolu ici: il s'agit de la cohérence du faisceau des relations linéaires entre un nombre fini de fonctions holomorphes (problème posé par CARTAN en 1944, mais que CARTAN n'avait pu résoudre).

L'ensemble des résultats démontrés dans ce Mémoire VII est condensé dans les théorèmes I, II, III, IV énoncés à la fin du Mémoire, et qui résolvent respectivement les problèmes (C_1), (C_2), (E) et (K). OKA énonce ces théorèmes pour un compact Δ, produit de disques compacts dans les plans de coordonnées.

Le *théorème I* dit que si l'on se donne des F_i holomorphes sur Δ, en nombre fini, et si une Φ holomorphe sur Δ appartient, en chaque point de Δ, à l'idéal ponctuel engendré par les F_i, alors Φ appartient à l'idéal engendré par les F_i dans l'anneau des fonctions holomorphes sur Δ. En termes de faisceaux, cela s'enonce comme suit: si on a un morphisme surjectif de faisceaux $\mathcal{O}^p \to J$ sur Δ, où $\mathcal{J}$ est un faisceau cohérent d'idéaux, le morphisme de sections $\Gamma(\Delta, \mathcal{O}^p) \to \Gamma(\Delta, \mathcal{J})$ est surjectif (ce qui résulte du théorème B appliqué au noyau de $\mathcal{O}^p \to \mathcal{J}$).

Le *théorème II* dit que si Δ est recouvert par des ouverts U_α dans chacun desquels on a une φ_α holomorphe, de façon qu'en tout point de $U_\alpha \cap U_\beta$ la différence $\varphi_\alpha - \varphi_\beta$ appartienne à l'idéal ponctuel engendré par les F_i, alors il existe une Φ holomorphe sur Δ telle que, en tout point de U_α, $\Phi - \varphi_\alpha$ appartienne à l'idéal ponctuel engendré par les F_i. Ceci, en termes de faisceaux, s'énonce comme suit: si $\mathcal{J}$ est un faisceau cohérent d'idéaux sur Δ, l'homomorphisme de sections $\Gamma(\Delta, \mathcal{O}) \to \Gamma(\Delta, \mathcal{O}/\mathcal{J})$ est surjectif (conséquence du théorème B appliqué à $\mathcal{J}$).

Le *théorème III* dit que si Δ est recouvert par des ouverts U_α, et si, dans chaque U_α, on a un idéal de $\mathcal{O}(U_\alpha)$ engendré par un nombre fini de fonctions holomorphes, de façon qu'en tout point de $U_\alpha \cap U_\beta$ les idéaux attachés à U_α et à U_β engendrent le même idéal ponctuel, alors il existe un système fini de fonctions holomorphes sur Δ qui engendre en tout point l'idéal ponctuel donné. – Ceci, en termes de faisceaux, s'énonce comme suit: si $\mathcal{J}$ est un faisceau cohérent d'idéaux sur Δ, il existe un nombre fini de sections $\in \Gamma(\Delta, \mathscr{I})$ qui engendrent ce faisceau (théorème A de CARTAN).

Enfin le *théorème IV* dit que si l'on se donne des F_i holomorphes en nombre fini sur Δ, le faisceau des relations linéaires (à coefficients holomorphes) entre les F_i est engendré par un nombre fini de sections de ce faisceau (cela résulte du théorème A parce que OKA a prouvé que ce faisceau est cohérent).

Au point de vue technique, il est bon d'attirer l'attention du lecteur sur l'important "théorème du reste" (nº 5). Il ressemble au lemme donné par H. CARTAN (Ann. E.N.S. 1944, p. 170) auquel on pourrait d'ailleurs le ramener, en observant que $F(x, y)$ est équivalente, dans le domaine considéré par OKA, à un polynôme en y de degré λ.

VIII. Fundamental Lemma

Lemme fondamental

Journal of the Mathematical Society of Japan 3 (1951), p. 204–214; 259–278

Introduction. The principal problems we have dealt with since Memoir I are the following: COUSIN problems, the problem of expansions and the problem of (different types of) convexity [1]. In Memoirs I–VI [2], we have seen, to put it in one word, that these problems can be solved affirmatively for univalent domains without points at infinity [3]. Furthermore, the author has verified, albeit without publishing this, that these results remain valid at least as far as domains without points at infinity and without interior ramification points [4].

We must therefore either introduce suitable points at infinity or allow points of ramification. Now, one will find that almost nothing is known about domains with interior ramification. For example, what happens to the local expansion of a function? We shall therefore deal first with the second problem.

Now, the fundamental idea of our researches is symbolised by Theorem II of Memoir I. We have actually used it in the form of Theorem I of Memoir II [5] because we could not solve problem (E). But the original form is indispensable

---

1) These problems are based on H. BEHNKE and P. THULLEN, Theorie der Funktionen mehrerer komplexer Veränderlichen, 1934. Let us explain them in precise form. Let $\mathfrak{D}, \mathfrak{D}_0$ be two domains over the space of n complex variables connected or not such that $\mathfrak{D}_0 \subseteqq \mathfrak{D}$ (i.e. such that $\mathfrak{D}_0$ is a "Teilbereich" of $\mathfrak{D}$). We shall say that $\mathfrak{D}_0$ is holomorph-convex with respect to $\mathfrak{D}$ if $\mathfrak{D}_0 \subsetneqq H$, H being the "Regularitätshülle of $\mathfrak{D}_0$, and if, in addition, for every domain Δ_0, connected or not, such that $\Delta_0 \Subset \mathfrak{D}_0$ (that is, $\Delta_0 \subset \mathfrak{D}_0$ and $\Delta_0 \ll \mathfrak{D}_0$), we can find a domain Δ, connected or not such that $\Delta_0 \subset \Delta \Subset \mathfrak{D}_0$ and such that, to every point P_0 of $\mathfrak{D}_0 - \Delta$, there corresponds a function f holomorphic on $\mathfrak{D}$ with $|f(P_0)| > \max |f(\Delta_0)|$. In particular, if $\mathfrak{D}_0$ has this property with respect to itself, we call it, with H. BEHNKE, holomorph-convex (regulärkonvex). The problems are then the following: COUSIN problems. Find a meromorphic (or holomorphic) function having given poles (or given zeros satisfying a certain additional condition). Problem of expansions. Let $\mathfrak{D}_0$ be a domain (connected or not) holomorph-convex with respect to $\mathfrak{D}$; for any function f holomorphic on $\mathfrak{D}_0$, find a series of holomorphic functions on $\mathfrak{D}$ which converges uniformly to f on any domain Δ_0, connected or not, such that $\Delta_0 \Subset \mathfrak{D}_0$. Problem of convexity. Is every pseudoconvex domain holomorph-convex? For univalent domains, one can replace "holomorph-convex" by "domain of holomorphy" because of the theorem of H. CARTAN and P. THULLEN.

2) The preceding Memoirs are: I. Rationally convex domains, 1936; II. Domains of holomorphy, 1937; III. The second COUSIN problem, 1939 (Journal of Science of Hiroshima University); IV. Domains of holomorphy and rationally convex domains, 1941; V. The CAUCHY integral, 1941 (Japanese Journal of Mathematics); VI. Pseudoconvex domains, 1942 (Tohôku Mathematical Journal); VII. On some arithmetical concepts, 1950 (Bulletin de la Société Mathematique de France).

3) More precisely, we obtained a necessary and sufficient condition for the second COUSIN problem; and the problem of convexity was only explained for two complex variables in order to reduce the ultimate repetition which is inevitable.

4) The author has written this out in detail in japanese and sent it to Prof. T. TAKAGI in 1943.

5) H. BEHNKE and K. STEIN have indicated several times that this theorem can be applied even to multivalent domains without ramification points.

in the study of domains ramified in the interior; this is the fundamental lemma of the title of the present Memoir VIII, which was written just to establish this lemma.

To establish this lemma for (finite) domains without points of ramification, it is obviously sufficient to solve problems (C_2) and (E) and to find, locally, a finite pseudobasis for a geometric ideal with indeterminate domains. Of these, we solved problems (C_2) and (E) in Memoir VII, and, more recently H. CARTAN has solved the last problem, based on Theorem IV of Memoir VII asserting that problem (K) is always solvable[6]. However, if we allow points of ramification, we meet the new difficulty that a holomorphic function on a characteristic variety is not necessarily the restriction of a holomorphic function in (ambient) space. As a consequence, this generates a class of problems (J) which contains, in a certain sense, geometric ideals and is more extensive.

In the present memoir, we shall solve this problem, again using Theorem IV of Memoir VII (see Theorem 2 below). We shall also establish the fundamental lemma and indicate briefly how it can be applied to the main problem mentioned at the beginning. And, in an appendix, we shall use the same theorem to obtain a necessary and sufficient condition that problem (J) be solvable for a given ideal.

Since we shall not admit points at infinity in the space of complex variables anywhere in this Memoir, we shall, in general, omit to specify this condition.

I. Holomorphic Ideals with Indeterminate Domains and Local Pseudobases

1. Generalities. Since the domains considered here are, without exception, *univalent*, we shall again omit this condition in general.

We introduced holomorphic ideals with indeterminate domain in No. 2 of Memoir VII. Let us briefly recall this notion. We consider, in the space $(x_1, x_2, \dots, x_n)$, an ordered pair (f, δ), where δ is a domain, *not necessarily connected*, and f is a function holomorphic on δ. Let (I) be a set of pairs (f, δ) (always ordered); instead of saying that $(f, \delta) \in (I)$, we shall also say that $f \in (I)$ on δ. (I) is said to be a holomorphic ideal with indeterminate domains if it satisfies the following conditions: 1°. if $(f, \delta) \in I$ and α is a function holomorphic

[6]) Let me explain briefly the march of researches on holomorphic ideals. It was W. RÜCKERT who transplanted the notion of "ideal" from the field of algebraic functions to that of analytic functions (1933, Math. Annalen, vol 107, pp. 259–281); and it was H. CARTAN who first noticed the essential difference and obtained an important result (1940, cited in Memoir VII). CARTAN also published the following two memoirs: Idéaux de fonctions analytiques de n variables complexes (Annales de l'École Normale Supérieure, (3), LXI); Idéaux et modules de fonctions analytiques de variables complexes (1950, Bulletin de la Société Mathématique de France).

Now, the author wrote Memoir VII without knowing of the existence of the first of these two papers of CARTAN, or of the existence of the memoir of RÜCKERT. Let us therefore compare our Memoir VII with these two, just mentioned. Memoir VII consists of two parts, the first of which shows that problems (C_1), (C_2) and (E) can be reduced to the single problem (K); this was already indicated by CARTAN, without proof, but with all the necessary preparation. In the second part, the author first proved the remainder theorem in order to solve problem (K); this theorem was already expounded and used by RÜCKERT.

on a (not necessarily connected) domain δ', then $\alpha f \in (I)$ on $\delta \cap \delta'$; 2°. if $(f, \delta) \in (I)$ and $(f', \delta') \in (I)$, then $f + f' \in (I)$ on $\delta \cap \delta'$. In the present memoir, we shall, in general, simply call this an *ideal.* The following topological property follows immediately from the definition: if (I) is an ideal and if $(f, \delta) \in (I)$ and $\delta' \subset \delta$, then $(f, \delta') \in (I)$. We can therefore tell whether or not $f \in (I)$ at a point P.

A point P of the space (x) will be called a *lacunary point* of the ideal (I) if no holomorphic function belongs to (I) at P. The set of lacunary points of (I) is clearly closed. A point P of the space (x) will be called a *zero* of (I) if any holomorphic function belonging to (I) at P vanishes at P. The set of zeros of (I) is clearly closed in the complement of the set of lacunary points of (I). Conversely, every closed set of points in the space (x) is obviously the set of zeros of a suitable ideal in this space.

A finite system of holomorphic functions F_i $(i = 1, 2, \ldots, p)$ on a domain $\mathfrak{D}$ of the space (x) will be called a (finite) pseudobasis of the ideal (I) if it has the following properties: 1°. Each F_i belongs to (I) at every point of $\mathfrak{D}$; 2°. For any function f belonging to (I) at an arbitrary point P of $\mathfrak{D}$, we have $f \equiv 0 \mod (F)$ at P. A system (F) is called a pseudobasis of the ideal (I) at a point P if it is a pseudobasis on a certain neighbourhood of P; every such pseudobasis is called a local pseudobasis. From our point of view, the main problem concerning ideals consists in finding local pseudobases for these ideals; we call this problem (J). In view of what we have seen about the zeros of an ideal, this problem is not always solvable. It is convenient to introduce the following term in studying this problem. Given two ideals (I_1) and (I_2) on the space (x), if they are so related that for any point P of a certain domain $\mathfrak{D}$, any function belonging to one of the ideals at P necessarily also belongs to the other at P, we shall say that these ideals are equivalent on $\mathfrak{D}$ and denote this by $(I_1) \sim (I_2)$. We say that $(I_1) \sim (I_2)$ at a point P if this is the case on a certain neighbourhood of P.

As for these problems (J), we have seen in Memoir VII that problem (K) is always solvable (Theorem IV). This result will be our starting point.

2. General Principles. Let us consider a system of simultaneous linear functional equations of the form

$$A_{i1} F_{i1} + A_{i2} F_{i2} + \ldots + A_{ip} F_{ip} = 0 \quad (i = 1, 2, \ldots, q)$$

in the space (x), with some identities of the form $A_{ij} = A_{kl}$ $(i, k = 1, 2, \ldots, q;\ j, l = 1, 2, \ldots, p)$. In this system, the functions F_{ij} are given holomorphic functions on a domain $\mathfrak{D}$, and the A_{ij} are unknown functions. Let $(A_{11}, A_{12}, \ldots, A_{qp})$ be a system of holomorphic functions on a domain δ, not necessarily connected, contained in $\mathfrak{D}$; if it satisfies the relation above identically, we call it a solution on δ. Consider the set (I) of all the (A_{11}, δ) such that to each of these (A_{11}, δ), there corresponds a solution of the above equation on δ of the form $(A_{11}, A_{12}, \ldots, A_{qp})$. (I) is clearly an ideal. We shall, in general, call such an ideal an *(L)-ideal. Theorem IV of Memoir VII* is then equivalent to saying that

Any (L)-ideal on a domain $\mathfrak{D}$ has a finite pseudobasis at every point of $\mathfrak{D}$.

Let us begin by analysing some general principles which follow immediately from this theorem.

1°. **H. Cartan's Corollary.** *Let* (I_1), (I_2) *be two holomorphic ideals with indeterminate domains in the space* (x) *and let* $(I)=(I_1)\cap(I_2)$; (I) *is clearly also such an ideal. For* (I) *to have a pseudobasis at a point* (x^0), *it is sufficient that* (I_1) *and* (I_2) *do.* (The proof is immediate.)

2°. **Corollary 1.** *Suppose given a holomorphic ideal with indeterminate domains* $(I)=\{(f,\delta)\}$ *in the space* (x) *and a point* (x^0). *We transform the ideal* (I) *with the help of functions* $F(x)$, $\Phi_1(x),\ldots,\Phi_p(x)$ *holomorphic on a neighbourhood* V *of* (x^0) *and form the set* $(J)=\{(\varphi,\delta')\}$ *as follows:*

$$\varphi=fF+A_1\Phi_1+\ldots+A_p\Phi_p,\qquad \delta'=V\cap\delta\cap\alpha_1\cap\ldots\cap\alpha_p$$

where A_i $(i=1,2,\ldots,p)$ *is an arbitrary holomorphic function on* α_i *(and* $(f,\delta)\in(I)$*).* (J) *is clearly an ideal. Suppose that* (J) *has a pseudobasis at the point* (x^0) *and that the functional equation*

$$A_0F+A_1\Phi_1+\ldots+A_p\Phi_p=0$$

has the following property: If $(A_0,A_1,\ldots,A_p)$ *is a solution of this equation at a point of* V, *then* A_0 *belongs necessarily to* (I) *at this point.*

Then (I) *has a pseudobasis at* (x^0).

In fact, if Ψ_i $(i=1,2,\ldots,q)$ is a pseudobasis of (J) in a neighbourhood V' of (x^0) $(V'\subseteqq V)$, consider the functional equation

$$B_1\Psi_1+\ldots+B_q\Psi_q=A_0F+A_1\Phi_1+\ldots+A_p\Phi_p;$$

we shall compare its (L)-ideal $\{(A_0,\delta)\}$, call it (K), with (I) on V'. By definition of (J), every function belonging to (I) at a point of V' belongs necessarily to (K). Conversely, by our second hypothesis, any function belonging to (K) at a point necessarily belongs to (I) at this point. Hence $(K)\sim(I)$. Hence, by the theorem, (I) has a pseudobasis at (x^0). Q.E.D.

3°. Let $(I)=\{(f,\delta)\}$ be an ideal in the space (x), and let $\Phi(x)$ be a holomorphic function on a domain V; let us consider the set $(J)=\{(\varphi,\delta')\}$ for which $\varphi=f+A\Phi$, $\delta'=V\cap\delta\cap\alpha$, A being an arbitrary holomorphic function on the domain α, connected or not. We call this the *adjoint* of (I); we define another ideal $\{(\varphi,\delta')\}$ by the condition $\varphi\Phi=f$, $\delta'=V\cap\delta$, and call it the *quotient* of (I) (both relative to Φ).

If the ideal (I) *has a pseudobasis at a point* (x^0) *in* V, *so do the adjoint and the quotient.*

This is obvious for the adjoint, and one verifies it immediately for the quotient using our theorem. Is there a converse to this statement?

Example. In the space of two complex variables (x, y), consider the ideal generated by the 2 elements $(x\,y, \Delta)$, $(1, \Delta')$, where Δ is the whole space (x, y) and Δ' is given by $|y|>0$. (I) does not have a pseudobasis at the origin. However, the adjoint of (I) with respect to (y, Δ) has the pseudobasis (y) on Δ, which is also a pseudobasis for the quotient of (I) with respect to (x, Δ). Thus the converse statement, either just for the adjoint, or just for the quotient, is not true. But we see that for (y, Δ), the quotient does not have a pseudobasis at the origin, nor does the adjoint with respect to (x, Δ). This leads us to the following fact:

Corollary 2. *Suppose given a holomorphic ideal with indeterminate domain,* (I), *in the space* (x) *and a function* $\Phi(x)$ *holomorphic in a neighbourhood of a point* (x^0). *If both the adjoint and the quotient of* (I) *with respect to* (Φ, V) *have pseudobases at* (x^0), *so then does the original ideal* (I).

In fact, let $(\Phi_1, \Phi_2, \dots, \Phi_p)$ be a pseudobasis of the adjoint on the neighbourhood V' of the point (x^0) $(V' \subseteqq V)$. If we have chosen V' sufficiently small, we have, identically, $\Phi_i = F_i + A_i \Phi$ $(i=1,2,\dots,p)$, where $(F_i) \in (I)$ on V', and the A_i are holomorphic functions on V'. Let f be a function belonging to (I) at a point (x') of V' but otherwise arbitrary. We have, identically near (x'),

$$f = \alpha_1 F_1 + \dots + \alpha_p F_p + (\alpha_1 A_1 + \dots + \alpha_p A_p)\Phi,$$

the α_i being holomorphic functions. Let us denote the quotient of (I) by (J) and $\alpha_1 A_1 + \dots + \alpha_p A_p$ by B. We must then have $B\Phi \in (I)$, that is to say, $B \in (J)$, at (x'). Conversely, if a function satisfies these two conditions at (x') we necessarily have $f \in (I)$ at (x'). Let now $(\Psi_1, \Psi_2, \dots, \Psi_q)$ be a pseudobasis of (J) on V' (again taken sufficiently small). The second of the necessary and sufficient conditions can be expressed by

$$\alpha_1 A_1 + \dots + \alpha_p A_p = \beta_1 \Psi_1 + \dots + \beta_q \Psi_q,$$

where the β_i $(i=1,2,\dots,q)$ are holomorphic functions. The ideal (I) is thus equivalent to an (L)-ideal, and so possesses a pseudobasis at (x^0). Q.E.D.

3. Geometric Ideals. Given a domain $\mathfrak{D}$ in the space (x) and a characteristic (or analytic) variety Σ in $\mathfrak{D}$,[7] let us consider the set (I) of pairs (f, δ) such that $\delta \subseteqq \mathfrak{D}$ and f is identically zero on $\Sigma \cap \delta$. Obviously, (I) is an ideal. We shall call it the geometric ideal with indeterminate domains (attached to Σ and defined in $\mathfrak{D}$).

H. Cartan's Theorem. *Every geometric ideal with indeterminate domains possesses a local pseudobasis.*

We shall prove this again, using *Corollary 1*. Let (x^0) be an arbitrary point of Σ; it is sufficient to show that (I) has a pseudobasis at (x^0). We know, thanks to WEIERSTRASS, that the part of Σ in a neighbourhood of (x^0) consists

[7]) A characteristic variety is a set of points which can be locally expressed as the set of common zeros of a finite number of holomorphic functions.

of a finite number of branches (elements). Even if we do not appeal to the fact that these branches are again characteristic varieties, we can define an ideal as above for each of these branches, and continue to call it a geometric ideal for the time being. Since (I) is the intersection of these ideals in a neighbourhood of (x^0), it is sufficient, *by Cartan's corollary*, to show that each of them has a pseudobasis at (x^0). This is evident if Σ is a point or a surface.

Let us therefore consider a space $(x_1, \dots, x_n, y_1, \dots, y_m)$ with $n>0$, $m>1$, a branch Σ of dimension n (always complex in this memoir) of a characteristic variety in the neighbourhood of a point (x^0, y^0) of Σ, and the corresponding geometric ideal (I). We shall show that (I) has a pseudobasis at (x^0, y^0). Thanks to WEIERSTRASS, we can choose the coordinates (x, y) and find a polycylinder $[(\gamma), (\gamma')]$ of the form (γ): $|x_i - x_i^0| < r$, $(i=1,2,\dots,n)$, (γ'): $|y_j - y_j^0| < \rho$ $(j=1,2,\dots,m)$ such that Σ and (I) are defined in $[(\gamma), (\gamma')]$ and the following conditions are satisfied: the projection[8] of Σ on the (y)-space is (γ'), and (I) contains, on $[(\gamma), (\gamma')]$, holomorphic functions of the form

$$F_i(x, y_i), \quad \Psi_j(x, y_1, y_j) = y_j \frac{\partial F_1(x, y_1)}{\partial y_1} - \Phi_j(x, y_1),$$
$$(i=1,2,\dots, m;\ j=2,\dots,m),$$

where $F_i(x, y_i)$ is a polynomial in y_i whose coefficient of highest degree is 1 and $\Phi_j(x, y_1)$ is a polynomial in y_1. We can further arrange for $F_1(x, y_1)$ to have the property that the projection of Σ onto the (x, y_1)-space coincides with $F_1(x, y_1) = 0$ and that the intersection of Σ and $\dfrac{\partial F_1(x, y_1)}{\partial y_1} = 0$, if it is non-empty, is of dimension $n-1$. (Consequently, F_1 does not have multiple factors.)

Let (x', y') be an arbitrary point of $[(\gamma), (\gamma')]$, and let $f(x, y)$ be a function belonging to (I) at (x', y'), but otherwise arbitrary. Because of *the remainder theorem* given in Memoir VII, we can find a holomorphic function $\varphi(x, y)$ such that $f \equiv \varphi \mod (F_2, F_3, \dots, F_m)$ at (x', y') and φ is a polynomial in $y_2, y_3, \dots, y_m$ whose degree has an upper bound independent of f and of (x', y'). We can therefore choose a positive integer λ independent of f and of (x', y') such that $\left(\dfrac{\partial F_1}{\partial y_1}\right)^{\lambda} f \equiv \psi \mod (F_2, \dots, F_m, \Psi)$ at (x', y'), ψ being a holomorphic function of (x, y_1). Since $\psi(x, y_1)$ belongs to (I), it is clearly divisible by $F_1(x, y_1)$ at (x', y'). Hence $\left(\dfrac{\partial F_1}{\partial y_1}\right)^{\lambda} f \equiv 0 \mod (F, \Psi)$ at (x', y').

As in Corollary 1, we now form $(J) = \{(\varphi, \delta')\}$ by transforming $(I) = \{(f, \delta)\}$ as follows:

$$\varphi = f\left(\frac{\partial F_1}{\partial y_1}\right)^{\lambda} + A_1 F_1 + \dots + A_{2m-1} \Psi_m, \qquad \delta' = \delta \cap \alpha_1 \cap \dots \cap \alpha_{2m-1}.$$

Now (J) has a pseudobase at (x^0, y^0), and we have only to examine the second condition. Consider the functional equation

$$A_0 \left(\frac{\partial F_1}{\partial y_1}\right)^{\lambda} + A_1 F_1 + \dots + A_{2m-1} \Psi_m = 0,$$

---

[8] The projection of the set of points (x', y') on the space (x) is the set of points (x').

and let $(A_0, A_1, \ldots, A_{2m-1})$ be any solution at an arbitrary point (x', y') in $[(\gamma), (\gamma')]$; then A_0 is necessarily zero on Σ since this is the case for $F_1, \ldots, \Psi_m$, but not for $\frac{\partial F_1}{\partial y_1}$; hence $A_0 \in (I)$ at (x', y'). Q.E.D.

It is now clear that the branch Σ is a characteristic variety since the common zeros of the functions of a pseudobasis coincides with Σ in a neighbourhood of (x^0, y^0).

4. Projections. Given a set of points (x', y') in the space $(x_1, \ldots, x_n, y_1, \ldots, y_m)$, the corresponding set of points (x') is called the projection of the given set on the space (x). Let $\mathfrak{D}$ be a domain in the space (x), $\mathfrak{D}'$ a domain in the space (y), not necessarily connected, and let (I) be a holomorphic ideal with indeterminate domains in the space (x, y), without lacunary points in $(\mathfrak{D}, \mathfrak{D}')$. Let Σ be the set of zeros of (I) in $(\mathfrak{D}, \mathfrak{D}')$. Suppose that for every (x) in $\mathfrak{D}$, the projection of Σ onto the (y)-space is $\Subset \mathfrak{D}'$. Consider, in the space (x), the set of pairs $J = \{(f, \delta')\}$ such that $\delta \subseteqq \mathfrak{D}$ and f belongs to (I) at every point of $(\delta, \mathfrak{D}')$; (J) is clearly an ideal, which we shall call *the projection of the ideal* (I) onto the space (x) with respect to $(\mathfrak{D}, \mathfrak{D}')$.

Theorem 1. *If the ideal* (I) *has a pseudobasis at every point of* $(\mathfrak{D}, \mathfrak{D}')$, *its projection* (J) *has a pseudobasis at every point of* $\mathfrak{D}$.

Let (x^0) be an arbitrary point of $\mathfrak{D}$. It is sufficient to show that (J) has a pseudobasis at (x^0). Thanks to WEIERSTRASS, the intersection of Σ and the variety $x_i = x_i^0$ $(i = 1, 2, \ldots, n)$ is obviously a finite number of points; let (x^0, y^0) be any one of these points (if they exist). We consider a polycylinder $\Delta' \Subset \mathfrak{D}'$ around (y^0), and a polycylinder Δ around (x^0) which is so small that $\Delta \Subset \mathfrak{D}$ and the projection of $\Sigma \cap (\Delta, \Delta')$ onto the space (y) is $\Subset \Delta'$. Let (K) be the projection of (I) onto the space (x) with respect to (Δ, Δ'). In (a sufficiently small) Δ, (J) is clearly the intersection of a finite number of ideals of the form (K), and, by virtue of *Cartan's corollary*, it is sufficient to justify the proposition for (K).

Let (K_1) be the projection of (I) onto the space $(x, y_1, \ldots, y_{m-1})$, with respect to (Δ, Δ'). If the proposition is true for this case, it is obviously also true for (K). We therefore suppose that $m = 1$, and denote y_1 by y; Δ' is then a circle. Under these circumstances, we shall verify that (K) has a pseudobasis at (x^0).

Since the ideal (I) has a pseudobasis at every point in a neighbourhood of (Δ, Δ'), it has a global pseudobasis in the neighbourhood of (Δ, Δ') *by Theorem III of Memoir VII*; we denote such a pseudobasis by $(F, \Phi_1, \ldots, \Phi_p)$. The set of common zeros of these functions necessarily coincides with Σ. We can therefore suppose that $F(x^0, y) \not\equiv 0$, because otherwise, one of the Φ_i $(i = 1, 2, \ldots, p)$ would have this property. We can further suppose that the equation $F(x^0, y) = 0$ has no root on the boundary of Δ' since Σ does not meet (x^0, boundary of Δ'). We reduce further the size of Δ so as to have $F = \omega F_1$ on a neighbourhood of (Δ, Δ'), where ω is holomorphic function, nowhere zero, and F_1 is a polynomial in y whose coefficients are holomorphic functions of (x) in a neighbourhood of Δ, and the coefficient of the highest power of y is 1. We can further assume that for any (x) in a neighbourhood of Δ, the equation $F_1=0$ has no roots outside Δ'. We suppose that F itself has these properties. Let λ be the degree of the polynomial F. By the *remainder theorem*, we can suppose that the Φ_i $(i=1,2,\ldots,p)$ are polynomials in y of degree $<\lambda$.

Let (x') be an arbitrary point of Δ, and let $\varphi(x,y)$ be a polynomial in y of degree $\leqq 2\lambda-2$ and belonging to (I) at every point in a neighbourhood of $[(x'),\Delta']$. If δ is a sufficiently small polycylinder around (x'), we have $\varphi\equiv 0 \bmod (F,\Phi)$ at every point in a neighbourhood of (δ,Δ'); consequently, by Theorem I of Memoir VII, φ can be expressed in the form

$$\varphi=C_0F+C_1\Phi_1+\ldots+C_p\Phi_p,$$

the C_i $(i=0,1,\ldots 2p)$ being functions holomorphic in a neighbourhood of (δ,Δ'). By the remainder theorem, we can choose the C_i $(i=1,2,\ldots,p)$ to be polynomials in y of degree $\leqq\lambda-1$. We assert then that C_0 is also a polynomial in y. In fact, we have $C_0=\Psi/F$, Ψ being a polynomial in y whose coefficients are holomorphic functions of (x) in a neighbourhood of δ; also, for any (x') in a neighbourhood of δ, the equation $F(x',y)=0$ has all its roots inside Δ', and C_0 is holomorphic in a neighbourhood of (δ,Δ'). On the other hand, if $a(x)$ is the coefficient of the highest power of y in Ψ, and $\eta(x)$ is an arbitrary root of the equation $\Psi=0$ in y, the only singularities of the analytic function $a(x)\cdot\eta(x)$ in the neighbourhood of δ are points of ramification. It follows from this that C_0 is a polynomial, consequently of degree $\leqq\lambda-2$.

We have thus seen that any function $\varphi(x,y)$ which is a polynomial in y of degree $\leqq 2\lambda-2$ and which belongs to (I) at any point in a neighbourhood of $[(x'),\Delta']$ can be represented in the above form with C_0, C_1, ..., C_p having the properties indicated. The converse is, obviously, also true. Let us set

$$F=y^\lambda+A_1y^{\lambda-1}+\ldots+A_\lambda,\quad \Phi_i=A_{i1}y^{\lambda-1}+\ldots+A_{i\lambda}\quad (i=1,2,\ldots,p),$$
$$C_0=u_2y^{\lambda-2}+\ldots+u_\lambda,\quad C_i=u_{i1}y^{\lambda-1}+\ldots+u_{i\lambda},\quad \varphi=B_0y^{2\lambda-2}+\ldots+B_{2\lambda-2};$$

the necessary and sufficient condition given above is then equivalent to saying that

$$B_0=u_2+\sum A_{i1}u_{i1},\ldots,B_{2\lambda-2}=u_\lambda A_\lambda+\sum u_{i\lambda}A_{i\lambda},\quad (i=1,2,\ldots,p).$$

As a special case of this, the necessary and sufficient condition that we have $f(x)\in(K)$ at x' is that we can find (u) satisfying, with f, the system of simultaneous functional equations

$$u_2+\sum u_{i1}A_{i1}=0,\ldots,f=u_\lambda A_\lambda+\sum u_{i\lambda}A_{i\lambda}\quad (i=1,2,\ldots,p),$$

in which the A_j, A_{ij} $(j=1,2,\ldots,\lambda)$ are given holomorphic functions on Δ and (x') is an arbitrary point of Δ. The ideal (K) is thus equivalent to an (L)-ideal and so possesses a pseudobasis at (x^0). Q.E.D.

II. Domains with Interior Ramification

5. Definitions. We shall follow the usage in the book of H. BEHNKE and P. THULLEN when we consider the following objects: domains without interior ramification points over the (finite) space of n complex variables $(x_1, x_2, \ldots, x_n)$; boundary points of such domains; points of ramification considered as boundary points; etc. We shall now define domains with interior ramification.

Let M be a ramification point of a domain $\mathfrak{D}$ [Translator's note: $\mathfrak{D}$ is an unramified domain over the space (x) and M is a boundary point of $\mathfrak{D}$.] We call M a *non-transcendental point of ramification* if it satisfies the following conditions: 1°. the order is finite[9]; 2°. Let $\underline{M}$ be the base-point (Grundpunkt) of M; there is a univalent domain $\underline{\delta}$ containing $\underline{M}$ such that if δ is the connected component of the part of $\mathfrak{D}$ over $\underline{\delta}$ which has M as a boundary point, then the set of base-points corresponding to boundary points of δ over $\underline{\delta}$ form a characteristic surface.

Let $\mathfrak{D}$ be an arbitrary domain as above. We consider a set of points, $\mathfrak{D}'$, obtained by adjoining to $\mathfrak{D}$ a part of its non-transcendental points of ramification, which could be empty or contain all of them, but which has the following property: if M is a point of ramification of $\mathfrak{D}$ belonging to $\mathfrak{D}'$, there exists a univalent domain $\underline{\delta}$ containing the base-point $\underline{M}$ of M such that, if δ denotes the connected component of the portion of $\mathfrak{D}$ over $\underline{\delta}$ which has M as a boundary point, then any non-transcendental point of ramification of δ belongs to $\mathfrak{D}'$. We shall, from now on, call such a $\mathfrak{D}'$ a *domain*. Any point of $\mathfrak{D}'$ which is not a point of ramification will be said to be *regular*. As for relations between two such domains, or the intersection of two such domains, it is sufficient to define them by means of the domains consisting of the regular points of the given domains.

Let $\mathfrak{D}$ be a domain and $\mathfrak{D}_0$ the set of regular points of $\mathfrak{D}$; $\mathfrak{D}$ will be called *pseudoconvex* if $\mathfrak{D}_0$ is.

Let $\mathfrak{D}$ be a domain and let P be any point of $\mathfrak{D}$; we shall generally denote the coordinates of P by (x). A function $f(P)$ is called *holomorphic* on $\mathfrak{D}$ if it satisfies the following conditions: 1°. to each point P of $\mathfrak{D}$, there corresponds one and only one value of f, which is finite; 2°. $f(P)$ is a continuous function of P; 3°. in the neighbourhood of any regular point of $\mathfrak{D}$, $f(P)$ is a holomorphic function of (x). A domain will be called a *domain of holomorphy* if there exists a function which is holomorphic on $\mathfrak{D}$ but is not holomorphic on any other domain containing $\mathfrak{D}$.

A set of points Δ which can be expressed in the form:

$$\Delta \Subset (R), \quad f_i(P) \in A_i \quad (i=1,2,\ldots,m)$$

will be called, with H. Cartan[10], a *polyhedral domain;* here (R) is a subdomain (Teilbereich, not necessarily connected) of a domain of holomorphy, $f_i(P)$ are holomorphic functions on (R) and the A_i are bounded, *closed, simply connected* plane domains. Domains of this type (which are closed, but not necessarily connected) were first considered by A. Weil[11]; it is for them that we established the fundamental lemma (Theorem II) in Memoir I; we are in the process of re-establishing this lemma in the present memoir.

---

[9]) Instead of taking the order to be the number m of sheets as in page 13 of the book of Behnke-Thullen, we shall define it to be $m-1$.

[10]) H. Cartan, 1950, No. 21.

[11]) A. Weil, 1932 (C.R.), 1935 (Math. Annalen).

6. Property (H). Let $\mathfrak{D}$ be a univalent domain in the space (x) and S a characteristic variety in $\mathfrak{D}$. A function u on S will be called holomorphic if it satisfies the following two conditions: 1°. to any regular point of S corresponds a well-determined value of u, in such a way that u is locally the restriction of a holomorphic function in the space (x); 2°. u is bounded on the regular points in a neighbourhood of an arbitrary point of S. u will be called holomorphic at a point of S if it is holomorphic in a neighbourhood of this point.

Let M be a point of S and u a function on S holomorphic at M; if there is a holomorphic function in a neighbourhood of M in the space (x) whose restriction is u, except perhaps at the singular points of S, we shall say that u has property (H). If every such u has property (H), we shall say that the point M has the property (H).[12] To study this property, we introduce the following lemma:

H. Cartan's Lemma. *(The theorem of three rings.) In the space* $(x_1, x_2, \ldots, x_n)$ $(n \geqq 3)$, *let us consider three univalent domains* $\Delta_1, \Delta_2, \Delta_3$ *of the following form:*

$$(\Delta_1) \qquad \rho_1 < |x_1| < r_1, \qquad |x_2| < r_2, \qquad |x_3| < r_3, \qquad (x_4, \ldots, x_n) \in \mathfrak{D}$$

$$(\Delta_2) \qquad |x_1| < r_1, \qquad \rho_2 < |x_2| < r_2, \qquad |x_3| < r_3, \qquad (x_4, \ldots, x_n) \in \mathfrak{D}$$

$$(\Delta_3) \qquad |x_1| < r_1, \qquad |x_2| < r_2, \qquad \rho_3 < |x_3| < r_3, \qquad (x_4, \ldots, x_n) \in \mathfrak{D},$$

where $\mathfrak{D}$ *is a univalent domain, the* r_i $(i = 1, 2, 3)$ *are positive numbers, and the* ρ_i *are positive numbers or zero. Suppose that we are given, for each circular permutation* (i, j, k) *of* $(1, 2, 3)$, *a holomorphic function* $g_i(x)$ *on* $\Delta_j \cap \Delta_k$, *in such a way that we have*

$$g_1 + g_2 + g_3 = 0$$

identically. Then, we can find functions $h_i(x)$ *holomorphic on* Δ_i $(i = 1, 2, 3)$ *such that we have, identically,*

$$g_1 = h_2 - h_3, \qquad g_2 = h_3 - h_1, \qquad g_3 = h_1 - h_2.$$ [13]

Lemma 1. *Let* $\mathfrak{D}$ *be a univalent domain in the space* $(x_1, x_2, \ldots, x_n)$ $(n \geqq 3)$, S *a characteristic surface,* u *a holomorphic function on* S *and* S_0 *a characteristic variety on* S *of (complex) dimension* $n-3$. *If* u *has property* (H) *at every point of* $S - S_0$, *it has this property everywhere on* S.

Let (x^0) be an arbitrary point of S_0; it is sufficient to show that u has property (H) at (x^0). We shall regard (x^0) as the origin, and restrict ourselves in what follows to a sufficiently small neighbourhood of the origin. Since S_0 is of dimension $(n-3)$, we can choose coordinates (x) and positive numbers ρ, r, r' so that, if we form, as in Cartan's lemma, the domains $\Delta_1, \Delta_2, \Delta_3$ with $\rho_1 = \rho_2 = \rho_3 = \rho$, $r_1 = r_2 = r_3 = r$, with the $\mathfrak{D}$ of that lemma being expressed by $|x_i| < r'$ $(i = 4, \ldots, n)$, then there are no points of S_0 in a neighbourhood of any Δ_j $(j = 1, 2, 3)$.

---

[12]) For example, the point $(0,0)$ on the surface $y^2 = x^3$ in (x, y)-space does not have property (H).
[13]) H. Cartan: Note sur le premier problème de Cousin, 1938, C.R.

Now u has property (H) at each point of S in a neighbourhood of Δ_1. Hence, by *Theorem II of Memoir VII* concerning problem (C_2), we can find a holomorphic function $F_1(x)$ in a neighbourhood of Δ_1 such that its restriction to S is u (it being always understood that this is only upto singular points of S).[14)] Similarly, we can find functions $F_2(x)$ and $F_3(x)$ for Δ_2 and Δ_3 respectively.

Now, $F_1 - F_2$ vanishes identically on S in $\Delta_1 \cap \Delta_2$. Let $F(x)=0$ be the equation of S, F being a holomorphic function without multiple factors on (C): $|x_i|<r$, $|x_j|<r'$ $(i=1,2,3;\ j=4,\dots,n)$; we therefore have

$$F_1 - F_2 = g_3 F$$

identically on $\Delta_1 \cap \Delta_2$, g_3 being a holomorphic function on $\Delta_1 \cap \Delta_2$. Similarly, on $\Delta_2 \cap \Delta_3$ and $\Delta_3 \cap \Delta_1$, we have respectively

$$F_2 - F_3 = g_1 F, \qquad F_3 - F_1 = g_2 F.$$

Since

$$g_1 + g_2 + g_3 = 0,$$

we can find, by virtue of *Cartan's lemma*, holomorphic functions $h_i(x)$ on Δ_i $(i=1,2,3)$, in such a way as to have, identically,

$$g_1 = h_2 - h_3, \quad g_2 = h_3 - h_1, \quad g_3 = h_1 - h_2.$$

Let us set

$$\Phi_i = F_i - h_i F \qquad (i=1,2,3);$$

each Φ_i is then a holomorphic function on Δ_i, and becomes equal to u on S; we have

$$\Phi_i - \Phi_j = (F_i - F_j) - (h_i - h_j)F = g_k F - g_k F = 0,$$

so that the Φ_i are just parts of the same holomorphic function $\Phi(x)$ on $\Delta_1 \cup \Delta_2 \cup \Delta_3$.

Now, thanks to Hartogs, $\Phi(x)$ must then remain holomorphic on the polycylinder (C). Let v be the restriction of $\Phi(x)$ to S; we have $v=u$ identically on the portion of S in $\Delta_1 \cup \Delta_2 \cup \Delta_3$. Now, we can regard each (irreducible) component of S in (C) as the position of the poles of a meromorphic function in (C), and, by Hartogs' theorem, necessarily contains points in $\Delta_1 \cup \Delta_2 \cup \Delta_3$. We therefore have $v=u$ identically on S, and u is thus the restriction of $\Phi(x)$ at the origin. Q.E.D.

7. (*W*)-functions. Let us consider a univalent domain $\mathfrak{D}$ in the space $(x_1, x_2, \dots, x_n)$ and a covering domain (Überlagerungsbereich) $\mathfrak{D}$ of $\underline{\mathfrak{D}}$, which is not necessarily connected. Suppose that $\mathfrak{D}$ has a finite number of sheets; then $\mathfrak{D}$ has a constant number ν of sheets over $\underline{\mathfrak{D}}$. Let P be an arbitrary point of $\mathfrak{D}$ having coordinates (x) and let $\eta_1(P), \eta_2(P), \dots, \eta_m(P)$ be holomorphic functions on $\mathfrak{D}$. Suppose that *for any pair of regular points of* $\mathfrak{D}$ *having the same coordinates, the analytic elements of* $\eta_1(P)$ *at these two points are different*. We

---

[14)] To apply the theorem concerning the closed polycylinder to the present case, it suffices to consider, as usual, $y = 1/x_1$.

consider the characteristic variety Σ in the space $(x_1, x_2, \ldots, x_n;\ y_1, y_2, \ldots, y_m)$ given by

$$y_i = \eta_i(P) \qquad (i = 1, 2, \ldots, m).$$

To a point P of $\mathfrak{D}$ corresponds a point M of Σ having the coordinates (x, η); and conversely, because of the above property of η_1, to any regular point of Σ, there corresponds one and only one point of $\mathfrak{D}$.

Let Δ be a univalent domain in the space (x, y) contained in the domain given by $(x) \in \underline{\mathfrak{D}}$ which contains points of Σ. Let us consider a holomorphic function $F(x, y)$ on Δ which has the following property: for any function u holomorphic on Σ at an arbitrary point M of $\Sigma \cap \Delta$, uF has property (H). We shall call F a *(W)-function*. It is clear that (W)-functions on Δ with respect to Σ form a holomorphic ideal with (determinate) domain Δ.

Let $P_1, P_2, \ldots, P_\nu$ be the points of $\mathfrak{D}$ over the point (x); let us form the function

$$F_1(x, y_1) = \prod [y_1 - \eta_1(P_i)] \qquad (i = 1, 2, \ldots, \nu).$$

We know, thanks to WEIERSTRASS, that to every holomorphic function $u(P)$ on $\mathfrak{D}$, there corresponds a polynomial $\Phi(x, y_1)$, whose coefficients are holomorphic functions of (x) on $\underline{\mathfrak{D}}$, such that $u = \dfrac{\Phi}{(\partial/\partial y_1) F_1}$ on $F_1 = 0$. The usual proof[15] applies whether $\mathfrak{D}$ is connected or not. In other words:

$\dfrac{\partial F_1}{\partial y_1}$ *is a (W)-function with respect to Σ on the domain* $(x) \in \underline{\mathfrak{D}}$.

Let us look at the function η_1. Let σ be the set of points of ramification of the domain $\mathfrak{D}$; σ is a characteristic surface on $\mathfrak{D}$ (i.e. the image of a characteristic variety on Σ of dimension $(n-1)$); we call σ the *ramification surface*. Let σ_0 be any (irreducible) component of σ and let P' be a point of σ_0 such that the set $\underline{\sigma_0}$ of base-points of the points of σ_0 (which we call the base-set of σ_0) is regular at the point corresponding to P', and such that no other component of σ passes through P'. By applying a pseudoconformal transformation in the neighbourhood of $\underline{P'}$ in (x)-space, $\underline{P'}$ being the base-point of P', we can suppose the characteristic surface $\underline{\sigma_0}$ as being given by $x_1 = 0$. Let $\mu - 1$ be the order of the ramification surface σ_0; in the neighbourhood of P', we have

$$\eta_1 = a_0 + a_1 t + a_2 t^2 + \ldots, \qquad t = (x_1)^{1/\mu},$$

the a_i $(i = 0, 1, \ldots)$ being holomorphic functions of $(x_2, x_3, \ldots, x_n)$. Two cases are possible, depending on the nature of a_1.

Case 1, when $a_1 \not\equiv 0$ on σ_0. – Let S be the characteristic surface in the space (x, y_1) given by $F_1(x, y_1) = 0$, let M' be the point on S corresponding to P' and let S_0 be the (irreducible) component of S in a sufficiently small neighbourhood of M' which corresponds to the sheets of $\mathfrak{D}$ on which P' lies. If $a_1 \neq 0$ at P', S_0 is clearly regular at M'. Hence, in this case, the set of singular points of S_0 is of dimension at most $(n-2)$; consequently, by *Lemma 1*, every point of S_0 has

[15]) See: W.F. OSGOOD, Lehrbuch der Funktionentheorie II, 1, 1929, pp. 116–117.

property (H). Any ramification surface having this property will be called a ramification surface *of the first kind* with respect to η_1.

Case 2, when $a_1 \equiv 0$ on σ_0. – Let T_0 be the set of points of S_0 corresponding to points of σ_0. In this case, the function $(x_1)^{1/\mu}$ shows that no point of T_0 has property (H) with respect to S_0. We say that a ramification surface with this property is *of the second kind* with respect to η_1.

Let P be a point of $\mathfrak{D}$; if there exists another point P' of $\mathfrak{D}$ having the same coordinates as $\underline{P}$ and such that $\eta_1(P)=\eta_1(P')$, we shall call P an *ambiguous point* with respect to η_1. Let τ be the set of points of $\mathfrak{D}$ which are ambiguous with respect to η_1; τ is clearly a characteristic surface on $\mathfrak{D}$, which we shall call the *surface of ambiguity*. Let M be the point of S corresponding to a point P of $\mathfrak{D}$ ambiguous with respect to η_1; there are at least two branches of S passing through M. M does not have property (H) with respect to S. In fact, consider a function u on S in the neighbourhood of M which is identically zero on one of the branches and 1 on the others. By definition, u is a holomorphic function on S, but there is no holomorphic function of (x, y_1) in the neighbourhood of M which equals u at the regular points of S.

We have thus seen that no point of S which corresponds either to a ramification point of the second kind or to an ambiguous point of $\mathfrak{D}$ with respect to η_1 can have property (H). Every other point of S obviously does have this property. From which, we obtain:

Let Δ be a bounded univalent domain in the space (x, y) which is contained, together with its boundary, in the domain $(x)\in\underline{\mathfrak{D}}$, and let $U(x, y)$ be a holomorphic function on Δ such that, if we set $u_0(P)=U[x_1, \ldots, x_n, \eta_1(P), \ldots, \eta_m(P)]$, then u_0 is identically zero on the ramification surface σ' of the second kind and on the surface τ of ambiguity with respect to η_1. We can then find a positive integer λ such that U^λ is a (W)-function in Δ with respect to Σ.

In fact, let $\mathfrak{D}_0$ be the domain in $\mathfrak{D}$ corresponding to $\Delta\cap\Sigma$; let $v_0(P)$ be the function obtained by substituting $y_i=\eta_i(P)$ $(i=1,2,\ldots,m)$ in $\partial F_1/\partial y_1$. We can find a positive integer λ depending in $\mathfrak{D}_0$ such that u_0^λ is divisible by v_0 on σ' and on τ, except perhaps for a characteristic variety (on $\mathfrak{D}$) of smaller dimension. Let M be an arbitrary point of Σ in Δ and u an arbitrary holomorphic function at M on Σ. Thanks to WEIERSTRASS, $u U^\lambda$ then has property (H) at every point of a neighbourhood of M, except perhaps on a variety of dimension $(n-2)$ on Σ; consequently without exception because of Lemma 1.

Let (x^0) be an arbitrary point of $\underline{\mathfrak{D}}$, and $(C)\Subset\underline{\mathfrak{D}}$ a polycylinder around (x^0); let R be a positive number such that $|\eta_i|<R$ $(i=1,2,\ldots,m)$ for any (x) in (C) and (Γ) the polycylinder of radius R around the origin in the space (y). We denote by Δ the polycylinder $[(C),(\Gamma)]$.

Let T be the set of common zeros of all (W)-functions on Δ with respect to Σ, and let S_0 be the set of singular points of Σ in Δ. We have $T\subseteqq S_0$.

First, we have $T\subseteqq\Sigma$. In fact, the geometric ideal (with indeterminate domain) attached to Σ on $(x)\in\underline{\mathfrak{D}}$ has a pseudobasis at every point in a neighbourhood of Δ by *Cartan's theorem*, consequently a pseudobasis $(\Phi_1, \Phi_2, \ldots, \Phi_p)$ on Δ *by Theorem III of Memoir VII*. The set of common zeros of the functions Φ_i $(i=1, 2, \ldots, p)$ is Σ in Δ. Now, each Φ_i is a (W)-function on Δ because it is identically zero on Σ. Hence $T \subseteqq \Sigma$.

We consider a non-singular transformation L of the space (x, y) of the form

$$x'_i = \sum A_{ik} x_k + \sum A_{i,n+l} y_l, \quad y'_j = \sum A_{n+j,k} x_k + \sum A_{n+j,n+l} y_l$$

$$(i, k = 1, 2, \ldots, n;\ j, l = 1, 2, \ldots, m),$$

$$|A_{pp} - 1| < \varepsilon, \quad |A_{pq}| < \varepsilon \quad (p \neq q,\ p, q = 1, 2, \ldots, n+m),$$

ε being a positive number which we shall choose later. The images of Σ, Δ in the space (x', y') will be denoted by the same letters. Let us consider the characteristic variety Σ_1 in $(x) \in \mathfrak{D}$ given by $F_i(x, y_i) = 0$ $(i = 1, 2, \ldots, m)$, where

$$F_i(x, y_i) = \prod [y_i - \eta_i(P_j)] \qquad (j = 1, 2, \ldots, \ldots, \nu),$$

and the P_j are the points of $\mathfrak{D}$ over the point (x). Σ consists of certain components of Σ_1. We consider a polycylinder $|x_i - x_i^0| < r_i$ $(i = 1, 2, \ldots, n)$, $\Subset \mathfrak{D}$ and $\Supset (C)$, and let $R' > R$ be a positive number such that for any (x) in this new polycylinder, we have $|\eta_i| < \rho < R'$, ρ being a fixed number. Then, there are no points of Σ_1 in the neighbourhood of any of the m sets of points $|x_i - x_i^0| < r_i$ $(i = 1, 2, \ldots, n)$, $|y_p| = R'$, p representing any one of $1, 2, \ldots, m$. We choose ε so small that the same is true of (x', y'); more precisely, the functions $F_j(x', y') = F_j(x, y_j)$ $(j = 1, 2, \ldots, m)$ are defined and holomorphic in the neighbourhood of $|x'_i - x'^0_i| < r_i$, $|y'_j| < R'$ $(i = 1, 2, \ldots, n;\ j = 1, 2, \ldots, m)$, (x'^0) being the image of (x^0), and, moreover, the functions $F_j(x', y')$ have no common zeros in the neighbourhood of any of the m sets $|x'_i - x'^0_i| < r_i$ $(i = 1, 2, \ldots, n)$, $|y'_p| = R'$ $(p = 1, 2, \ldots, m)$. Thanks to WEIERSTRASS, under these circumstances, we can easily solve for (y') from the system of simultaneous equations $F_j = 0$ $(j = 1, 2, \ldots, m)$ if $|x'_i - x'^0_i| < r_i$ $(i = 1, 2, \ldots, n)$. Since Σ is the union of (irreducible) components of Σ_1, we can therefore also solve for (y') from the equations representing Σ. Let $\mathfrak{D}'$ be the multivalent domain (not necessarily connected) over $|x'_i - x'^0_i| < r_i$ $(i = 1, 2, \ldots, n)$ corresponding to Σ and $y'_j = \eta'_j$ $(j = 1, 2, \ldots, m)$ the equations of Σ, the η'_j being holomorphic functions on $\mathfrak{D}'$. We take ε so small that Δ is $\Subset$ in the polycylinder

$$|x'_i - x'^0_i| < r_i, \quad |y'_j| < R' \qquad (i = 1, 2, \ldots, n;\ j = 1, 2, \ldots, m).$$

Let us consider $(A_{11}, A_{12}, \ldots, A_{n+m,n+m})$ as a point in the polycylinder of radius ε (in the space of $(n+m)^2$ complex variables; ε is as indicated above). It is obvious that η'_1 has the same property as η_1 except perhaps for points (A) in a set of the first category. Let M be an arbitrary regular point of Σ in Δ; it is easy to see that the corresponding point of $\mathfrak{D}'$ is a regular point of $\mathfrak{D}'$ and is not an ambiguous point of η'_1, again excepting points (A) in a set of the first category. We can therefore find (x', y') fulfilling these conditions.

In view of CARTAN's theorem and Theorem III of Memoir VII, we can find a holomorphic function $\Psi(x)$ on Δ such that the image on $\mathfrak{D}'$ of the restriction of Ψ to Σ is identically zero on the ramification surface of the second kind and on the surface of ambiguity with respect to η_1', while Ψ does not vanish at M. In view of what we have seen above, Ψ^λ is a (W)-function on Δ relative to Σ, λ being a certain positive integer. Hence $T \subseteqq S_0$.

8. Nullstellensatz. We have the following lemma:

Lemma of Hilbert-Rückert. – *Let $F_i(x)$ $(i=1,2,\ldots,p)$, $f(x)$ be holomorphic functions at a point (x^0). If $f(x)$ vanishes identically on the set of common zeros of the F_i, we can find a positive integer λ such that $f^\lambda \equiv 0 \bmod (F)$.*[16)]

If F_i $(i=1,2,\ldots,p)$ and f are holomorphic on a polycylinder (C) around (x^0) and are related in the way indicated above on (C), then the lemma is valid at every point of (C); consequently, *by Theorem I of Memoir VII, the lemma is valid on any polycylinder $(C_0) \Subset (C)$.*

If we apply this lemma to the preceding lemma, we obtain

Lemma 2. *Let $F(x,y)$ be a holomorphic function on Δ vanishing identically on the set of singular points S_0 of Σ, and let Δ' be a domain $\Subset \Delta$. Then, one can find a positive integer λ such that F^λ is a (W)-function on Δ' with respect to Σ.*

Now, the essential character of our researches consists in actually *finding* solutions. Hence to show how one can find the number λ in the Nullstellensatz, we shall prove it directly.

1°. *Let $F_1, F_2, \ldots, F_p$ be holomorphic functions at a point (x^0, y^0) in the space $(x_1,\ldots,x_n;\ y_1,\ldots,y_m)$ $(n \geqq 0,\ m>0)$ none of which is identically zero, and let Σ be the set of common zeros of these functions. We suppose that the (irreducible) components of Σ are n-dimensional. For any function $f(x,y)$ vanishing identically on Σ in the neighbourhood of (x^0,y^0), we can find a positive integer μ and a holomorphic function $H(x,y)$ which is not identically zero on any component of Σ in such a way that we have $Hf^\mu \equiv 0 \bmod (F)$ in a neighbourhood of (x^0,y^0).*

Suppose that (x^0,y^0) is a point of Σ. In what follows, we shall place ourselves in a neighbourhood of (x^0,y^0) on which the given conditions are valid. We choose the coordinates and positive numbers r, ρ, ρ' $(\rho > \rho')$ suitably and consider the polycylinder $\Delta = [(\gamma), (\gamma')]$, (γ) being given by $|x_i - x_i^0| < r$ $(i=1,2,\ldots,n)$ and (γ') by $|y_j - y_j^0| < \rho$ $(j=1,2,\ldots,m)$; we may do this in such a way that, when (x) lies in (γ), Σ has no point on $|y_q - y_q^0| \geqq \rho'$ for any q among $1,2,\ldots,m$. Because of the results of WEIERSTRASS, we can then find m functions $\Phi_j(x,y_j)$ with the following properties: $\Phi_j(x,y_j)$ is a polynomial in y_j without multiple factors whose coefficients are holomorphic functions of (x) in (γ), the coefficient of the highest power of y_j is 1, and the projection of Σ in the (x,y_j)-space is given by $\Phi_j = 0$ for $(x) \in (\gamma)$.

We shall show that $\Phi_j^\nu \equiv 0 \bmod (F)$ at (x^0,y^0) for a certain positive integer ν independent of j. Consider the ideal (I) with indeterminate domains generated by (F_k, Δ) $(k=1,2,\ldots,p)$, and its projection (J) on the space (x,y_j) relative to Δ.

---

16) See: S. BOCHNER and W. MARTIN, Several complex variables, 1948, Chapter X.

Let Σ' be the projection of Σ on the space (x, y_j); then, for $(x) \in (\gamma)$, Σ' is a characteristic surface. Now, by *Theorem 1*, (J) has a pseudobasis at every point of $[(x) \in (\gamma), |y_j - y_j^0| < \rho]$; Σ' is necessarily the set of zeros of (J). Since Φ_j is identically zero on Σ', we can find an integer ν' such that $\Phi_j^{\nu'} \in (J)$ at (x^0, y_j^0). Then $\Phi_j^{\nu'} \in (I)$ at (x^0, y^0). Let ν be the largest of the ν' for $j = 1, 2, \ldots, m$; we have $\Phi_j^{\nu} \equiv 0 \mod (F)$ at (x^0, y^0).

Let us consider the characteristic variety T given by $\Phi_j = 0$ $(j = 1, 2, \ldots, m)$ in Δ. Σ consists of (irreducible) components of T in Δ; let Σ' be the union of the components of T which do not belong to Σ. Let us consider the geometric ideal (K) with indeterminate domain attached to T in Δ. It is easy to see that $(\Phi_1, \Phi_2, \ldots, \Phi_m)$ *is a pseudobasis of* (K) *on* Δ (e.g. by using the remainder theorem)[17].

Let φ be a holomorphic function on Δ vanishing identically on Σ' without vanishing identically on any component of Σ. We then have $f\varphi \equiv 0 \mod (\Phi)$ at (x^0, y^0). Now $\Phi_j^{\nu} \equiv 0 \mod (F)$ at (x^0, y^0). Hence $(f\varphi)^{\mu} \equiv 0 \mod (F)$ at (x^0, y^0), where $\mu = m(\nu - 1) + 1$; this justifies the proposition.

2°. We shall now verify the Nullstellensatz formulated above. Suppose that none of the $F_i(x)$ $(i = 1, 2, \ldots, p)$ is identically zero. We first show that if the theorem holds in the case in which Σ consists of components of the same dimension, then it is always true. For example, let $\Sigma = \Sigma_1 \cup \Sigma_2$, where Σ_1 is of dimension λ $(0 < \lambda < m)$ and Σ_2 is of dimension $(\lambda - 1)$. Assuming that Σ_2 is given, in a neighbourhood of (x^0), as the set of zeros of holomorphic functions $\Phi_1(x), \Phi_2(x), \ldots, \Phi_q(x)$, we consider

$$F_1(x), F_2(x), \ldots, F_p(x), \qquad y\Phi_1(x), y\Phi_2(x), \ldots, y\Phi_q(x)$$

in the neighbourhood of the point $(x^0, 0)$ in the space $(x_1, \ldots, x_n, y)$; the set of common zeros of these functions clearly consists of components of dimension λ. $f(x)$ continues to satisfy the vanishing condition and the proposition is true for the new set of functions by hypothesis. Consequently, if we set $y = 0$, it remains true for (F). The same reasoning applies to the general case.

Suppose therefore that the components of Σ are of dimension r $(r < n)$. By the preceding proposition, we have $Hf^{\mu} \equiv 0 \mod (F)$ at (x^0) $(H, \mu$ having the meaning indicated there). Now, if $f^{\nu} \equiv 0 \mod (H, F)$ at (x^0) for a certain positive integer ν, we necessarily have $f^{\nu+\mu} \equiv 0 \mod (F)$ at (x^0). Let Σ' be the intersection of Σ and $H = 0$; by our condition on H, Σ' consists of components of dimension $(r - 1)$ (if it is non-empty). Hence, if the proposition is true for $r - 1$,

---

[17]) In fact, let $f(x, y)$ be a holomorphic function vanishing identically on T in a neighbourhood of a point (a, b) of T. We consider a polycylinder δ around (a) in the space (x) and a polycylinder δ' around (b) in the space (y); we denote its component in the y_j-plane by δ'_j. We suppose that the above condition on f is satisfied on (δ, δ') and that, for any (x) in δ, $\Phi_j(x, y_j) = 0$ has λ_j roots in δ'_j. We then have

$$f = f_0 + \varphi \Phi_1 \tag{1}$$

identically on (δ, δ'), where f_0, φ are holomorphic functions and f_0 is a polynomial in y_1 of degree $< \lambda_1$. Let us choose (x') in δ such that $\Phi_1(x', y_1) = 0$ has λ_1 roots $\eta_1, \ldots, \eta_{\lambda_1}$ in δ'_1, and points y'_k $(k = 2, \ldots, m)$ in δ'_k such that $\Phi_k(x', y'_k) = 0$. Substituting these values in (1), we have $f' = f'_0 + \varphi' \Phi'_1$. Now, if we set $y_1 = \eta_l$ $(l = 1, \ldots, \lambda_1)$, we have $\Phi'_1 = 0$, $f' = 0$, and consequently $f'_0 = 0$. Since f'_0 is a polynomial in y_1 of degree $< \lambda_1$, this means that $f'_0 \equiv 0$. Hence, all the coefficients of f_0 have to be identically zero on $\Phi_k(x, y_k) = 0$, $(k = 2, \ldots, m)$.

it is true for r. The proposition is obviously true for $r=0$ (by 1°). It is therefore always true.

9. (*Z*)-Ideals. We return to the domain $\mathfrak{D}$ over the space (x) and the corresponding variety Σ as at the beginning of No. 7. If we attempt to define a *distribution* ($\mathfrak{z}$) *of zeros* on $\mathfrak{D}$, extending the procedure of the *second Cousin problem* on univalent domains, we immediately run into a new phenomenon, namely:

In general, a characteristic surface on a domain with interior ramification cannot be represented, even locally, as the set of zeros of a single function holomorphic on the domain.

The example which the author has constructed is not very simple and will be given in a later memoir.

Let σ be an irreducible characteristic surface in a domain $\Delta \subseteqq \mathfrak{D}$, and let u be a holomorphic function on Δ vanishing identically on σ; we shall begin by defining the *order of vanishing* of u on σ.

Case 1. σ is not a ramification surface of Δ. Let P_0 be a point of σ which is not a ramification point of $\mathfrak{D}$; in the neighbourhood of P_0, $u(P)$ is a holomorphic function of (x), (x) being the coordinates of P; consequently, we know what the order λ of vanishing of u on σ is; λ does not depend on P_0 and will be called the order of vanishing of u on σ. (If $u \equiv 0$, then $\lambda = \infty$; if u is not identically zero on σ, then $\lambda = 0$.)

Case 2. σ is a ramification surface of Δ. Let P_0 be a point of σ which does not belong to any other component of the ramification surface of. In the neighbourhood of P_0, σ can be represented by an equation $f(x)=0$ where f is a holomorphic function of (x) without multiple factors. Let $\mu - 1$ be the order of the ramification surface σ. Then, if $u \not\equiv 0$, we can find a positive integer λ such that $u/f^{\lambda/\mu}$ is finite in a neighbourhood of P_0, but this is no longer the case for $u/f^{(\lambda+1)/\mu}$; λ does not depend on P_0, and we shall call it the order of vanishing of u on σ. (If $u \equiv 0$, the order is infinite; if u is not identically zero on σ, the order is zero.)

Let σ be a characteristic surface in a domain $\Delta \subseteqq \mathfrak{D}$, let $\sigma_1, \sigma_2, \ldots$ be the (irreducible) components of σ (in Δ); and let $\lambda_1, \lambda_2, \ldots$ be positive integers. We define a distribution ($\mathfrak{z}$) of zeros in Δ by means of $\{(\sigma_i, \lambda_i)\}$ $(i=1,2,\ldots)$ as follows: let u be a holomorphic function on Δ; we say that u *has at least the zeros* ($\mathfrak{z}$) if the order of vanishing of u on σ_i is at least λ_i $(i=1,2,\ldots)$.

Let P_0 be a point on σ and let $(u_1, u_2, \ldots, u_p)$ be a finite system of holomorphic functions in a neighbourhood of P_0. We shall say that the system (u) *defines the zeros* ($\mathfrak{z}$) *at* P_0 if the set of common zeros of these functions is σ upto a set of points of lower dimension and if the smallest order of vanishing of these functions on σ_i is λ_i $(i=1,2,\ldots)$.

Next, let u be a holomorphic function on $\mathfrak{D}$ at a point P_0 of $\mathfrak{D}$; let M_0 be the point of Σ corresponding to P_0. If there exists a function $U(x, y)$, holomorphic at M_0 in the space (x, y), such that the image (on $\mathfrak{D}$) of the restriction of U to Σ is u, i.e. if $U(x, \eta) = u$, we shall say that u *has property* (H) at P_0 with respect to Σ.

Let us now consider a distribution $(\mathfrak{z})$ of zeros on $\mathfrak{D}$; *suppose that* $(\mathfrak{z})$ *is everywhere locally defined by a finite system of holomorphic functions on* $\mathfrak{D}$ *which have property* (H) *with respect to* Σ. We consider, in the space (x, y), the set (I) of pairs (f, δ) such that $\delta \subsetneqq$ the domain $(x) \in \underline{\mathfrak{D}}$ and such that, if (f', δ') is the image (on $\mathfrak{D}$) of the restriction of (f, δ) to Σ, then f' has at least the zeros $(\mathfrak{z})$ in δ'. (I) is clearly an ideal; we shall call such an ideal a *(Z)-ideal.*

Theorem 2. *The (Z)-ideal defined above has a pseudobasis at every point of* $[(x) \in \underline{\mathfrak{D}}]$.

1°. We shall start with the *special case* in which there exists a (W)-function $W(x, y)$ on the set $(x) \in \underline{\mathfrak{D}}$ with respect to Σ which is such that if Z is the set of zeros of the ideal (I), then $W(x, y)$ is not identically zero on any component of T.

Let $F_1(x, y), F_2(x, y), \ldots, F_p(x, y)$ be a system of holomorphic functions at an arbitrary point M_0 of T which defines $(\mathfrak{z})$ at the image P_0 of M_0,[18] and let τ be the image of T. It is easy to see that we can choose constants c_i $(i = 1, 2, \ldots, p)$ such that the image F' of the restriction to Σ of the function $F = \sum c_i F_i$ has the same order of vanishing on τ as $(\mathfrak{z})$, while it is not identically zero on any other component of $\underline{\tau}$, where $\underline{\tau}$ is the base-set of τ. Let τ' be the union of the components of $F' = 0$ not contained in τ, and let $\Phi(x)$ be a holomorphic function of (x) such that it vanishes to at least the order of vanishing of F' on τ', while it is not zero outside $\underline{\tau}'$.

Using these functions W, F, Φ, we transform the ideal $I = \{(f, \delta)\}$ as in *Corollary 1*, and form $(J) = \{(\varphi, \delta')\}$ as follows:

$$\varphi = f\Phi W + A_0 F + A_1 \Psi_1 + \ldots + A_r \Psi_r, \qquad \delta' = V \cap \delta \cap \alpha_0 \cap \alpha_1 \cap \ldots \cap \alpha_r,$$

here V is a (univalent) neighbourhood of M_0 on which F, Φ, W are holomorphic, and (Ψ) is a pseudobasis on V of the geometric ideal (with indeterminate domains) attached to Σ; (Ψ) exists by *Cartan's theorem.* We shall denote the image on $\mathfrak{D}$ of the restriction of a function $\psi(x, y)$ by ψ'. The above relation then becomes

$$\varphi' = f'\Phi W' + A_0' F'$$

on $\mathfrak{D}$. Thus φ'/F' is holomorphic (on the domain in $\mathfrak{D}$ corresponding to δ') and has property (H) with respect to Σ. We therefore have $\varphi \equiv 0 \bmod (F, \Psi_1, \ldots, \Psi_r)$ at M_0. Thus (J) has a local pseudobasis at M_0, and we have only to examine the second condition in Corollary 1.

Let us therefore consider the functional equation

$$B\Phi W + A_0 F + A_1 \Psi_1 + \ldots + A_r \Psi_r = 0$$

on V. Let $(B, A_0, \ldots, A_r)$ be any solution of this equation at an arbitrary point (x', y') of V; it is sufficient to verify that $B \in (I)$ at (x', y'). This is obvious if (x', y')

[18]) That is, a system such that if F_i' is the image on $\mathfrak{D}$ of the restriction of F_i $(i = 1, 2, \ldots, p)$, then (F') defines $(\mathfrak{z})$ in a neighbourhood of P_0.

is not a point of Σ. We therefore suppose that (x, y) is a point of Σ and set $Y_i = 0$ $(i=1,2,\ldots,r)$; we then have the identity

$$B' \Phi W + A'_0 F' = 0.$$

Now, the order of vanishing of the function $\Phi W'$ on τ is zero. Hence B' vanishes on τ at least to the order of vanishing of F', and consequently, $B \in (I)$ at (x', y') by definition. The proposition is thus true in the special case.

The general case can be reduced to the above case using *Theorem 1*. We now explain how this can be done.

2°. Let $\underline{\Delta}$ be a polycylinder such that $\underline{\Delta} \Subset \underline{\mathfrak{D}}$, and let Δ be any connected component of the portion of $\mathfrak{D}$ above $\underline{\Delta}$. By CARTAN's corollary, it is sufficient to verify the result on Δ. Let σ be the ramification surface of $\mathfrak{D}$ and $\underline{\sigma}$ its base-set. Only finitely many of the (irreducible) components of $\underline{\sigma}$ in $\underline{\mathfrak{D}}$ meet $\underline{\Delta}$; we denote them by $\underline{\sigma_1}, \underline{\sigma_2}, \ldots, \underline{\sigma_q}$. By COUSIN's theorem, we can find a holomorphic function $f(x)$ on $\underline{\Delta}$ vanishing to order 1 on $\underline{\sigma_1}$ and being nowhere zero outside. Let r be the least common multiple of $s_1+1, s_2+1, \ldots,$ where $s_1, s_2, \ldots$ are the orders of the ramification surfaces of Δ situated above $\underline{\sigma'_1}$; consider

$$S(x) = (f(x))^{\frac{1}{r}}.$$

Let Δ' be an arbitrary connected component of the intersection of Δ with the domain of existence of $S(x)$. Consider, in the space $(x_1, \ldots, x_1, y_1, \ldots, y_m, z)$, the characteristic variety Σ':

$$y_i = \eta_i(P), \quad z = S(P), \quad (P \in \Delta', \underline{P} = (x); i = 1, 2, \ldots, m)$$

here $\underline{P}$ is the base-point of P.

The system of holomorphic functions in the space (x, y) which defines a distribution of zeros on $\mathfrak{D}$ also defines a distribution of zeros on Δ' since we can consider the functions in the space (x, y, z); we shall denote this distribution by $(\mathfrak{z}')$. Let (I') be the (Z)-ideal in the space (x, y, z) defined by (Σ', Δ') and $(\mathfrak{z}')$. It is easy to see that for $(x) \in \underline{\Delta}$, the ideal (I) *is the projection of* (I') on the space (x, y). Hence, by Theorem 1, it is sufficient to prove that (I') has a pseudobasis at every point of the domain $(x) \in \underline{\Delta}$. Now, if we choose suitably the constant c and form $\eta'_1 = \eta_1 + cS$, then this function has Δ' contained in its domain of existence and does not have a ramification surface of the second kind over $\underline{\sigma_1}$. In other words, we may assume, without loss of generality, that Δ does not have a ramification surface of the second kind over $\underline{\sigma_1}$ (relative to η_1). By reasoning similarly with $\underline{\sigma_2}, \underline{\sigma_3}, \ldots, \sigma_q$ successively, we can suppose that η_1 has no ramification surface of the second kind.

3°. Let us consider again the linear transformation L of the space (x, y) considered at the end of No. 7. If we choose ε sufficiently small, Δ is transformed into a domain Δ' over the space (x'). It is easy to see that the relation is *biunique and pseudoconformal.* Hence, we can study the ideal (I) using Δ' instead of using Δ; since there is no ramification surface of the second kind, we can therefore suppose *that the ramification surface of* Δ *and the surface of ambiguity with respect to* η_1 *do not have a common component.*

Let P_1, P_2 be a pair of regular points of $\mathfrak{D}$ over the same base-point (x) such that there is no point of intersection of the two surfaces mentioned above over

(x); we ask to find a holomorphic function $S(P)$ on $\mathfrak{D}$ such that $S(P_1) \neq S(P_2)$. We once again understand by $\underline{\mathfrak{D}}$ a polycylinder $\Subset$ in the earlier domain. Suppose that the same point M on Σ correponds to the two points P_1, P_2; if this is not the case, one of the functions η_i $(i=1,2,\ldots,m)$ fulfills the required condition. By hypothesis, M cannot be situated on the image in Σ of the ramification surface. The derivatives of η_1 with respect to x_i $(i=1,2,\ldots,n)$ are holomorphic on $\mathfrak{D}$ excepting the ramification surface σ; and on σ, it is obvious that these derivatives have at most poles[19]. Now, $\mathfrak{D}$ being a part of the domain of existence of η_1, there is certainly, among the derivatives of η_1, one, say $\xi(P)$, such that $\xi(P_1) \neq \xi(P_2)$. Let (Γ) be a polycylinder around the origin such that for $(x) \in \underline{\mathfrak{D}}$, the projection of Σ on the space (y) is $\Subset (\Gamma)$. By *Cartan's theorem* and *Theorem III of Memoir VII*, we can find a function $f(x,y)$ holomorphic on the polycylinder $[\underline{\mathfrak{D}}, (\Gamma)]$ such that the image f' of the restriction of f to Σ is identically zero on σ and $f(M) \neq 0$. We choose a sufficiently large positive integer λ and set $f'^{\lambda} \xi = S$; S is then holomorphic on $\mathfrak{D}$ and we have $S(P_1) \neq S(P_2)$.

The ramification surface of Δ and the surface of ambiguity with respect to η_1 do not have a common component; further, η_1 has no ramification surface of the second kind in Δ. Hence, by the preceding considerations, we can choose a holomorphic function $S(P)$ on Δ and construct the characteristic variety Σ' in the space (x,y,z) given by $y_i = \eta_i(P)$, $z = S(P)$ $(P \in \Delta,\ \underline{P} = (x), i = 1,2,\ldots,m)$ in such a way that the set of singular points of Σ' is of dimension at most $(n-2)$.

Because of Theorem 1, we can suppose that this is already the case we are considering, more precisely that the portion of Σ corresponding to Δ has no singularities upto a set of dimension at most $(n-2)$. By Lemma 2, we can find a (W)-function with respect to Σ which is not identically zero on any component of T, T being the set of zeros of the ideal (I) in the neighbourhood of any point of the portion of Σ corresponding to Δ. This being the special case already considered, (I) has a pseudobasis at every point of this portion of Σ.

Q.E.D.

III. The Fundamental Lemma and its Applications

10. Fundamental Lemma. Let us consider a polyhedral domain Δ over the (finite) space of n complex variables $(x_1, x_2, \ldots, x_n)$ such that $\Delta \Subset (R)$, where (R) is a portion (Teilbereich; not necessarily connected) of the domain of existence of an analytic function $f_1(x)$; suppose that Δ is of the form

$$x_i \in A_i, \quad f_j(P) \in B_j, \quad P \in (R) \quad (i=1,\ldots,n; j=1,\ldots,m),$$

where the $f_j(P)$ are holomorphic functions on (R), A_i, B_j are closed, bounded, simply connected univalent domains in the plane, and P has coordinates (x). Consider the closed cylindrical domain (A,B) in the space $(x_1, x_2, \ldots, x_n, y_1, y_2, \ldots, y_m)$ and the characteristic variety Σ: $y_j = f_j(P)$ $(j=1,\ldots,m)$. Δ then corresponds to the part of Σ over (A,B) in such a way that the boundaries

[19]) A pole of a function on $\mathfrak{D}$ is a point of $\mathfrak{D}$ where the function is not holomorphic, but can be represented locally as a quotient of two holomorphic functions on $\mathfrak{D}$.

correspond to each other. The theorems of Memoir VII (which we shall simply note VII in what follows) apply to (A, B) because a neighbourhood of each A_i, B_j can be transformed conformally to a circle.

Let us consider a set Δ_0 of points of Δ satisfying

$$x_i \in A_i^0, \quad f_j(P) \in B_j^0 \quad (i=1,\ldots,n; j=1,\ldots,m),$$

where the A_i^0, B_j^0 are closed simply connected domains such that $A_i \supset A_i^0$, $B_j \supset B_j^0$. In what follows, we look upon Δ as being fixed and Δ_0 as varying arbitrarily.

Let u be an arbitrary holomorphic function on (R) in a neighbourhood of Δ_0; we can regard this as a holomorphic function on Σ. Let $U(x, y)$ be a holomorphic function of (x, y) in a neighbourhood of Σ which vanishes identically on the set S_0 of singular points of Σ but not on Σ; $U(x, y)$ exists by *Cartan's theorem* and *Theorem III of VII*. Consequently, by *Lemma 2*, we can choose a positive integer λ so that U^λ is a (W)-function with respect to Σ in a certain neighbourhood of (A, B). Since the geometric ideal with indeterminate domain attached to Σ possesses a pseudobasis in the neighbourhood of (A, B) (by CARTAN's theorem and Theorem III of VII), we can find, in view of Theorem II of VII, a holomorphic function $F(x, y)$ in a neighbourhood of (A^0, B^0) such that $F = U^\lambda u$ on Σ.

Let u_0 be the image (on (R)) of the restriction (to Σ) of U^λ, and let $(I) = \{(f, \delta)\}$ be the (Z)-ideal defined in a neighbourhood of (A, B) as follows: if (f', δ') is the image on (R) of the restriction of (f, δ) to Σ, then f' is divisible by u_0 in δ'. By *Theorem 2*, and *Theorem III of VII*, (I) has a pseudobasis on a neighbourhood of (A, B), which we shall denote by $(\Phi_1, \Phi_2, \ldots, \Phi_\mu)$; the image of the restriction of each Φ_i is divisible by u_0 in a neighbourhood of Δ. Since $F \in (I)$ in a neighbourhood of (A^0, B^0), we have $F \equiv 0 \bmod (\Phi)$ at every point in a neighbourhood of (A^0, B^0); consequently, by *Theorem I of VII*, this is also globally the case.

Let us briefly formulate what we have seen.

Fundamental Lemma. *Let us consider a polyhedral domain Δ given in the form described above as a portion of a multivalent domain (R), not necessarily connected, over (finite) space (x), the corresponding polycylinder (A, B) in the space (x, y), and the characteristic variety Σ as above. One can then find a holomorphic function u_0 in a neighbourhood of Δ on (R) and a finite system of holomorphic functions $\Phi_i(x, y)$ $(i=1,2,\ldots,\mu)$ in a neighbourhood of (A, B), so that the image on (R) of the restriction to Σ of each $\Phi_i(x, y)$ is divisible by u_0, and which have the following property: let $\Delta_0 (\Delta_0 \subset \Delta)$ be a polyhedral domain of the form given above, but otherwise arbitrary; let (A^0, B^0) be the corresponding closed cylindrical domain and let u be an arbitrary holomorphic function in the neighbourhood of Δ_0 on (R); there then exists a holomorphic function $F(x, y)$ on a neighbourhood of (A^0, B^0) such that $F = u \cdot u_0$ on Σ and such that $F \equiv 0 \bmod (\Phi)$ globally.*

We shall show briefly how one can apply this result to the main problems we have been considering since Memoir I.

11. The Cousin Problems. Let us take for A_1 a closed circle; let Δ_1 be the part of Δ for which $X \leqq \varepsilon$, where X is the real part of x_1 and ε is a sufficiently small positive number, and let Δ_2 be the part where $X \geqq -\varepsilon$. Suppose that $\Delta_1 \supset \{0\}$, $\Delta_2 \supset \{0\}$ and let $\Delta_0 = \Delta_1 \cap \Delta_2$.

Theorem 3. *With this configuration, given a holomorphic function u in a neighbourhood of Δ_0 on (R), we can find functions u_1, u_2, holomorphic on neighbourhoods of Δ_1 and of Δ_2 on (R) respectively, such that we have $u_1 - u_2 = u$ identically.*

In fact, the part of A_1 for which $-\varepsilon \leqq X \leqq \varepsilon$ is simply connected; we denote it by A_1^0. Let $A_k^0 = A_k$, $B_j^0 = B_j$ $(k = 2, \ldots, n; j = 1, \ldots, m)$. The polyhedral domain corresponding to (A^0, B^0) will then be the Δ_0 indicated above. The lemma holds for these Δ_0, u and we have, identically,

$$F = c_1 \Phi_1 + c_2 \Phi_2 + \ldots + c_\mu \Phi_\mu$$

in a neighbourhood of (A^0, B^0), c_i $(i = 1, 2, \ldots, \mu)$ being holomorphic functions. Let $\mathfrak{D}'$ be the part of (A, B) for which $X \leqq \varepsilon$ and $\mathfrak{D}''$ the part $X \geqq -\varepsilon$. By Cousin's theorem, we can find holomorphic functions c_i', c_i'' in neighbourhoods of $\mathfrak{D}'$ and $\mathfrak{D}''$ respectively such that we have

$$c_i' - c_i'' = c_i \qquad (i = 1, 2, \ldots, \mu)$$

identically. We set

$$F' = \sum c_i' \Phi_i, \qquad F'' = \sum c_i'' \Phi_i;$$

F' and F'' are then functions holomorphic, respectively in a neighbourhood of $\mathfrak{D}'$ and in one of $\mathfrak{D}''$, and we have

$$F' - F'' = F.$$

We denote the images on (R) of the restrictions of F' and F'' to Σ by f' and f'' respectively; we then have

$$f' - f'' = u_0 u.$$

Set

$$f' = u_0 u_1, \qquad f'' = u_0 u_2.$$

Because of the properties of Φ_i, u_1, u_2 are then holomorphic functions on neighbourhoods of Δ_1, Δ_2 on (R) respectively, and are such that $u_1 - u_2 = u$.

Q.E.D.

12. Expansion of Holomorphic Functions.

Theorem 4. *Under the circumstances of the fundamental lemma, given a holomorphic function u(P) in a neighbourhood of Δ_0 on (R) and a positive number ε, we can find a holomorphic function v(P) in a neighbourhood of Δ on (R) such that we have $|u(P) - v(P)| < \varepsilon$ in a neighbourhood of Δ_0.*

In fact, the function $F(x, y)$ of the fundamental lemma can be expressed in the form $F = \sum c_k \Phi_k$ $(k = 1, 2, \ldots, \mu)$ in the neighbourhood of (A^0, B^0), the c_k being holomorphic functions. Since A_i^0, B_j^0 $(i=1,\ldots,n;\ j=1,\ldots,m)$ are simply connected, for any positive number ε', we can find entire functions $c'_k(x,y)$ such that

$$|c_k(x,y)-c'_k(x,y)|<\varepsilon'$$

in a neighbourhood of (A^0, B^0). Let us set $G(x,y)=\sum c'_k \Phi_k$ and denote by $g(P)$ the image on (R) of its restriction to Σ; $g(P)$ is then holomorphic in a neighbourhood of Δ and divisible by u_0. Hence, if we set

$$g=u_0\, v,$$

v is holomorphic in a neighbourhood of Δ. Since the image on (R) of the restriction of F to Σ is $u_0 u$, it is clear that if we choose ε' sufficiently small, we have $|u-v|<\varepsilon$ in a neighbourhood of Δ_0. Q.E.D.

IV. Appendix

13. A Condition for the Solvability of Problem (J). – Let (I) be a holomorphic ideal with indeterminate domains in the space (x) and let Σ be the set of its zeros. We shall say that (I) has *the property* (R) at the point (x^0) if Σ can be expressed as the common zeros of a finite number of functions of (I) in a neighbourhood of (x^0).

A set $(\mathfrak{F})$ of holomorphic ideals with indeterminate domain in the space (x) will be said to form an (A)-family at a point (x^0) if it has the following two properties:

1°. If an ideal (I) belongs to $(\mathfrak{F})$, and if $\Phi(x)$ is a holomorphic function at (x^0), then the adjoint and the quotient of (I) with respect to Φ on a sufficiently small neighbourhood of (x^0) belong again to $(\mathfrak{F})$.

2°. Every ideal (I) in $(\mathfrak{F})$ has property (R) at (x^0).

Theorem 5. *For a holomorphic ideal* (I) *with indeterminate domain in the space* (x) *to have a pseudobasis at* (x^0), *it is necessary and sufficient that there exist an* (A)*-family at* (x^0) *containing* (I).

In fact, the condition is necessary because the set of ideals having a pseudobasis at (x^0) obviously forms an (A)-family at (x^0).

Suppose, therefore, that there exists an (A)-family $(\mathfrak{F})$ at (x^0) containing the ideal (I); we shall show that (I) has a pseudobasis at (x^0); we shall omit the term "at (x^0)" in what follows.

Let Σ be the set of zeros of (I); its dimension will be assumed to be smaller than n. By *Cartan's theorem*, the geometric ideal with indeterminate domains attached to Σ has a pseudobasis, which we shall denote by $(F_1, F_2, \ldots, F_p)$. Since Σ is given as the common zeros of a finite number of functions of (I), we can find because of the *lemma of Hilbert-Rückert*, a positive integer λ such that $F_1^\lambda \in (I)$.

Let us consider the adjoint and the quotient of (I) with respect to F_1 (on a sufficiently small neighbourhood of (x^0)); according to *Corollary 2*, if both have pseudobases, so also does (I). Now, the adjoint contains F_1 and the quotient contains $F_1^{\lambda-1}$. If $\lambda-1>1$, we apply the same procedure to the quotient, and so on. We proceed similarly with $F_2, F_3, \ldots, F_p$ successively, and arrive at a system $[(J_1),(J_2),\ldots,(J_q)]$ of ideals such that we can attain any (J_j) $(j=1,2,\ldots,q)$ from (I) by a finite number of operations of taking adjoints and quotients with respect to one of the functions (F_i) $(i=1,2,\ldots,q)$; further, if each (J_j) has a pseudobasis, so also does (I). In addition, each (J_j) contains (F).

Let (J) be an arbitrary ideal of this system and let Σ' be the set of its zeros. Two cases are possible: either $\Sigma'=\Sigma$, in which case (J) has (F) as a pseudobasis; or $\Sigma'\neq\Sigma$, in which case $\Sigma'\subset\Sigma$ as is easily seen. Let us consider the subsystem (S) consisting of all the (J) such that $\Sigma'\subset\Sigma$.

Let us consider, in general, a characteristic variety T passing through (x^0). According to WEIERSTRASS, the portion of T in a neighbourhood of (x^0) consists of a finite number of branches. If the number of branches of dimension i is v_i $(i=n-1,n-2,\ldots,0)$, we make correspond to T the n-tuple $\alpha=(v_{n-1},v_{n-2},\ldots,v_0)$. Consider the set A of all the α (for a given n) and order it as follows: let $\alpha'=(v'_{n-1},v'_{n-2},\ldots,v'_0)$ be an element of A different from α; we shall say that $\alpha<\alpha'$ if either $v_{n-1}<v'_{n-1}$ or $v_{n-1}=v'_{n-1},\ldots,v_i=v'_i$, $v_{i-1}<v'_{i-1}$ $(i=n-1,n-2,\ldots,1)$.

We make correspond to the ideal (I) the element α of the set A associated to the set Σ of zeros of (I) at (x^0); and, to the system (S) if it is not empty, we make correspond the largest β of the elements of A similarly attached to the ideals of (S). Since $\Sigma'\subset\Sigma$, we have $\beta<\alpha$.

Now, each ideal (J) of (S) belongs to $(\mathfrak{F})$. We can therefore apply to it the same procedure and obtain a system of ideals which corresponds to the system (S) attached to (I). Let (S_1) be the union of these systems when (J) runs over (S_1). If (S_1) is non-empty, the element γ of A attached to (S_1) in the same way satisfies the inequality $\gamma<\beta<\alpha$. Consequently, we can only continue a finite number of times. (I) therefore possesses a pseudobasis at (x^0). Q.E.D.

Commentaire de H. Cartan

Ce Mémoire VIII, de lecture très difficile, est consacré à l'étude des fonctions holomorphes sur les "domaines intérieurement ramifiés". Il s'agit, en réalité, de l'étude des espaces analytiques *normaux* (i.e. dont l'anneau des germes de fonctions holomorphes en chaque point est intègre et intégralement clos), et plus généralement de la "normalisation" d'un espace analytique réduit.

Cette étude soulève des problèmes de nature locale; le théorème essentiel est le suivant (cf. Séminaire H. CARTAN, 1953/54, exposé 11): étant donné un espace analytique réduit Σ, de faisceau structural $\mathcal{O}(\Sigma)$, le faisceau $\tilde{\mathcal{O}}(\Sigma)$ des clôtures intégrales $\tilde{\mathcal{O}}_x(\Sigma)$ des anneaux locaux $\mathcal{O}_x(\Sigma)$ est un faisceau cohérent sur Σ. C'est ce que, en fait, démontre OKA sans que ce résultat soit clairement énoncé. Comme conséquence immédiate, l'ensemble des points $x\in\Sigma$ où Σ n'est pas normal est un sous-ensemble analytique de Σ.

La lecture de ce Mémoire VIII est encore compliquée par le fait que OKA mélange l'étude de ces problèmes de nature locale à des considérations globales, qui faisaient déjà l'objet du Mémoire VII. C'est la raison pour laquelle la partie I du présent Mémoire est consacrée à l'approfondissement technique de notions relatives aux faisceaux cohérents d'idéaux, notamment le théorème 1 qui a servi d'une manière essentielle dans la preuve du théorème 2 de la Partie II du Mémoire. OKA donne aussi une démonstration originale de la cohérence du faisceau d'idéaux attaché à un sous-ensemble analytique de $\mathbb{C}^n$ (qu'il appelle "théorème de H. CARTAN"), et donne divers critères de cohérence, notamment le "corollaire 2". Il donnera aussi un critère de cohérence dans l'Appendice.

Mais le but essentiel d'OKA est l'étude des "domaines intérieurement ramifiés". Un tel domaine D est, par définition, un revêtement ramifié à un nombre fini ν de feuillets d'un domaine $\underline{D}$ de $\mathbb{C}^n$. On peut le considérer comme l'image d'un sous-ensemble analytique Σ de $\mathbb{C}^{n+m}$ par la projection p: $\mathbb{C}^{n+m} \to \mathbb{C}^n$, p définissant une bijection de l'ensemble des points réguliers de Σ sur un ouvert dense de D. OKA définit alors ce qu'il entend par *fonction holomorphe* sur D (c'est une fonction continue qui est holomorphe aux points réguliers de D); en transportant cette définition à Σ, on obtient les fonctions holomorphes dans l'ouvert des points réguliers de Σ et qui ont une limite en chaque point singulier a lorsqu'on reste dans une composante irréductible de Σ au point a. Les germes de fonctions holomorphes en $a \in \Sigma$ ne sont autres que les éléments de la clôture intégrale de l'anneau $\mathcal{O}_a(\Sigma)$ induit par les germes de fonctions holomorphes de l'espace ambiant C^{n+m}. Autrement dit, lorsqu'on aura prouvé l'existence de l'espace *normalisé* $\tilde{\Sigma} \xrightarrow{q} \Sigma$ ($\tilde{\Sigma}$ étant considéré comme sous-ensemble analytique de $\mathbb{C}^{n+m+p}$, et q étant induit par la projection canonique $\mathbb{C}^{n+m+p} \to \mathbb{C}^{n+m}$, - tout ceci étant vrai au moins localement), alors les fonctions holomorphes sur D s'identifient aux fonctions holomorphes sur $\tilde{\Sigma}$ (c'est-à-dire induites localement par des fonctions holomorphes de l'espace ambiant $\mathbb{C}^{n+m+p}$). Bien sûr, une fonction holomorphe sur D, considérée comme fonction sur Σ (ou plutôt sur l'ensemble des composantes irréductibles aux points de Σ) n'est pas toujours induite localement par une fonction holomorphe de l'espace ambiant $\mathbb{C}^{n+m}$. Dans la terminologie d'OKA, les germes de fonctions holomorphes en un point $a \in \Sigma$ qui sont induits par des germes de fonctions holomorphes dans $\mathbb{C}^{n+m}$ sont dits posséder la "propriété (H)". Le point a possède la propriété (H) si tout germe de fonction holomorphe en a possède la propriété (H); cela revient à dire que Σ, muni de la structure analytique induite par l'espace ambiant, est *normal* au point a.

Dans cette situation de revêtement ramifié p: $\Sigma \to \underline{D}$, OKA introduit la notion de *fonction* (W): c'est une F holomorphe dans l'espace ambiant $\mathbb{C}^{n+m}$ telle que la multiplication par F transforme toute fonction holomorphe sur Σ en une fonction possédant la propriété (H). Naturellement, cette notion peut se définir soit globalement, soit localement. L'existence locale de telles fonctions (W) est prouvée. Ces fonctions (W) sont ce que H. CARTAN appelle *dénominateurs universels* pour le sous-ensemble analytique Σ de $\mathbb{C}^{n+m}$ (cf. Séminaire H. CARTAN, 1953/54, exposé 9). Si une fonction holomorphe F de l'espace ambiant s'annule aux points singuliers de Σ sans être identiquement nulle dans un ouvert non vide de Σ, il existe une puissance F^λ qui est (localement) un dénominateur universel (lemme 2); de plus, étant donné un point $a \in \Sigma$, *il existe* un système fini de fonctions u_i holomorphes dans l'espace ambiant au voisinage de a, telles que chaque u_i soit un dénominateur universel en tout point de Σ assez voisin de a, et telles que l'annulation de toutes les u_i définisse exactement l'ensemble des points singuliers de Σ voisins de a.

Nous espérons que les indications qui précèdent aideront le lecteur dans l'étude de ce Mémoire VIII. Le cœur du Mémoire est constitué par le théorème 2, qui concerne ce que Oka appelle les "idéaux (Z)". Le commentateur avoue que cette notion d'idéal (Z) lui demeure un peu obscure, mais il croit que le théorème 2 fournit essentiellement le résultat annoncé au début de ce commentaire, à savoir que le faisceau des clôtures intégrales $\tilde{\mathcal{O}}_a(\Sigma)$ est un faisceau cohérent sur l'espace analytique Σ (muni de la structure analytique induite par l'espace ambiant). La démonstration se fait en deux temps: on examine d'abord le "cas spécial" où, au voisinage de $a \in \Sigma$, le sous-ensemble S des points singuliers de Σ est de codimension $\geqq 2$ dans Σ. On ramène ensuite le cas général à ce cas spécial en considérant und revêtement ramifié $\Sigma' \to \underline{D}$ qui se factorise par $\Sigma' \xrightarrow{q} \Sigma \xrightarrow{p} \underline{D}$; Σ' est un sous-ensemble analytique d'un espace $\mathbb{C}^{n+m+r}$, tel que l'ensemble des points singuliers de Σ' soit de codimension $\geqq 2$ dans Σ' au point considéré. On applique alors le cas spécial à Σ' et on obtient un faisceau cohérent sur Σ'. Il reste à en déduire la cohérence du faisceau étudié sur Σ, et pour cela on est ramené à utiliser le théorème 1 de la Partie I du présent Mémoire, qui a été établi juste pour servir ici.

En fait, on peut observer ce qui suit: une fois démontré le cas spécial où l'ensemble des points singuliers de Σ est de codimension $\geqq 2$ au voisinage du point $a \in \Sigma$, on conclut que si Σ est normal au point a, cette condition est remplie, donc le faisceau des $\tilde{\mathcal{O}}_x(\Sigma)$ est cohérent au voisinage de a, ce qui entraîne $\tilde{\mathcal{O}}_x(\Sigma) = \mathcal{O}_x(\Sigma)$ pour x voisin de a; autrement dit, Σ est normal en tout point d'un voisinage de a. Sachant ainsi que l'ensemble des points où Σ est normal est ouvert on peut définir le normalisé $\tilde{\Sigma}$ de Σ, muni de l'application canonique $\tilde{\Sigma} \to \Sigma$ (qui est un homéomorphisme sur un ouvert dense). Il reste alors à montrer que le faisceau structural de $\tilde{\Sigma}$, considéré comme faisceau sur Σ, est cohérent, ce qui se fait grâce au théorème 1 de la Partie I.

Le "lemme fondamental" de la Partie III et ses applications sont alors des conséquences faciles de ce qui précède.

IX. Unramified Domains Without Points at Infinity

Domaines finis sans point critique intérieur

Japanese Journal of Mathematics **27** (1953), p. 97–155

Introduction. 1. This is the ninth in a series of memoirs [1], the first of which was published in 1936. We shall first cast an eye over the terrain in which we find ourselves.

The general theory of analytic continuation in a single variable is like an open field; despite many efforts [2] one has not found any facts which could not have been predicted by formal logic. The case of several variables, on the other hand, seems to us like mountainous country, very precipitous.

In 1902, FABRY remarked that the radii of convergence of a double series are not arbitrary; from this, we were led by HARTOGS in 1906 to the very fundamental and really curious fact that every domain of holomorphy is pseudoconvex.

From then on until 1932, each new problem in the field gave rise to other problems; the lines of this accumulation of difficulties, as well as the flow of ideas, were drawn with great emphasis in the following work:

BEHNKE-THULLEN, Theorie der Funktionen mehrerer komplexer Veränderlichen, 1934.

The main problems dealt with in this book are: the inverse HARTOGS problem, first and second Cousin problems, problem of expansion of functions [3].

In 1935, WEIL took the first step in the opposite direction, that is, in the direction of solving these problems; thanks to this, the last three of the above problems could be solved for rationally convex domains [4].

It was just at this time, and to study these problems, that we began our researches. The following memoirs of BEHNKE-STEIN give a vivid picture of the change in the status of these problems:

---

[1] The previous memoirs are: I. Rationally convex domains, 1936; II. Domains of holomorphy, 1937; III. The second COUSIN problem, 1939 (Journal of Science of Hiroshima University); IV. Domains of holomorphy and rationally convex domains, 1941; V. The CAUCHY integral, 1941 (Japanese Journal of Mathematics); VI. Pseudoconvex domains, 1942 (Tohoku Mathematical Journal); VII. On some arithmetical notions, 1950 (Bulletin de la Société Mathématique de France); VIII. Fundamental lemma, 1951 (Journal of Mathematical Society of Japan). (We have recalled the road which we have followed.)

[2] See e.g. the beautiful memoirs of DENJOY (C.R. Paris).

[3] See pages 54, 69, 79 of the book. It is truly thanks to this book that we were able to begin our researches.

[4] Before him, these problems could only be solved on cylindrical domains. See: A. WEIL, Sur les séries de polynômes de deux variables complexes, 1932 (C.R., Paris). A. WEIL, L'intégrale de Cauchy et les fonctions de plusieurs variables, 1935 (Math. Annalen).

Analytische Funktionen mehrerer Veränderlichen zu vorgegebenen Null- und Polstellenflächen, 1937 [5].

Die Konvexität in der Funktionentheorie mehrerer komplexer Veränderlichen, 1940 [6].

Die Singularitäten der analytischen Funktionen mehrerer Veränderlichen, 1952 [7].

2. In the present memoir, we shall deal with the problems indicated above, as well as the arithmetical problems introduced in Memoir VII, for pseudoconvex domains without interior ramification and without points at infinity; the essential part of this memoir is not very different from what we have expounded in japanese in 1943 [8].

We shall see in the memoir following this one that when one permits interior points of ramification, one meets a problem which seems to me to be extremely difficult (see also No. 23 below). It is to prepare the methods and to illuminate the nature of this difficulty that we have decided to publish the present memoir separately [9].

This memoir consists of three chapters. In Chapter I, we shall add a quantitative complement to the lemma given in Memoir VIII. In Chapter II, we shall obtain a second preparatory lemma. In Chapter III, we shall treat the above problems using these lemmas. (For the precise form, see Nos. 1, 7, 24.)

Chapter I. Complement to the Fundamental Lemma

1. The Problem. We want to solve *the inverse Hartogs problem* without invoking *the Weil integral* [10]. This gives rise to the following problem concerning the lemma established in the preceding memoir.

In the lemma of Memoir VIII, we found that to each holomorphic function u on a neighbourhood of Δ_0 on (R), there corresponds a holomorphic function $F(x, y)$ on a neighbourhood of (A_0, B_0) such that $F = u u_0$ on Σ and $F \equiv 0 \bmod (\Phi)$ globally. Let now

$$F = A_1 \Phi_1 + A_2 \Phi_2 + \ldots + A_\mu \Phi_\mu$$

identically, the A_i $(i = 1, 2, \ldots, \mu)$ being holomorphic functions in a neighbourhood of (A_0, B_0); suppose that

$$|u| < M \text{ on } V_0,$$

where V_0 is a domain (not necessarily connected) such that $\Delta_0 \Subset V_0 \subseteqq (R)$, and M is a positive number. We ask if it is possible to choose the A_i $(i = 1, 2, \ldots, \mu)$

---

[5]) (Jahresbericht der Deutschen Mathematiker-Vereinigung). See also: H. Cartan, Note sur le premier problème de Cousin, 1938 (C.R., Paris).

[6]) (Mitteilungen der Mathematischen Gesellschaft in Hamburg, VIII.)

[7]) (Nieuw Archief voor Wiskunde, Amsterdam.)

[8]),[9]) See the note in the introduction to Memoir VIII. In that manuscript, one finds already problems (C_1), (C_2) (explicitly), and problem (E) (implicitly).

[10]) See: Memoir VI and B, Chapter III below.

with the bound

$$|A_i| < KM,$$

K being a positive constant independent of u.

We shall solve this by inspecting successively the quantitative relations in the theorems which went into the lemma.

2. The Remainder Theorem. Let us start with the remainder theorem formulated in No. 5 of Memoir VII (in what follows, we shall omit the word "Memoir") as follows:

Consider a domain of the form $[\mathfrak{D},(C)]$ in the space $(x_1, x_2, \ldots, x_n, y)$, where $\mathfrak{D}$ is a univalent (finite) domain in the space (x) and (C) is a circle in the y-plane. Let us consider a holomorphic function $F(x,y)$ in $[\mathfrak{D},(C)]$ such that, for any point (x^0) of $\mathfrak{D}$, the equation $F(x^0,y)=0$ has λ roots in (C), λ being a finite integer independent of (x^0). Then, any holomorphic function $f(x,y)$ in $[\mathfrak{D},(C)]$ can be put in the form

$$f(x,y) = f_0(x,y) + \varphi(x,y)\,F(x,y),$$

where f_0 and φ are holomorphic on $[\mathfrak{D},(C)]$; f_0 is a polynomial in y of degree at most $\lambda - 1$ (identically zero if $\lambda = 0$). Further, such a decomposition is unique.

This theorem, due to W. RÜCKERT[11], has been examined quantitatively by H. CARTAN[12]; we shall simply formulate his result:

Let (C_0) be a circle, concentric with (C) and contained in (C); suppose that all λ roots of $F(x^0,y)$ remain in (C_0) for an arbitrary point (x^0) of $\mathfrak{D}$. Then, if $|f|$ has an upper bound M on $[\mathfrak{D},(C)]$, we have

$$|f_0| < KM, \qquad |\varphi| < KM$$

on $[\mathfrak{D},(C_0)]$, K being a positive constant independent of f.

3. The Local Problem (C_1). We now inspect problem (C_1) quantitatively. Let us first consider the problem locally; we shall prove the following:

Given holomorphic functions $F_1, F_2, \ldots, F_p$ in a *polycylinder* (C), $|x_i| < R$ $(i = 1, 2, \ldots, n)$ in the space (x),[13] there correspond a positive constant r smaller than R and a positive constant K which have the following property: Let $f(x)$ be a holomorphic function on (C) such that $f \equiv 0 \bmod (F)$ at every point of (C), and such that $|f| < M$ on (C) (M is a positive number), one can then find holomorphic functions $A_j(x)$ $(j = 1, 2, \ldots, p)$ on the polycylinder (γ), $|x_i| < r$ $(i = 1, 2, \ldots, n)$ with the bounds

$$|A_j(x)| < KM \text{ on } (\gamma)$$

in such a way that we have $f = A_1 F_1 + A_2 F_2 + \ldots + A_p F_p$ identically.

---

[11]) Math. Annalen, 1933.

[12]) H. CARTAN: Idéaux de fonctions analytiques de n variables complexes, 1944 (Annales de l'École Normale, pages 192–194).

[13]) By "polycylinder" in the space (x), we mean a set of points of the form $|x_i - x_i^0| < r_i$ $(i = 1, 2, \ldots, n)$, (x^0) being a definite point and r_i, positive numbers.

We shall generalise problem (C_1) as follows:

Suppose given a system of simultaneous functional equations

(a) $$f_i = A_{i1}F_{i1} + A_{i2}F_{i2} + \dots + A_{ip}F_{ip} \quad (i=1,2,\dots,q),$$

where f_i, f_{ij} ($j=1,2,\dots,p$) stand for given holomorphic functions on a domain $\mathfrak{D}$ over the space (x), and the A_{ij} stand for the unknown functions, together with a certain number of identities of the form

$$A_{ij} = A_{kl} \quad (i \neq k)$$

($i, k=1,\dots,q$; $j, l=1,\dots,p$). If this system of equations has a solution (A) at every point of $\mathfrak{D}$,[14] find a solution on $\mathfrak{D}$.

Let us rearrange the form of the functional equation. Suppose given, for example,

$$f_1 = A_{11}F_{11} + A_{12}F_{12}, \quad f_2 = A_{21}F_{21} + A_{22}F_{22} \quad \text{with} \quad A_{12} = A_{21}.$$

Let us set

$$\Phi_{11} = F_{11}, \quad \Phi_{12} = F_{12}, \quad \Phi_{13} = 0,$$
$$\Phi_{21} = 0, \quad \Phi_{22} = F_{21}, \quad \Phi_{23} = F_{22};$$

the given functional equation then takes the form

$$f_1 = B_1\Phi_{11} + B_2\Phi_{12} + B_3\Phi_{13}$$
$$f_2 = B_1\Phi_{21} + B_2\Phi_{22} + B_3\Phi_{23}.$$

By applying this method of arranging the functional equation (a) (together with the identities indicated), we can always put the system in the form

$$f_i = B_1\Phi_{i1} + B_2\Phi_{i2} + \dots + B_r\Phi_r \quad (i=1,2,\dots,q),$$

where $(\Phi_{i1}, \Phi_{i2}, \dots, \Phi_{ir})$ is equal, upto order, to $(F_{i1}, F_{i2}, \dots, F_{ip}, 0, \dots, 0)$.[15]

We shall justify the proposition for the problem so generalised. Since it is obviously true for $(0, q)$, it is sufficient to justify it for $(n, q+1)$ and $(n+1, 1)$ assuming it to be true for all (n', q') such that $n \geqq n' > 0$, $q \geqq q' > 0$. We begin with the case of $(n, q+1)$.

Consider the simultaneous functional equations

(1) $$f_i = A_1F_{i1} + A_2F_{i2} + \dots + A_pF_{ip} \quad (i=1,2,\dots,q+1)$$

in the space $(x_1, x_2, \dots, x_n)$, where f_i, F_{ij} ($j=1,\dots,p$) are given holomorphic functions on the polycylinder (C), $|x_k| < R$ ($k=1,2,\dots,n$), and the A_j stand for the unknown functions; we suppose that the equation (1) has solutions at every point of (C). We are to find a solution on the polycylinder (γ), $|x_k| < r$ ($k=1,2,\dots,n$) with the bounds $|A_j(x)| < KM$, where $r(<R)$, K are positive con-

---

14) Let us remark that in the equation (a) (with the given identities), even if $f_i \equiv 0 \mod (F_{i1}, F_{i2}, \dots, F_{ip})$ at every point of $\mathfrak{D}$, one does not necessarily have local solutions; for example the system $x_1 = A_{11}x_1 + A_{12}x_2$, $x_2 = A_{21}x_1 + A_{22}x_2$, with $A_{12} = A_{21}$.

15) Just by this method of arrangement, H. CARTAN has greatly shortened our proof that problem (K) is always solvable (VII, No. 6). See: H. CARTAN, Idéaux et modules de fonctions analytiques de plusieurs varables, 1950 (Bulletin de la Société Mathématique de France).

stants independent of the f_i, and M is the upper bound of $|f_i(x)|$ on (C), under the hypothesis indicated.

The proposition being true for $(n, 1)$ by hypothesis, there exist positive constants $R_1 (<R)$ and K_1 corresponding to the functional equation

$$f_1 = A_1 F_{11} + A_2 F_{12} + \ldots + A_p F_{1p}, \tag{2}$$

(K_1 is independent of f_1) such that we can find holomorphic functions $A_i^0(x)$ $(i=1, 2, \ldots, p)$ on the polycylinder (C_1), $|x_k| < R_1$ $(k=1, 2, \ldots, n)$ with the bounds

$$|A_i^0| < K_1 M$$

in such a way that these functions satisfy (2) identically. Let us set

$$B_i = A_i - A_i^0 \quad (i=1, 2, \ldots, p);$$

the equation (2) then becomes

$$B_1 F_{11} + B_2 F_{12} + \ldots + B_p F_{1p} = 0. \tag{3}$$

There exists a formula for the solutions of the functional equation (3) on any polycylinder Δ such that $\Delta \Subset (C_1)$ (Theorems IV and III of VII); this formula has the form

$$B_i = \sum C_{ij} \Pi_{ij} \quad (i=1, \ldots, p; j=1, \ldots, r')$$

with some identities of the form $C_{ij} = C_{kl}$ $(i \neq k;\ i, k = 1, \ldots, p;\ j, l = 1, \ldots, r')$. If we apply the same method of rearrangement as above, we obtain a formula for the solutions having the form

$$B_i = \sum C_i \Pi_{ij} \quad (i=1, \ldots, p; j=1, \ldots, r).$$

If we substitute this formula, and the relation $B_i = A_i - A_i^0$ in

$$f_i = \sum A_k F_{ik} \quad (i=2, \ldots, q+1; k=1, \ldots, p), \tag{4}$$

we obtain a functional equation of the form

$$\varphi_i = \sum C_j \Phi_{ij} \quad (i=2, \ldots, q+1; j=1, \ldots, r), \tag{5}$$

where $\varphi_i = f_i - \sum A_k^0 F_{ik}$ $(k=1, \ldots, p)$, and Φ_{ij} are well determined holomorphic functions on (C_1) independent of (f). This being evidently in the case (n, q), there exists a solution of (5) having the required properties. The same is therefore true for the given equation (1).

Let us next consider, in the space $(x_1, x_2, \ldots, x_n, y)$, a polycylinder (C), $|x_i| < R$, $|y| < R$ $(i=1, 2, \ldots, n)$ and the functional equation

$$f = A_1 F_1 + A_2 F_2 + \ldots + A_p F_p, \tag{6}$$

where f and F_j $(j=1, 2, \ldots, p)$ represent given holomorphic functions on (C) such that $f \equiv 0 \bmod (F)$ at every point of (C), and the A_j stand for the unknown functions. We shall justify the proposition for this equation under the hypothesis indicated earlier. We suppose that one, at least, of the functions F_j is not identically zero (if not, there is no problem); say $F_1(x, y) \neq 0$ identically to fix our ideas. Thanks to WEIERSTRASS, we can regard $F_1(x, y)$ as being a polynomial in y such that the coefficient of the highest power of y is 1 (by changing (x, y) and R). Let us choose three positive numbers ρ, ρ', ρ'' such that $\rho<R$, $\rho''<\rho'<R$ in such a way that for all (x') in the polycylinder Δ, $|x_i|<\rho$ $(i=1,2,\ldots,n)$, the equation $F_1(x', y)=0$ has the same number of roots in the circle (C''), $|y|<\rho''$, and there are no roots on $\rho''\leqq|y|<\rho'$. Let λ be the number of roots. We can always suppose that F_1 is of degree λ in y by placing ourselves on $[\Delta,(C')]$, (C') being the circle $|y|<\rho'$.

By the remainder theorem, the function $f(x, y)$ can then be put in the form $f=f_0+\varphi F_1$, where f_0 and φ are holomorphic functions on $[\Delta,(C')]$ and moreover, f_0 is a polynomial in y of degree at most $\lambda-1$ (identically zero if $\lambda=0$); in addition, if $|f|<M$ on $[\Delta,(C')]$, we have $|f_0|<KM$, $|\varphi|<KM$ on $[\Delta,(C'')]$, K being a constant independent of f. We can therefore suppose that the function f in equation (6) is a polynomial in y of degree at most $\lambda-1$ (we suppose, in what follows, that $\lambda>0$ since, otherwise, there is no problem). Similarly, we can suppose that the functions F_k $(k=2,3,\ldots,p)$ in equation (6) are polynomials in y of degree at most $\lambda-1$.

Under these conditions, we consider equation (6) on $[\Delta,(C'')]$. Let $A_j(x, y)$ $(j=1,2,\ldots,p)$ be a solution of (6); for the moment, we shall call it a special solution if all the functions A_j are polynomials in y of degree at most $\lambda-1$. The degree of $A_1(x, y)$ in a special solution is necessarily at most $\lambda-2$ (if $\lambda=1$, $A_1=0$).

Let (x^0) be an arbitrary point of Δ. The equation (6) has a local solution at (x^0, y'), y' being any point of (C'); consequently, *by Theorem* I of VII, the equation (6) has a solution for a certain neighbourhood of $[(x^0),(C'')]$. Because of the conditions imposed on equation (6), in particular on F_1, it follows from this that for any point (x^0) of Δ, the equation (6) has a local special solution (one has only to apply *the remainder theorem* to the preceding solution).
Let us set

$$f=\varphi_1 y^{\lambda-1}+\varphi_2 y^{\lambda-2}+\ldots+\varphi_\lambda,$$

$$F_1=y^{\lambda-1}+\Phi_1 y^{\lambda-1}+\Phi_2 y^{\lambda-2}+\ldots+\Phi_\lambda,\qquad A_1=B_2 y^{\lambda-2}+\ldots+B_\lambda,$$

$$F_i=\Phi_{i1} y^{\lambda-1}+\Phi_{i2} y^{\lambda-2}+\ldots+\Phi_{i\lambda},\qquad A_i=B_{i1} y^{\lambda-1}+B_{i2} y^{\lambda-2}+\ldots+B_{i\lambda}$$

$(i=2,3,\ldots,p)$; then, for (A) to be a special solution of the equation (6), it is necessary and sufficient that (B) satisfy the following simultaneous functional equations, defined on Δ:

$$(7)\qquad 0=B_2+\sum B_{i1}\Phi_{i1},\quad 0=(B_2\Phi_1+B_3)+\sum(B_{i1}\Phi_{i2}+B_{i2}\Phi_{i1}),\ldots,$$
$$\varphi_\lambda=B_\lambda\Phi_\lambda+\sum B_{i\lambda}\Phi_{i\lambda}\quad(i=2,3,\ldots,p).$$

The equations (7) have a local solution at every point of Δ because equation (6) has a special local solution at any point of Δ. Now, this is in the case (n,q); consequently, by hypothesis, there exists a solution at the origin satisfying the required condition. The same is therefore true for equation (6).

The proposition in question is therefore true. In a word, we have seen that *the generalised problem* (C_1) *always has local solutions satisfying the quantitative condition.*

4. The Global Problem (C_1). We now inspect problem (C_1) quantitatively and globally. For this, it is necessary to continue to treat it in generalised form.

Let us recall the geometric configuration of No. 4 in VII.

Let us take a closed circle $(\bar{C}_i)$ in each x_i-plane ($i=1,\dots,n$) and let $(\bar{C})$ be the closed polycylinder having the (C_i) as components. Denote by E_0 an arbitrary point of $(\bar{C})$.

We separate the variable x_n into its real and imaginary parts, $x_n = X + iY$, and consider a line of the form $x_i = x_i^0$ ($i=1,\dots,n-1$), $Y = Y^0$, the x_i^0 being complex numbers, Y^0, real. Let E_1 be the intersection of this line with the closed polycylinder $(\bar{C})$; E_1 is a segment which may reduce to a point.

We consider similarly $E_2, E_3, \dots$ by raising each time the real dimension by 1, terminating with E_{2n}. E_2, for example, is a closed cylindrical set with a point of $(\bar{C}_i)$ as component in the x_i-plane for $i=1,2,\dots,n-1$, and $(\bar{C}_n)$ as component in the x_n-plane; E_{2n} stands for the polycylinder $(\bar{C})$.

We have just seen that the problem (C_1) (generalised and quantitative) is solvable in the neighbourhood of E_0; let us therefore pass to the *case of* E_1.

We consider an arbitrary one of the sets E_1 and denote it by the same letter; we may suppose that it is a real segment.

The component of E_1 in $(x_1,\dots,x_{n-1})$ space is a point which we denote by Q and its component in the x_n-plane is a segment which we denote by l; l is horizontal, its left extremity will be denoted by m_0, its right extremity by m_q; we introduce $q-1$ points $m_1,\dots,m_{q-1}$ on the segment from left to right. Let l_i be the closed segment $[m_{i-1}, m_i]$ ($i=1,\dots,q$). Let M_i be the cylindrical set (Q, m_i) in the space (x) and L_i the cylindrical set (Q, l_i). We set

$$L_1 \cup L_3 \cup L_5 \cup \dots = \Delta_1, \qquad L_2 \cup L_4 \cup L_6 \cup \dots = \Delta_2.$$

We then have

$$\Delta = \Delta_1 \cup \Delta_2 = E_1, \qquad \Delta_0 = \Delta_1 \cap \Delta_2 = M_1 \cup M_2 \cup \dots \cup M_{q-1}.$$

In a polycylinder (C') concentric with (C) and containing (C), we now consider the functional equation

$$f_i = A_1 F_{i1} + A_2 F_{i2} + \dots + A_p F_{ip} \qquad (i=1,2,\dots,p'), \tag{1}$$

assuming that it has a solution at every point of (C') (the notation having the usual meaning). Since this problem (C_1) is solvable (quantitatively) at every point of E_1, we can solve it in the neighbourhood of each l_h if we choose the l_h ($h=1,2,\dots,q$) sufficiently small; more precisely, we can find a circle α_h in the x_n-plane around the midpoint of l_h which contains l_h and a polycylinder (β) in the space $(x_1,\dots,x_{n-1})$ with centre Q which are independent of the f_i, in such a way that there are holomorphic functions A_j ($j=1,2,\dots,p$) in (β, α_h) satisfying the identities (1) and the condition

$$|A_j| < K_1 M,$$

where K_1 is a positive constant independent of the f_i and M is an upper bound for the $|f_i|$ on (C').

Let $\alpha_h \cap \alpha_{h+1} = \gamma_h$ $(h<q)$; we consider

$$(\beta, \alpha_1) \cup (\beta, \alpha_3) \cup \ldots = V_1, \qquad (\beta, \alpha_2) \cup (\beta, \alpha_4) \cup \ldots = V_2,$$

and denote the A_j corresponding to V_1 by A'_j, that corresponding to V_2 by A''_j. Set $A'_j - A''_j = B_j$; we then have

$$B_1 F_{i1} + B_2 F_{i2} + \ldots + B_p F_{ip} = 0 \qquad (i = 1, 2, \ldots, p') \tag{2}$$

identically on $V_0 = V_1 \cap V_2 = (\beta, \gamma_1) \cup (\beta, \gamma_2) \cup \ldots \cup (\beta, \gamma_{q-1})$.

We have a formula

$$B_j = \sum C_k \Phi_{jk} \qquad (j = 1, \ldots, p; k = 1, \ldots, p'') \tag{3}$$

for the solutions of the functional equation (2) on any polycylinder (C'') concentric with (C) and such that $(C) \Subset (C'') \Subset (C')$ (Theorems IV and III of VII; the Φ_{jk} are holomorphic functions). We suppose that $V_0 \subset (C'')$.

If we return to the functions $B_j = A'_j - A''_j$, the equations (3) give us again a problem (C_1) on V_0 because the B_j are holomorphic functions on V_0 such that this equation has a solution at every point of V_0.

We consider a circle δ_h around the point m_h $(h = 1, 2, \ldots, q-1)$ in the x_n-plane and a polycylinder (β') with centre Q in the space $(x_1, x_2, \ldots, x_{n-1})$ which are such that $\delta_h \Subset \gamma_h$, $(\beta') \Subset (\beta)$ and such that for any holomorphic functions B_j having the property indicated and satisfying $|B_j| < M$ on V_0, we can find holomorphic functions C_k which satisfy the identities (3) and the conditions

$$|C_k| < K_2 M$$

on the closed polycylinder (β', δ_h), K_2 being a positive constant independent of B_j (and, naturally, of M).

Let us consider the *following Cousin integral*[16]:

$$I_k(x) = \frac{1}{(2\pi i)^n} \sum_h \int_{d_h} \frac{C_k(x_1, \ldots, x_{n-1}, t)}{t - x_n} dt \qquad (h = 1, 2, \ldots, q-1),$$

where d_h is the vertical diameter of δ_h on which the integration is carried out from bottom to top (and i is the imaginary unit).

Let r be a positive number smaller than half the length of any d_h, and let λ be the set of points of the x_n-plane whose distances from l are smaller than r; we partition λ by the diameters d_h as follows: let λ_1 be the part of λ which is left of d_1, λ_2 the part between d_1 and d_2, and so on. Let

$$\mu_1 = \lambda_1 \cup \lambda_3 \cup \ldots, \qquad \mu_2 = \lambda_2 \cup \lambda_4 \cup \ldots.$$

In this geometric configuration, the integral $I_k(x)$ gives a holomorphic function on each of the domains (connected or not) (β', μ_g) $(g = 1, 2)$, which we denote by $\varphi_k^{(g)}(x)$; they remain holomorphic at every point of $[(\beta'), (d_h \cap \lambda)]$ $(h = 1, 2, \ldots, q-1)$ and we have

$$\varphi_k^{(1)}(x) - \varphi_k^{(2)}(x) = C_k(x)$$

---

[16]) We are calling COUSIN integral any integral of the same general form as that used by COUSIN (see: Acta Math. 1895), and which, in our context, has the required property and plays the same role.

identically. Further, we clearly have

$$|\varphi_k^{(1)}(x)| < K_3 M, \quad |\varphi_k^{(2)}(x)| < K_3 M,$$

where K_3 stands for a positive constant and M is an upper bound for $|B_j|$ on V_0.

We now set

$$\psi_i^{(g)}(x) = A_1^{(g)} F_{i1} + A_2^{(g)} F_{i2} + \ldots + A_p^{(g)} F_{ip} \quad (i = 1, 2, \ldots, p'),$$

$$-\sum_j [\sum_k \varphi_k^{(g)} \Phi_{jk}] F_{ij} = a_1^{(g)} F_{i1} + a_2^{(g)} F_{i2} + \ldots + a_p^{(g)} F_{ip} \quad (j = 1, \ldots, p; k = 1, \ldots, p'')$$

on (β, μ_g) respectively ($g = 1, 2$; we have set $A_j^{(1)} = A'_j$, $A_j^{(2)} = A''_j$).

We then find that $\psi_i^{(g)} = f_i$, $a_i^{(1)} = a_i^{(2)}$ identically; consequently

$$f_i = a_1 F_{i1} + a_2 F_{i2} + \ldots + a_p F_{ip} \quad (i = 1, 2, \ldots, p')$$

identically; further,

$$|a_j| < K_4 M \quad (j = 1, 2, \ldots, p),$$

where K_4 stands for a positive constant independent of the f_i, and M is the same number as above. Let r' be the smaller of the numbers r and the radii of (β').

We have thus seen that, given a generalised problem (C_1) in the polycylinder (C'), we can find positive constants K_4 and r' independent of (f), and having the properties indicated; in other words, the generalised quantitative problem (C_1) is always solvable for E_1. Repeating the same mode of reasoning, we arrive at the following result:

Given holomorphic functions $F_{i1}, F_{i2}, \ldots, F_{ip}$, and f_i in a polycylinder (C) in the space (x) such that the functional equation $f_i = A_1 F_{i1} + A_2 F_{i2} + \ldots + A_p F_{ip}$ ($i = 1, 2, \ldots, p'$) has a solution (A) at every point of (C), and given a polycylinder (C_0) concentric with (C) such that $(C_0) \Subset (C)$, we can always find a constant K independent of the (f_i) in such a way that, if M is an upper bound of $|f_i|$ on (C), we can choose holomorphic functions A_j ($j = 1, 2, \ldots, p$) on (C_0) satisfying the functional equation as well as the condition $|A_j| < KM$.

5. Problem (C_2). Between problems (C_1) and (C_2), there is the relation which we have given in No. 1 of VII; hence the following result is obtained immediately from the above result:

Consider two concentric polycylinders (C), (C_0) in the space (x) having radii R_i, R_i^0 ($i = 1, 2, \ldots, n$) respectively, where $R_i > R_i^0$. Let r be a positive number with $r < R_i - R_i^0$, and let (x^0) be an arbitrary point of (C_0). We describe the polycylinder (γ) with centre (x^0) and radius r. Let $F_j(x)$ ($j = 1, 2, \ldots, p$) be a system of holomorphic functions on (C). Under these circumstances, there exists a positive constant K with the following property: Given a holomorphic function $\varphi(x)$ on (γ) such that $|\varphi| < M$, M being a positive number independent of (γ), and such that for any pair $((\gamma_1), (\gamma_2))$ of these (γ), the corresponding functions φ_1, φ_2

satisfy $\varphi_1 \equiv \varphi_2 \operatorname{mod}(F)$ *at every point of* $(\gamma_1) \cap (\gamma_2)$, *we can find a holomorphic function* $\Phi(x)$ *on* (C_0) *such that* $|\Phi| < KM$ *and* $\Phi \equiv \varphi \operatorname{mod}(F)$ *at every point of* (γ).

We have treated problems (C_1), (C_2) on polycylinders, but the results hold on univalent cylindrical domains; to verify this, it is enough to apply the usual procedure of passing to higher dimensional spaces.

6. (*W*)-functions: Complement to the Lemma. We shall examine (*W*)-functions quantitatively. Let us recall the contents of No. 7 of VIII. Consider a domain $\mathfrak{D}$ (not necessarily connected) over the space $(x_1, x_2, \ldots, x_n)$ which may contain non-transcendental points of ramification as interior points. Let $\eta_1(P)$, $\eta_2(P), \ldots, \eta_m(P)$ be holomorphic functions on $\mathfrak{D}$. Let us suppose that for any pair of regular points of $\mathfrak{D}$ having the same coordinates, the analytic elements of $\eta_1(P)$ at these points are different. We consider the characteristic variety Σ in the space $(x_1, x_2, \ldots, x_n; y_1, y_2, \ldots, y_m)$ given by

$$y_j = \eta_j(P), \qquad P \in \mathfrak{D} \quad (j = 1, 2, \ldots, m).$$

1°. Let Δ be a domain *(not necessarily connected)* such that $\Delta \Subset \mathfrak{D}$ and such that Δ is a *covering surface* (Überlagerungsbereich) of $\underline{\Delta}$, where $\underline{\Delta}$ is the projection of Δ on the space (x) (Grundbereich). Δ then has a constant number ν of sheets. Let $P_1, P_2, \ldots, P_\nu$ be the points of Δ over the point (x), (x) being any point of $\underline{\Delta}$; we form the function

$$F_1(x, y_1) = \prod [y_1 - \eta_1(P_i)] \qquad (i = 1, 2, \ldots, \nu).$$

Thanks to Weierstrass, we know that to every holomorphic function $u(P)$ on Δ, there corresponds, in a well-defined way, a polynomial $\Phi(x, y_1)$ in y_1, whose coefficients are holomorphic functions of (x) in $\underline{\Delta}$, so that $u \dfrac{\partial F_1}{\partial y_1} = \Phi$ on $F_1 = 0$. If we inspect the proof[17] of this result, we obtain immediately the following:

For any domain A (not necessarily connected) such that $A \Subset \underline{\Delta}$, *there exists a constant K independent of u such that the coefficients of* y_1 *in the polynomial* $\Phi(x, y_1)$ *are smaller than KM in absolute value on A, M being an upper bound for* $|u|$ *on* Δ.

2°. Let (x^0, y^0) be any point on the characteristic variety Σ. Let us consider a nonsingular linear transformation $\mathfrak{L}$ of the form

$$x_i' = \sum A_{i,k}(x_k - x_k^0) + \sum A_{i,n+l}(y_l - y_l^0),$$
$$y_j' = \sum A_{n+j,k}(x_k - x_k^0) + \sum A_{n+j,n+l}(y_l - y_l^0)$$
$$(i, k = 1, \ldots, n; j, l = 1, \ldots, m).$$

We constructed the function $\dfrac{\partial F_1}{\partial y_1}$ in the space (x, y) with respect to Σ; we wish to construct, by the same method, a function $W(x, y)$ in the space (x', y')

[17]) See: W.F. Osgood, Lehrbuch der Funktionentheorie II_1, 1929, pages 116, 117.

with respect to the image of Σ in a neighbourhood of the origin in the new space. If we choose $\mathfrak{L}$ suitably, this is possible, and, in addition, W has the same property as $\frac{\partial F_1}{\partial y_1}$. As for these functions W, if we examine our earlier reasoning, we easily find that:

We can choose $n+1$ transformations $\mathfrak{L}_i$ $(i=1,2,\ldots,n+1)$ such that the corresponding functions W_i exist and have the property indicated above, and moreover, the set of common zeros of these functions is contained in the set S_0 of singular points of Σ in the neighbourhood of (x^0, y^0).

3°. Suppose that (x^0, y^0) belongs to S_0. Let $U(x, y)$ be a holomorphic function in a neighbourhood of (x^0, y^0) which vanishes identically on S_0. By the "*Nullstellensatz*" (No. 8 in VIII), we find that $U^\lambda \equiv 0 \bmod (W_1, W_2, \ldots, W_{n+1})$ in a neighbourhood of (x^0, y^0), λ being a positive integer.

4°. Let us now consider the problem posed in No. 1 above. (See No. 10 of VIII.) Because of the results of No. 4 onwards, we find easily (using the Borel lemma) that *the answer is affirmative, as long as the integer λ is chosen sufficiently large.*

Chapter II. Pseudoconvex Domains; a Second Lemma

7. Generalities. Let us recall what we said in the introduction to Memoir VI. In 1906, F. HARTOGS made what I think is an epoch-making discovery in the theory of analytic functions of several variables; namely that every domain of holomorphy is subject to a very curious restriction which we call being *pseudo-convex*[18].

The same restriction was found successively in different parts of the theory; we shall rapidly recall them. In 1910, E.E. LEVI showed that *domains of meromorphy* are also pseudoconvex[19]; G. JULIA indicated in 1926 that so also are *domains of normality* for holomorphic functions[20]. In 1931, W. SAXER added the statement that the same is true (in a certain sense) for *families of meromorphic functions*[21]. And finally, in 1934, the author indicated the corresponding result for *families of characteristic surfaces*[22].

These theorems were studied, and diverse consequences deduced, by the discoverers themselves, and other mathematicians, especially by HARTOGS and E.E. LEVI[23].

The present chapter is divided into three parts A, B, C. In A, we shall expound, abstractly, what we owe to HARTOGS and to E.E. LEVI.

[18]) F. HARTOGS, Einige Folgerungen aus der Cauchyschen Integralformel bei Funktionen mehrerer Veränderlichen (Münch. Berichte).

[19]) E.E. LEVI, Studii sui punti singolari essenziali delle funzioni analitiche di due o più variabili complesse (Annali di Matematica).

[20]) G. JULIA, Sur les familles de fonctions analytiques de plusieurs variables (Acta Mathematica).

[21]) W. SAXER, Sur les familles de fonctions méromorphes de plusieurs variables (C.R., Paris).

[22]) K. OKA, Note sur les familles de fonctions analytiques multiformes etc. (Journal of Science of the Hiroshima Univ.).

[23]) See the book of BEHNKE-THULLEN and the developments indicated in the introduction. (See also the above note of the author.)

In B, we shall extend the results in Memoir VI concerning pseudoconvex functions. And in C, we shall treat a new problem which is peculiar to multivalent domains[24].

A. Pseudoconvex Domains

8. Domains. The domains which we shall consider from here on in the present memoir *contain neither points at infinity nor interior points of ramification*, with a single exception in No. 24. We shall not, therefore, indicate these conditions. Domains of exactly this kind are explained in *the book of Behnke-Thullen;* what we shall do is imply to introduce and add some terminology and notation which will be useful in what follows.

We consider the space of n complex variables $x_1, x_2, \ldots, x_n$; and over this space (x), we consider a point P; a point is a thing (Ding) which has well-defined coordinates (x). The point (x) in the space (x) is called the *projection* of the point P (on the space (x)); and is denoted by $\underline{P}$. We shall sometimes say that P lies over $\underline{P}$. If E is a set of points P, the set of projections $\underline{P}$ is called the projection of E and is denoted by $\underline{E}$.

In general, we shall use, for sets of points over the space (x), the terms and the notation of the theory of abstract sets[25].

We shall call a set of points over the space (x) a domain if it possesses a system (S) of subsets, called neighbourhoods (Umgebungen) which satisfies "neighbourhood postulates" ("Umgebungspostulate") and, in addition, is connected in a sense which respects (S).

Consider a domain $\mathfrak{D}$ over the space (x). $\mathfrak{D}$ has, by definition, a definite system (S) of neighbourhoods. We can therefore draw continuous curves on $\mathfrak{D}$. Let P_1, P_2 be an arbitrary pair of points of $\mathfrak{D}$; we denote by $d(P_1, P_2)$ the lower bound of the lengths of rectifiable curves on $\mathfrak{D}$ joining P_1 and P_2; $d(P_1, P_2)$ satisfies the distance postulates ("Entfernungspostulate"). We shall define the *distance* (Entfernung) of points P_1, P_2 on $\mathfrak{D}$ as this $d(P_1, P_2)$. By introducing this, we can use the terminology of "metric spaces" for domains.

We consider relations between two domains over the same space (x). Let $\mathfrak{D}$, $\mathfrak{D}_0$ be two domains such that every point of $\mathfrak{D}_0$ belongs to $\mathfrak{D}$. We consider the correspondence which associates to any point of $\mathfrak{D}_0$ the same point in $\mathfrak{D}$; if this correspondence is bicontinuous we say that $\mathfrak{D}_0$ is a *subdomain* of $\mathfrak{D}$ and denote it $\mathfrak{D}_0 \subseteqq \mathfrak{D}$; we also say that $\mathfrak{D}$ is an *extension* of $\mathfrak{D}_0$. If, in addition, $\mathfrak{D}_0 = \mathfrak{D}$ as sets, we say that the two domains $\mathfrak{D}_0$, $\mathfrak{D}$ are *equal* (or *identical*) and denote this by $\mathfrak{D}_0 = \mathfrak{D}$.

---

[24]) Very recently, the author has received several memoirs of P. Lelong and of F. Norguet on the subject of the present chapter, as well as on the inverse Hartogs problem. But the author is now devoting himself to the inspection of his own solution (naturally subjective) of the problem mentioned in the Introduction; he will therefore speak (objectively) of these papers on a later occasion. As far as the author is concerned, since the time of Memoir VI (1942) (in which the essential difficulty was the discovery of the use of integral equations), the main difficulties lie in this problem and the lemma expounded in Memoir VIII (for which Memoir VII was written).

[25]) See: F. Hausdorff, Mengenlehre.

Let $\mathfrak{D}_1, \mathfrak{D}_2$ be two domains. If we can establish a biunique and bicontinuous correspondence between the points of $\mathfrak{D}_1$ and those of $\mathfrak{D}_2$ in such a way that corresponding points have the same projection (on the space (x)), we say that these two domains are *equivalent*, and denote this by $\mathfrak{D}_1 \sim \mathfrak{D}_2$.

To define pseudoconvex domains, it is necessary to distinguish clearly between the two relations $\mathfrak{D}_1 = \mathfrak{D}_2$ and $\mathfrak{D}_1 \sim \mathfrak{D}_2$.

Let $\mathfrak{D}_0, \mathfrak{D}$ be two domains. If we can establish a correspondence between all the points of $\mathfrak{D}_0$ and a portion of the points of $\mathfrak{D}$ such that to any point of $\mathfrak{D}_0$ there corresponds a point of $\mathfrak{D}$ uniquely and continuously, and, moreover, corresponding points have the same projection, we shall say, with BEHNKE and THULLEN, that $\mathfrak{D}_0$ is *contained in* $\mathfrak{D}$, and denote it by $\mathfrak{D}_0 < \mathfrak{D}$. (Remark that this contains the case when $\mathfrak{D}_0 \sim \mathfrak{D}$.) And similarly for other notions. (See the book of BEHNKE-THULLEN.)

If $\mathfrak{D}_0 \subseteqq \mathfrak{D}$ and, further, if the set of points of $\mathfrak{D}_0$ is compact in the domain $\mathfrak{D}$; more precisely, if from any infinite sequence of points of $\mathfrak{D}_0$, we can choose a subsequence of points converging to a point of $\mathfrak{D}$; we shall say that the subdomain is *completely interior* to $\mathfrak{D}$ and shall denote it by $\mathfrak{D}_0 \Subset \mathfrak{D}$.

We have restricted these explanations to (connected) domains, but they apply, by definition, also to *(not necessarily connected) domains.*

Let E be a set of points over the space (x), and let $\mathfrak{D}$ be a domain over the same space; let us denote by E' the set of points of $\mathfrak{D}$. The notations $E = \mathfrak{D}$ and $E \subseteqq \mathfrak{D}$ stand, respectively for the relations $E = E'$ and $E \subseteqq E'$. The symbol $E \Subset \mathfrak{D}$ means that the set of points E is contained in the set $\mathfrak{D}$, and that, in addition, E is compact in the domain $\mathfrak{D}$; we shall also say that E is *completely interior* to $\mathfrak{D}$.

The Intersection of a Family of Domains. We consider, with BEHNKE and THULLEN, an abstract set $M = \{m, n, p, \ldots\}$ and a set of domains $\{\mathfrak{D}_m, \mathfrak{D}_n, \mathfrak{D}_p, \ldots\}$ over the space (x). From this, we construct a set δ of points over the same space as follows: if there exists a polycylinder γ_0 with centre (x^0) such that each domain $\mathfrak{D}_m$ of the set contains a subdomain V_m equivalent to γ_0, we denote the point of V_m over (x^0) by P_m and say that $Q = \{P_m\}$ is a point of δ (having coordinates (x^0)). If $Q = \{P_m\}$, $Q' = \{P'_m\}$ is an arbitrary pair of points of δ, we define $Q = Q'$ if and only if $P_m = P'_m$ for every m in M.

We shall attach a system (S) of neighbourhoods to the set of points (δ) thus defined over the space as follows: Let Q be any point of δ, and let us consider a polycylinder (γ) with centre (x^0) and contained in γ_0. Let (x') be any point of γ and P'_m the point of V_m over (x'). Consider the point $Q' = \{P'_m\}$ of δ; if we let (x') run over γ, the point Q' describes a set v; v is a univalent subset of δ. We define $(S) = \{v\}$. (S) satisfies the neighbourhood postulates. We shall denote by the same letter δ the set δ together with this neighbourhood system. Then δ is a domain (not necessarily connected) over the space (x).

This (not necessarily connected) domain δ clearly has the following properties:

1°. $\delta < \mathfrak{D}_m$ for any $m \in M$.

2°. Let δ' be a (not necessarily connected) domain over the space (x) having the first property. We then necessarily have $\delta' < \delta$.

Upto equivalence, there is only one domain (connected or not) over the space (x) having these two properties. We call δ the *intersection* of the set of domains $\{\mathfrak{D}_m, \mathfrak{D}_n, \mathfrak{D}_p, \ldots\}$ and denote it by $\delta \sim \mathfrak{D}_m \cap \mathfrak{D}_n \cap \mathfrak{D}_p \cap \ldots$. [The same applies also to not necessarily connected domains.]

We define the *kernel* of a sequence of domains in the same way. As for *convergence* of a sequence of domains, we must adhere, without modification, to the definition in the book.

Boundary Points. Concerning boundary points, there is nothing to modify or to add. But, in order to be able to define pseudoconvex domains, let us look for a moment at what the book explains about them.

Let $\mathfrak{D}$ be a domain over the space (x). Let us consider a sequence of points P_i $(i=1, 2, \ldots)$ in $\mathfrak{D}$; suppose that this sequence has no points of accumulation in $\mathfrak{D}$ and that it has the following properties: 1°. The sequence of projections $\underline{P}_i$ $(i=1, 2, \ldots)$ converges to a point (x^0). 2°. For every positive number ρ, we can find a positive integer m such that any two points P_i, P_j with $i \geqq m$, $j \geqq m$ can be joined by a continuous curve lying in $\mathfrak{D}$ whose projection remains in the polycylinder γ with centre (x^0) and radius ρ. We then say that $R=(P_1, P_2, \ldots)$ defines a *boundary point* of $\mathfrak{D}$ having coordinates (x^0). Let $R'=(P_1', P_2', \ldots)$ be another boundary point. We say that $R=R'$ if R' has the same coordinates as R and, further, R and R' are related in the following way: for any positive number ρ, we can find a positive integer m such that P_i, P_i', where $i \geqq m$, can be joined by a continuous curve in $\gamma \cap \mathfrak{D}$, γ being the polycylinder defined above. We shall call any univalent subdomain of $\mathfrak{D}$ containing almost all points of the sequence P_i $(i=1, 2, \ldots)$ a *neighbourhood* of the boundary point R.

Rotations. From now on, we shall mean by a *rotation* of the space (x) about the point (ξ) any linear transformation of the form $x_i'-\xi_i=\sum a_{ij}(x_j-\xi_j)$ $(i, j=1, 2, \ldots, n)$, the a_{ij} being constants, which preserves (euclidean) distances, but is otherwise arbitrary.

9. Pseudoconvex Domains. Let us begin by defining the continuity theorem. Let $\mathfrak{D}$ be a domain over the space $(z_1, z_2, \ldots, z_n)$ where $n \geqq 1$ (but we shall explain the notion first when $n>1$). Let M be a boundary point of $\mathfrak{D}$. We shall say that M satisfies the *continuity theorem* if, whenever the following condition (C) is fulfilled, M has the property (P) given below (after condition (C)). We shall say that $\mathfrak{D}$ satisfies the continuity theorem if this is the case for any boundary point of $\mathfrak{D}$.

Condition (C). We distinguish one of the variables $(z_1, z_2, \ldots, z_n)$ $(n \geqq 2)$ and denote it by y; we denote the other variables by $(x_1, x_2, \ldots, x_{n-1})$. Let (ξ, η) be the coordinates of M. The condition is that there then corresponds to M a positive number ρ such that, if we denote by A the set of points (x, y) satisfying $x_i=\xi_i$ $(i=1,2,\ldots,n-1)$, $0<|y-\eta|<\rho$, there exists a set of points A^* contained in a single neighbourhood of M such that $\underline{A}^*=A$.

For any positive number δ' such that $\delta'<\rho$, there then corresponds a positive number r such that, if we denote by E the set of points $|x_i-\xi_i|<r$ (i

$=1, 2, \ldots, n-1)$, $|y-\eta|=\delta'$, there exists a set E^* of points contained in a single univalent subdomain of $\mathfrak{D}$ and satisfying $\underline{E}^*=E$, $A^*\cap E^*\supset 0$.[26]

Property (P). To the point M, there corresponds a pair of positive numbers (δ, δ') where $\delta \leqq r$ (r being the number corresponding to δ' introduced above) such that for any point (x') in the polycylinder $|x_i'-\xi_i|<\delta$ $(i=1, 2, \ldots, n-1)$ there exists a point y' of the circumference $|y-\eta|=\delta'$ fulfilling the following condition: Let P be the point of E^* over (x', y'), l the (closed) line segment in the y-plane joining the points y', η and L the segment $((x'), l)$ in the space (x, y). Then, if we describe a segment on $\mathfrak{D}$ starting at the point P in such a way that the projection lies always on L, then we necessarily meet a boundary point of $\mathfrak{D}$ (either at the final point of L or on the way).

The case $n=1$ can be regarded as a special case where the domain $\mathfrak{D}$ is independent of $(x_1, x_2, \ldots, x_{n-1})$.

Let now $\mathfrak{D}$ be a domain over the space $(x_1, x_2, \ldots, x_n)$ $(n \geqq 1)$. We shall call $\mathfrak{D}$ *pseudoconvex* if every boundary point M of $\mathfrak{D}$ satisfies the continuity theorem and if, in addition, this property of M holds after an arbitrary bi-unique pseudoconformal transformation of the space (x) in the neighbourhood of the projection $\underline{M}$.

In the case $n=1$, any domain is pseudoconvex.

10. Other Definitions. We shall now define pseudoconvex domains in different ways. Let us start with the continuity theorem. We call the continuity theorem of the preceding section the *continuity theorem* (A), and shall consider continuity theorems (B) and (C) in what follows.

Let $\mathfrak{D}$ be a domain over the space $(z_1, z_2, \ldots, z_n)$ where $n>1$, and let M be a boundary point of $\mathfrak{D}$. Let $(x_1, x_2, y_1, y_2, \ldots, y_{n-2})$ be any permutation of $(z_1, z_2, \ldots, z_n)$ and let (ξ, η) be the coordinates of M. We take, in the space (x), a point (a) different from (ξ); we describe the hypersphere S with centre (a) whose boundary passes through (ξ), and a hypersphere σ with centre (ξ). Let β be the part of σ exterior to S; β is a simply connected domain. Let $B=(\beta, \eta)$ in the space (x, y). We shall say that M satisfies the *continuity theorem* (B) if, under the above geometric circumstances, there never exists a set of points of $\mathfrak{D}$ contained in a single neighbourhood of M whose projection is B, whatever be (a), σ. We describe $\mathfrak{D}$ (over the space (z)) by the same expression if this is the case for every boundary point of $\mathfrak{D}$.

In the case $n=1$, we consider that every domain satisfies the continuity theorem (B).

Let, again, $\mathfrak{D}$ be a domain over the space $(z_1, z_2, \ldots, z_n)$ where $n>1$. Let y be one of the $z_1, z_2, \ldots, z_n$, and $x_1, x_2, \ldots, x_{n-1}$, the others. We consider the following two domains in the space (x, y):

$$|x_i-x_i^0|<r, \quad \rho'<|y-y^0|<\rho \quad (i=1, 2, \ldots, n-1);$$
$$|x_i-x_i^0|<r' \,(<r), \quad |y-y^0|<\rho \quad (i=1, 2, \ldots, n-1),$$

[26]) It follows from this that any point of E^* lying over $x_i=\xi_i$ $(i=1, 2, \ldots, n-1)$ belongs to A^*.

(where (x^0, y^0) is a definite point and r, r', ρ, ρ' are positive numbers); we denote the union of these two domains by Δ; Δ is a simply connected domain. Let us denote the polycylinder $|x_i - x_i^0| < r$, $|y - y^0| < \rho$ $(i = 1, 2, \ldots, n-1)$ by C. The domain $\mathfrak{D}$ (over the space (z)) will be said to satisfy the *continuity theorem* (C) if, with the above geometric configuration, $\mathfrak{D}$ has the following property: whenever there exists a domain Δ^* such that $\Delta^* \sim \Delta$, $\Delta^* \subseteqq \mathfrak{D}$, there necessarily exists a domain C^* such that $C^* \sim C$, $\Delta^* \subset C^* \subseteqq \mathfrak{D}$.

We consider that any domain over the plane of one complex variable satisfies the continuity theorem (C). Let us remark that only the continuity theorem (C) is *global*.

Let us now consider a domain $\mathfrak{D}$ over the space $(x_1, x_2, \ldots, x_n)$ $(n \geqq 1)$; we shall say that $\mathfrak{D}$ is (B)-*pseudoconvex* if an arbitrary boundary point M of $\mathfrak{D}$ satisfies the continuity theorem (B) and if, further, this property of M remains invariant under any biunique pseudoconformal transformation of the space (x) in the neighbourhood of $\underline{M}$.

Let, again, $\mathfrak{D}$ be a domain over the space $(x_1, x_2, \ldots, x_n)$ $(n \geqq 1)$, and let M be a boundary point of $\mathfrak{D}$. Consider a polycylinder γ around $\underline{M}$ with a certain radius ρ, and let δ be the connected component of the portion of $\mathfrak{D}$ over γ having M as a boundary point. With this geometric configuration, we say that $\mathfrak{D}$ is (C)-*pseudoconvex* if for any M, there is a positive number ρ_0 such that, for any $\rho \leqq \rho_0$, the subdomain δ corresponding to ρ satisfies the continuity theorem (C), and, in addition, this property holds after any biunique pseudoconformal transformation of the polycylinder γ.

From the definition, it follows that any domain over the x_1-plane is (B)- and (C)-pseudoconvex at the same time.

11. Equivalence of the Definitions. We shall show that these three kinds of pseudoconvexity define one and the same class of domains, which we shall call, from now on, simply "pseudoconvex domains". Since this is evident over the plane of one complex variable, we shall place ourselves in what follows, over the space $(x_1, x_2, \ldots, x_n)$ where $n > 1$.

1°. It is easy to see that *every* (C)-*pseudoconvex domain is* (A)-*pseudoconvex* (i.e. pseudoconvex in the sense of No. 9).

2°. We shall show that *every* (A)-*pseudoconvex domain is* (B)-*pseudoconvex.* We have only to verify that every (A)-pseudoconvex domain satisfies the continuity theorem (B).

Let $\mathfrak{D}$ be an (A)-pseudoconvex domain over the space (x), and let M be a boundary point of $\mathfrak{D}$. We denote two of the $x_1, x_2, \ldots, x_n$ again by x_1, x_2 and the others, if they are present, by $y_1, y_2, \ldots, y_{n-2}$. We consider the configuration indicated in the definition and suppose that, on the contrary, there is a set B^* over the space (x, y), contained in a single neighbourhood of M with $\underline{B}^* = B$.

We transform the space (x) into the space (x') by translations and rotations and continue to denote the images of $S, \sigma, \ldots$ by the same letters; we can do this so that the images of (a), (ξ) are $(0, 0)$ and $(R, 0)$ respectively, R being the radius of S. In the space (x'), the only point on $x'_1 = R$ to lie on S is $(R, 0)$, the others are exterior to S. Moreover, the points (x') satisfying $\Re(x'_1) > R$, $\Re(x'_1)$

being the real part of x_1', lie outside S. With this configuration, the existence of B^* clearly contradicts the domain $\mathfrak{D}$ over the space (x, y) satisfying the continuity theorem (A). $\mathfrak{D}$ therefore necessarily satisfies the continuity theorem (B).

3°. It only remains to show that *every* (B)-*pseudoconvex domain is* (C)-*pseudoconvex*. To do this, we shall first verify that any connected component of the intersection of two (B)-pseudoconvex domains is again (B)-pseudoconvex.

In fact, let us consider two (B)-pseudoconvex domains $\mathfrak{D}_1, \mathfrak{D}_2$ over the space (x); we choose arbitrarily a connected component $\mathfrak{D}_0$ of the intersection $\mathfrak{D}_1 \cap \mathfrak{D}_2$. Let M be a boundary point of $\mathfrak{D}_0$. We transform a neighbourhood of the point $\underline{M}$ in the space (x) by a biunique pseudoconformal transformation and denote the images of (x), M, $\mathfrak{D}_0$, $\mathfrak{D}_1$, $\mathfrak{D}_2$ by (x'), M', $\mathfrak{D}_0'$, $\mathfrak{D}_1'$, $\mathfrak{D}_2'$ respectively (where, of course, $\mathfrak{D}_0'$, $\mathfrak{D}_1'$, $\mathfrak{D}_2'$ are only defined over a neighbourhood of $\underline{M'}$). Let us suppose, contrary to the statement, the existence of the set of points indicated in the definition with respect to $\mathfrak{D}_0'$ and M'. Since $\mathfrak{D}_1'$, $\mathfrak{D}_2'$ are (B)-pseudoconvex, it would follow immediately from this that M' is an interior point of $\mathfrak{D}_0'$.

In view of what we have just seen, in order to justify the proposition, it is sufficient to affirm that every (B)-pseudoconvex domain satisfies the continuity theorem (C). We shall verify this; let $\mathfrak{D}$ be a (B)-pseudoconvex domain over the space (x). We again denote by $(x_1, x_2, \ldots, x_{n-1}, y)$ a permutation of $(x_1, x_2, \ldots, x_n)$, we consider the geometric configuration indicated in the continuity theorem (C) and suppose that there is a domain Δ^* such that $\Delta^* \sim \Delta$, $\Delta^* \subsetneqq \mathfrak{D}$. It is sufficient to show that there exists a domain C^* such that $C^* \sim C$, $\Delta^* \subset C^* \subsetneqq \mathfrak{D}$. To simplify the notation, we suppose that (x^0, y^0) is $(0, 0)$.

Let (x') be an arbitrary point of the polycylinder $|x_i| < r$ $(i = 1, 2, \ldots, n-1)$, let y' be an arbitrary point of the circumference $\Sigma'\colon |y| = \rho''$, where $\rho'' = \frac{1}{2}(\rho + \rho')$, and let L be the line segment in the space (x, y) with initial point (x', y') and final point $(x', 0)$. Let P be the point of Δ^* over (x', y'). We describe a line segment continuously in the domain $\mathfrak{D}$ starting at P such that its projection always lies on L. Two cases are possible; either we meet a boundary point M of $\mathfrak{D}$ (along the way or at the final point of L), or we do not meet any boundary point. In the first case, we denote the coordinates of M by (x', y'').

If (x') lies in the polycylinder $|x_i| < r'$ $(i = 1, 2, \ldots, n-1)$, the point M never exists, whatever be y'. We denote this polycylinder by γ_0.

Let us examine the polycylinder γ_1:

$$|x_1| < r, \quad |x_j| < r' \qquad (j = 2, 3, \ldots, n-1)$$

for this property of the existence of M. Suppose that there is a point (a) in γ_1 such that to the point (a, b) corresponds M with coordinates (a, b'), where b is a suitable point on $\Sigma'(|y| = \rho'')$. Let us consider the expression

$$d(x, y) = \left(\left|\frac{K}{x_1}\right|^2 + |y|^2\right)^{1/2},$$

where K is a positive number, and the square root is the non-negative determination. Let Σ be the circle in the space (x) given by $|x_1| < r$, $x_i = a_i$ $(i = 2, 3, \ldots, n-1)$. Let (x', y') be an arbitrary point on (Σ, Σ'); if to (x', y') corresponds the point M, (x', y''), we shall associate to (x', y') the number $d = d(x', y'')$; if there is no corresponding point M, we associate nothing. Since for any point (x') of γ_0, only the second case is possible, the set $\{d\}$ is bounded; let d_0 be its upper bound.

For K sufficiently large, and for r_1 such that $0<r_1<r$ and sufficiently close to r, we have

$$d(a, b') = \left(\left|\frac{K}{a_1}\right|^2 + |b'|^2\right)^{1/2} > \left[\left(\frac{K}{r_1}\right)^2 + (\rho'')^2\right]^{1/2} = d_1.$$

Hence, for this K, we have $d_0 > d_1$. For every point (x', y') of (Σ, Σ') satisfying $r_1 < |x'_1| < r$, the associated number d, even if it is defined, necessarily satisfies $d < d_1\ (< d_0)$. From this, it follows easily that there exists a point (ξ, η) on (Σ, Σ') to which corresponds a boundary point M_0 with coordinates (ξ, η') such that we effectively have $d(\xi, \eta') = d_0$ $(\xi_i = a_i,\ i = 2, 3, \ldots, n-1)$.

We now transform the space (x) by

$$X_1 = \frac{K}{x_1}, \qquad X_2 = x_2, \ldots, X_{n-1}, \qquad Y = y;$$

let $\mathfrak{D}'$, M'_0 be respectively the images of $\mathfrak{D}$, M_0 over the space (X). We take the two points $(0, 0)$, and Q: $\left(\frac{K}{\xi_1}, \eta'\right)$ and describe the hypersphere S around $(0, 0)$ so that its boundary passes through Q; we also take a sufficiently small hypersphere σ with centre Q and denote by β the part of σ exterior to S. Let B be the set of points of the space (X, Y) given by $(X_1, Y) \in \beta$, $X_i = \xi_i$ $(i = 2, 3, \ldots, n-1)$. We then find, without difficulty, that there exists a set B^* over the space (X, Y) contained in a single neighbourhood of the point M'_0 and such that $\underline{B^*} = B$. Since M'_0 is a boundary point of $\mathfrak{D}'$ and $\mathfrak{D}'$ is (B)-pseudoconvex, this is absurd. Therefore, there cannot correspond a boundary point of $\mathfrak{D}$ to any point (x', y') of the set (γ_1, Σ').

We now go over to $|x_1| < r$, $|x_2| < r$, $|x_k| < r'$ $(k = 3, 4, \ldots, n-1)$ and reason exactly in the same way, and so on. And we find, finally, that there exists a domain C^* such that $C^* \sim C$, $\Delta^* \subset C^* \subseteq \mathfrak{D}$. We have thus seen that:

The three kinds of pseudoconvex domains (A), (B), (C), *are but one and the same.*

We shall apply the term pseudoconvex domain also to domains which are not connected. Along the way, we have also seen the following:

The intersection of two pseudoconvex domains (over the same space) is again pseudoconvex.

Every pseudoconvex domain satisfies the continuity theorem (C).

12. The Kernel of a Sequence of Pseudoconvex Domains. *Given a sequence of pseudoconvex domains over the space* (x), *the kernel (connected or not), if it exists, is also pseudoconvex.*

In fact, let $\mathfrak{D}_1, \mathfrak{D}_2, \ldots, \mathfrak{D}_p, \ldots$ be the sequence of pseudoconvex domains over the space $(x_1, x_2, \ldots, x_n)$; we can suppose that $n \geqq 2$. As criterion for pseudoconvexity, we adopt the continuity theorem (C).

Consider $E_1 \sim \mathfrak{D}_1$; $E_2 \sim \mathfrak{D}_1 \cap \mathfrak{D}_2, \ldots, E_p \sim \mathfrak{D}_1 \cap \mathfrak{D}_2 \cap \ldots \cap \mathfrak{D}_p$; each of these domains is then pseudoconvex.

Consider $\delta_1 \sim \mathfrak{D}_1 \cap \mathfrak{D}_2 \cap \ldots \cap \mathfrak{D}_p \cap \ldots$. Since $E_p > E_{p+1}$ $(p=1, 2, \ldots)$ and E_p is pseudoconvex, we find immediately, using the criterion we have adopted, that δ_1 is also pseudoconvex.

Setting $\delta_p = \mathfrak{D}_p \cap \mathfrak{D}_{p+1} \cap \ldots$, we consider the sequence $\delta_1, \delta_2, \ldots, \delta_p, \ldots$; we then have $\delta_p < \delta_{p+1}$. Consider a domain δ_0 (connected or not) having the following two properties: 1°. $\delta_0 > \delta_p$ $(p=1, 2, \ldots)$; 2°. let δ be a (not necessarily connected) domain having the first property; then $\delta > \delta_0$. By reasoning analogous to the case of intersections, one can check that δ_0 effectively exists if one at least of the δ_p $(p=1, 2, \ldots)$ is non-empty. When δ_0 exists, it is unique upto equivalence. It is this δ_0 which one calls the *kernel* of the sequence $\mathfrak{D}_p$ $(p=1, 2, \ldots)$.

Suppose that δ_0 exists. Let us again denote by $(x_1, x_2, \ldots, x_{n-1}, y)$ a permutation of $(x_1, x_2, \ldots, x_n)$.

We describe in the space (x, y) a polycylinder C of the form $|x_i - x_i^0| < r$, $|y - y^0| < \rho$ $(i=1, 2, \ldots, n-1)$. Let $\varDelta$ be the set of points of C satisfying at least one of the two conditions $\rho' < |y - y^0| < \rho$, $|x_i - x_i^0| < r' (< r)$ $(i=1, 2, \ldots, n-1)$. With this configuration, suppose that there is a domain $\varDelta^*$ such that $\varDelta^* \sim \varDelta$, $\varDelta^* \subseteqq \delta_0$.

Now by the very reasoning by which we verified the existence of δ_0, it follows that to each point P of δ_0 there is a polycylinder γ around P and a positive integer m such that, if P_p is the point of δ_p corresponding to P in δ $(p \geqq m$; this is under the correspondence implicit in the first property of $\delta_0)$ then every δ_p, where $p \geqq m$, contains a subdomain equivalent to γ and containing P_p; let us denote by m again the smallest of these positive integers.

Suppose that $\varDelta^* \Subset \delta_0$. By what we have just seen and the Borel lemma, we then find that there is a domain C^* such that $C^* \sim C$, $\varDelta^* \subset C^* \subseteqq \delta_0$. From this, the same result without the hypothesis $\varDelta^* \Subset \delta_0$ then follows. Thus the kernel δ_0 of the sequence $\mathfrak{D}_p$ $(p=1, 2, \ldots)$ satisfies the continuity theorem (C).

Let now γ be an arbitrary polycylinder in the space (x); let us transform γ by a biunique pseudo-conformal transformation; the kernel of the sequence of images of $\gamma \cap \mathfrak{D}_p$ $(p=1, 2, \ldots)$ is the image of $\gamma \cap \delta_0$. Since the image of each $\gamma \cap \mathfrak{D}_p$ is pseudoconvex, it follows from what we have seen above that the image of $\gamma \cap \delta_0$ still satisfies the continuity theorem (C). δ_0 is therefore pseudoconvex.

B. Pseudoconvex Functions

13. Definition. Real functions satisfying the differential condition of E.E. Levi do not allow addition, that is to say, even if $\mathfrak{L}(\varphi_1) \geqq 0$, $\mathfrak{L}(\varphi_2) \geqq 0$, we do not necessarily have $\mathfrak{L}(\varphi_1 + \varphi_2) \geqq 0$. (Consequently, we cannot, for example, take arithmetical means.) It is to avoid this inconvenience that we conceived of pseudoconvex functions in Memoir VI. Since we only did this for two variables there, we shall extend the notion.

Consider a characteristic variety L of one complex dimension in the space $(x_1, x_2, \ldots, x_n)$ of the form

$$x_i = f_i(u) \qquad (i = 1, 2, \ldots, n),$$

where u stands for one of the variables $x_1, x_2, \ldots, x_n$ and the $f_i(u)$ represent polynomials in u of degree at most 1; in what follows, we shall call L a *characteristic line*. Let $\mathfrak{D}$ be a domain over the space (x) [without points at infinity or interior ramification], let P be an arbitrary point of $\mathfrak{D}$ and (x), the coordinates of P. Let $\varphi(P)$ be a *real, singlevalued* function of P (which can take the value $-\infty$). We shall call $\varphi(P)$ a *pseudoconvex*[27] *function* of P on the domain $\mathfrak{D}$ if this function satisfies the *following conditions:* 1°. $e^{\varphi(P)}$ is finite and upper semicontinuous (with respect to P). 2°. Let P_0 be an arbitrary point of $\mathfrak{D}$ having coordinates (x^0), and let L be a characteristic line passing through P_0, but otherwise arbitrary; in a neighbourhood of x_j^0, the restriction of $\varphi(P)$ to L is then a subharmonic function with respect to x_j, where x_j is any one of the variables $x_1, x_2, \ldots, x_n$ which is not constant on L.

As for the notion of subharmonic functions, this is well-known and we shall not explain it[28]. Let us only recall that we have agreed to count the *constant* $-\infty$ as a subharmonic function. However, it is convenient to make a simple remark about the criterion for a function to be subharmonic.

Since subharmonic functions can be analysed locally, it is sufficient to have a *local criterion.* Let $\psi(z)$ be a real singlevalued function of the complex variable (which may take the value $-\infty$) in the neighbourhood of a point z_0, such that $e^{\psi(z)}$ is finite and upper semicontinuous. For the function to be subharmonic (with respect to z in the neighbourhood of z_0) it is necessary and sufficient that it satisfy the following two conditions:

1°. $\psi(z)$ *does not have a strict relative maximum* (that is to say, there is no circle of the form $|z-z'|<\rho$ where $\psi(z)<\psi(z')$, z' being fixed. 2°. Let $u(z)$ be a harmonic function of z (that is, with respect to the real and imaginary parts of z) defined for $|z|<\infty$, but otherwise arbitrary; then the *same property holds for* $\psi(z)+u(z)$.

From the properties of subharmonic functions one immediately obtains corresponding properties of pseudoconvex functions; these are as follows:

1°. Given a pseudoconvex function $\varphi(x)$ and a positive constant c, $c\varphi(x)$ is pseudoconvex.

2°. If $\varphi_1(x)$, $\varphi_2(x)$ are pseudoconvex functions, so is $\varphi_1(x)+\varphi_2(x)$.

3°. Under the same condition, the supremum $\max[\varphi_1(x), \varphi_2(x)]$ is also pseudoconvex.

4°. Let $\varphi_1(P), \varphi_2(P), \ldots, \varphi_p(P), \ldots$ be a sequence of pseudoconvex functions on a domain $\mathfrak{D}$ over the space (x) such that the sequence $e^{\varphi_p(P)}$ converges to

---

27) Plurisousharmonique (plurisubharmonic) according to P. LELONG.

28) See, for example: F. RIESZ, Sur les fonctions subharmoniques et leur rapport avec la théorie du potential, I, 1926; II, 1930 (Acta Mathematica).
T. RADÓ, Remarques sur les fonctions subharmoniques, 1928 (C.R., Paris).

$e^{\varphi(P)}$. If the convergence is uniform on every domain completely interior to $\mathfrak{D}$, or if the sequence is decreasing, then the limit $\varphi(P)$ is also pseudoconvex.

5°. Every convex increasing function (excepting the constant $+\infty$) of a pseudoconvex function is again pseudoconvex.

14. Hartogs Radii. We want now to collect together examples of pseudoconvex functions. Now, in the present case, we recall immediately an idea due to HARTOGS, namely *radii of holomorphy.*

We shall define them abstractly. Let $\mathfrak{D}$ be a domain over the space $(x_1, x_2, \ldots, x_n)$, let P be a point of $\mathfrak{D}$ and let (x^0) be the coordinates of P. We consider a polycylinder C, $|x_i - x_i^0| < r_i$ $(i = 1, 2, \ldots, n)$ around the point P such that there exists a domain C^* satisfying the conditions $C^* \sim C$, $C^* \subseteqq \mathfrak{D}$, $P \in C^*$. We shall denote the upper bound of the $\{r_n\}$ by $R(P)$, and call it the *Hartogs radius* of the domain $\mathfrak{D}$ (over the space (x)) with respect to x_n. Similarly for the other x_i.

Let $\mathfrak{D}$ be a pseudoconvex domain over the space $(x_1, x_2, \ldots, x_n)$. The Hartogs radius $R(P)$ of $\mathfrak{D}$ with respect to x_n is a function such that $-\log R(P)$ is pseudoconvex.

To keep the picture clearer, we suppose that $\mathfrak{D}$ is *bounded*. There is no loss of generality because the intersection of two pseudoconvex domains is again pseudoconvex and the limit of a decreasing sequence of pseudoconvex functions is also pseudoconvex.

The function $\log 1/R(P)$ is real, single valued, bounded on any set completely interior to $\mathfrak{D}$ and is clearly upper semicontinuous. To show that it is pseudoconvex, it suffices to consider the second condition.

Let P_0 be an arbitrary point of $\mathfrak{D}$ having the coordinates (x^0), which we suppose to be (0) to simplify the notation. Let L be an arbitrary characteristic line passing through P_0. We distinguish between two cases according to the position of L relative to the coordinate axes.

We place ourselves first in the general situation in which L can be expressed parametrically by a suitable one of the variables $x_1, x_2, \ldots, x_{n-1}$, say x_1, in the following form:

$$(1) \qquad x_i = A_i x_1 \quad (i = 2, 3, \ldots, n),$$

the A_i being constants. We make the transformation

$$X_1 = x_1, \quad X_i = x_i - A_i x_1 \quad (i = 2, 3, \ldots, n)$$

in the space (x), and continue to denote the images of $\mathfrak{D}$, P and L by the same letters; the equation of L then reduces to $X_i = 0$ $(i = 2, 3, \ldots, n)$. On the other hand, let $R'(P)$ be the HARTOGS radius of $\mathfrak{D}$ over the space (X) with respect to X_n. If we regard x_1 as being fixed, the transformation $X_n = x_n - A_n x_1$ represents a translation of the x_n-plane. We have, therefore,

$$R'(P) = R(P).$$

The first case is thus reduced to the special case in which L is expressed in the following form:

(2) $$x_i=0 \quad (i=2,3,\dots,n).$$

In this case, the restriction of $R(P)$ to L is a function of the single variable x_1 which we denote by $R(x_1)$. I say that $[R(x_1)]^{-1}$ cannot have a strict relative maximum in the neighbourhood of the origin. This is obvious because $\mathfrak{D}$ satisfies *the continuity theorem* (C).

Let, next, $u(x_1)$ be a harmonic function of x_1 on $|x_1|<\infty$, but otherwise arbitrary, and let $f(x_1)$ be an entire function having $-u(x_1)$ as real part. We consider the transformation

$$X_j=x_j, \quad X_n=x_n e^{f(x_1)} \quad (j=1,2,\dots,n-1);$$

it is biunique on the space (x). Let us continue to denote the images of $\mathfrak{D}$, L by the same letters. $\mathfrak{D}$ remains pseudoconvex over the space (X). The restriction to L of the HARTOGS radius of $\mathfrak{D}$ over the space (X) with respect to X_n is given by

$$R(x_1)e^{-u(x_1)}.$$

Hence the above property of $-\log R(x_1)$ is conserved by the function $-\log R(x_1)+u(x_1)$. Hence, by the criterion mentioned, $-\log R(x_1)$ is subharmonic.

It only remains to examine the special case when L is expressed in the form

(3) $$x_i=0 \quad (i=1,2,\dots,n-1).$$

The restriction of $R(P)$ to L is then a function of the single variable x_n, which we denote by $R(x_n)$.

In the plane x_n, let us draw the circle γ with centre the origin and radius $\frac{1}{3}R(0)$. Let x_n' be an arbitrary point of γ; the circle C, $|x_n-x_n'|<R(x_n')$ contains the circle γ, and on its circumference, there is at least one point ξ_n such that there is a boundary point M of $\mathfrak{D}$ over the point $(0,\dots,0,\xi_n)$, which can attained by starting at P_0 and following the broken line on $\mathfrak{D}$, the vertices of whose projection on the space (x) are, in order, (0), $(0,\dots,0,x_n')$, $(0,\dots,0,\xi_n)$. We let x_n' describe the circle γ and consider the set $\{\xi\}$. We clearly have

$$-\log R(x_n)=\max(-\log|x_n-\xi|)$$

on γ, where the term on the right is the upper bound when ξ describes the set indicated. The restriction $-\log R(x_n)$ is therefore subharmonic.

We shall modify the HARTOGS radii a little, since that will be more convenient for our purposes.

Let us consider a domain $\mathfrak{D}$ over the space (x). Let P be any point of $\mathfrak{D}$. Consider a hypersphere around the point $\underline{P}$ in the space (x) such that there is a domain S^* over the space (x) satisfying $S^*\sim S$, $S^*\subseteqq\mathfrak{D}$, $P\in S^*$, and let r be the radius of S. We denote by $d(P)$ the upper bound of all these r and call it the *(euclidean) boundary distance* on the domain $\mathfrak{D}$.

Let $d(P)$ be the euclidean boundary distance on a pseudoconvex domain $\mathfrak{D}$ over the space (x); $-\log d(P)$ *is pseudoconvex in* $\mathfrak{D}$.

In fact, we can suppose that $\mathfrak{D}$ is bounded. Let P_0 be an arbitrary point of $\mathfrak{D}$ and (x^0) its coordinates. Let $R(P)$ be the *Hartogs radius* of $\mathfrak{D}$ with respect to x_n; we obviously have

$$d(P_0) \leqq R(P_0).$$

Let T be an arbitrary rotation about the point P_0 which changes the coordinates of P in $\mathfrak{D}$ from (x) to (X), and let $R_T(P)$ be the HARTOGS radius of $\mathfrak{D}$ over (X) with respect to X_n. We have

$$d(P_0) \leqq R_T(P_0).$$

On the other hand, let M be one of the boundary points of $\mathfrak{D}$ whose distances (Entfernungen) from the point P_0 on $\mathfrak{D}$ are the shortest, and let (ξ) be the coordinates of M; it is obvious that the distance between the points (x^0), (ξ) is $d(P_0)$. We can find a rotation T_0 about (x^0) such that, if (ξ') is the image of (ξ), we have

$$\xi'_j = x_j^0, \qquad \xi'_n = x_n^0 + d(P_0) \quad (j=1,2,\ldots,n-1);$$

we then have

$$d(P_0) = R_{T_0}(P_0).$$

We therefore have

$$d(P) = \min[R_T(P)]$$

(the term on the right means the lower bound when T runs over the set indicated). Since $-\log R_T(P)$ is pseudoconvex, so is $-\log d(P)$.

The function $-\log d(P)$ has the following properties:

1°. It is *continuous* on $\mathfrak{D}$, except when $\mathfrak{D}$ coincides with the space (x), in which case it reduces to the constant $-\infty$.

2°. *When P tends to a boundary point of $\mathfrak{D}$, it tends to $+\infty$.*

15. Differential Condition. Let us consider a real (single valued) function $\varphi(x)$ which is continuous and has continuous partial derivatives up to second order with respect to the real and imaginary parts of the x_i $(i=1,2,\ldots,n)$ in a portion of the space (x). We shall look for the condition that $\varphi(x)$ be pseudoconvex.

We take a point (x^0) (in the given portion of space) and consider (in a neighbourhood of this point) a characteristic variety λ of one (complex) dimension passing through this point and having the form

$$x_j = f_j(t) \qquad (j=1,2,\ldots,n),$$

where the $f_j(t)$ are holomorphic functions of t in a neighbourhood of the origin (and such that $f_j(0) = x_j^0$). We separate t, x_j into real and imaginary parts and set

$$t = u + iv, \qquad x_j = u_j + iv_j$$

(i being the imaginary unit). We shall denote $\varphi(x)$ by

$$\varphi(u_1, u_2, \ldots, u_n; v_1, v_2, \ldots, v_n) = \varphi(u_j, v_j).$$

Under these circumstances we calculate, in a neighbourhood of (x^0),

$$\Delta\Phi(u,v)=\frac{\partial^2\Phi}{\partial u^2}+\frac{\partial^2\Phi}{\partial v^2}$$

for the restriction of $\varphi(x)$ to λ:

$$\Phi(u,v)=\varphi[u_j(u,v),v_j(u,v)].$$

If we set

$$\frac{\partial u_j}{\partial u}=\frac{\partial v_j}{\partial v}=\alpha_j,\qquad \frac{\partial v_j}{\partial u}=-\frac{\partial u_j}{\partial v}=\beta_j,$$

we have

$$\Delta\Phi=\sum_j\sum_k\left[\left(\frac{\partial^2\varphi}{\partial u_j\partial u_k}+\frac{\partial^2\varphi}{\partial v_j\partial v_k}\right)(\alpha_j\alpha_k+\beta_j\beta_k)+\left(\frac{\partial^2\varphi}{\partial u_j\partial v_k}-\frac{\partial^2\varphi}{\partial u_k\partial v_j}\right)(\alpha_j\beta_k-\alpha_k\beta_j)\right]$$

$(j, k=1, 2, \ldots, n)$. *We shall denote the right hand term in this identity by* $W(\varphi;\alpha,\beta)$.

Since, for $\Phi(t)$ to be subharmonic, it is necessary and sufficient that we have $\Delta\Phi\geqq 0$ everywhere, it follows immediately from the above calculation that the following holds:

Let $\varphi(x)$ be a real continuous function having continuous partial derivatives up to second order (with respect to the real and imaginary parts of the complex variables). For $\varphi(x)$ to be pseudoconvex, it is necessary and sufficient that $W(\varphi;\alpha,\beta)\geqq 0$ everywhere.

From the computation above, it also follows immediately that

Under the same conditions, the property that $\varphi(x)$ is pseudoconvex is preserved by every biunique pseudoconformal transformation.

16. The Main Property. Let $\varphi(x_1,x_2,\ldots,x_n,y)$ be a real continuous function in the neighbourhood of a point (x^0,y) of the space (x,y). We separate x_j, y $(j=1,2,\ldots,n)$ into their real and imaginary parts and set

$$x_j=u_j+iv_j,\qquad y=y_1+iy_2,$$

and denote $\varphi(x,y)$ by $\varphi(u,v,y_1,y_2)$. Let us suppose that $\varphi(u,v,y_1,y_2)$ has continuous partial derivatives up to second order, and that $W(\varphi;\alpha,\beta)$ is everywhere positive for all systems of real values of (α,β) except for $(0,0)$; we shall, in what follows, denote this simply by

$$W(\varphi;\alpha,\beta)>0$$

(if there is no confusion).

Suppose, further, that

$$\left(\frac{\partial\varphi}{\partial y_1}\right)^2+\left(\frac{\partial\varphi}{\partial y_2}\right)^2>0\quad\text{at}\quad(x^0,y^0).$$

Under these circumstances, we propose to find a characteristic surface σ passing through (x^0, y^0) in such a way that it does not leave the portion of the space where

$$\varphi(x, y) > \varphi(x^0, y^0)$$

in the neighbourhood of (x^0, y^0), except at this point itself.

To simplify the notation, suppose that (x^0, y^0) is the origin and that $\varphi(x^0, y^0)$ is zero. Let us consider a characteristic surface σ of the form

$$y = f(x) = \sum_j a_j x_j + \sum_j \sum_k b_{jk} x_j x_k \quad \text{with} \quad b_{jk} = b_{kj}$$

$(j, k = 1, 2, \ldots, n)$. We separate a_j, b_{jk} and $f(x)$ into real and imaginary parts:

$$a_j = \alpha_j + i\beta_j, \qquad b_{jk} = \gamma_{jk} + i\delta_{jk}, \qquad f(x) = P(u, v) + iQ(u, v);$$

then

$$\gamma_{jk} = \gamma_{kj}, \qquad \delta_{jk} = \delta_{kj}.$$

If we substitute $y_1 = P(u, v)$, $y_2 = Q(u, v)$ in $\varphi(x, y) = \varphi(u, v, y_1, y_2)$, we have

$$\Phi(x) = \Phi(u, v) = \varphi(u, v, P(u, v), Q(u, v)).$$

Since the function $\Phi(u, v)$ has partial derivatives up to second order, we can expand it as follows:

$$\begin{aligned}\Phi(u, v) = &\sum_j \left(\frac{\partial \Phi}{\partial u_j} u_j + \frac{\partial \Phi}{\partial v_j} v_j \right) \\ &+ \tfrac{1}{2} \sum_j \sum_k \left[\frac{\partial^2 \Phi}{\partial u_j \partial u_k} u_j u_k + 2 \frac{\partial^2 \Phi}{\partial u_j \partial v_k} u_j v_k + \frac{\partial^2 \Phi}{\partial v_j \partial v_k} v_j v_k \right] + \varepsilon,\end{aligned}$$

$$\varepsilon = \tfrac{1}{2} \sum_j \sum_k [\xi_{jk} u_j u_k + 2\eta_{jk} u_j v_k + \zeta_{jk} v_j v_k] \qquad (j, k = 1, 2, \ldots, n),$$

where the partial derivatives stand for their values at the origin (for the moment; we shall use a similar convention also in what follows). As for ξ_{jk}, η_{jk}, ζ_{jk}, we have, for example,

$$\xi_{jk} = \frac{\partial^2 \Phi(\theta u, \theta v)}{\partial u_j \partial u_k} - \frac{\partial^2 \Phi(0, 0)}{\partial u_j \partial u_k}, \qquad 0 < \theta < 1;$$

since the partial derivatives are continuous, ξ_{jk} tends to 0 when (u, v) tends to $(0, 0)$; the same is true for η_{jk}, ζ_{jk}.

Suppose now that we can make the $(\alpha, \beta, \gamma, \delta)$ of the equation of σ satisfy the conditions

$$\frac{\partial \Phi}{\partial u_j} = \frac{\partial \Phi}{\partial v_j} = 0, \qquad \frac{\partial^2 \Phi}{\partial u_j \partial u_k} = \frac{\partial^2 \Phi}{\partial v_j \partial v_k}, \qquad \frac{\partial^2 \Phi}{\partial u_j \partial v_k} = -\frac{\partial^2 \Phi}{\partial u_k \partial v_j} \quad (j, k = 1, 2, \ldots, n).$$

We then have

$$\begin{aligned}\Phi(u, v) &= \tfrac{1}{4} \sum_j \sum_k \left[\left(\frac{\partial^2 \Phi}{\partial u_j \partial u_k} + \frac{\partial^2 \Phi}{\partial v_j \partial v_k} \right) (u_j u_k + v_j v_k) \right. \\ &\qquad \left. + \left(\frac{\partial^2 \Phi}{\partial u_j \partial v_k} - \frac{\partial^2 \Phi}{\partial u_k \partial v_j} \right) (u_j v_k - u_k v_j) \right] + \varepsilon \\ &= \tfrac{1}{4} W(\Phi; u, v) + \varepsilon.\end{aligned}$$

Now, by hypothesis, we have $W(\varphi; \alpha, \beta) > 0$; hence, by the computation of the preceding section, we easily have $W(\Phi; u, v) > 0$. Because of the form of ε, we can therefore choose a positive number ρ such that we have

$$\Phi(u, v) > 0 \quad \text{for} \quad |x_j| < \rho \qquad (j = 1, 2, \ldots, n),$$

except at the origin. This means that σ remains in the portion $\varphi(x, y) > 0$ of space, except at the origin. It is therefore sufficient to choose a system of $(\alpha, \beta, \gamma, \delta)$ satisfying the conditions indicated.

First, by a simple calculation, the condition $\dfrac{\partial \Phi}{\partial u_j} = \dfrac{\partial \Phi}{\partial v_j} = 0$ becomes

$$\frac{\partial \Phi}{\partial u_j} = \frac{\partial \varphi}{\partial u_j} + \frac{\partial \varphi}{\partial y_1} \alpha_j + \frac{\partial \varphi}{\partial y_2} \beta_j = 0, \qquad \frac{\partial \Phi}{\partial v_j} = \frac{\partial \varphi}{\partial v_j} - \frac{\partial \varphi}{\partial y_1} \beta_j + \frac{\partial \varphi}{\partial y_2} \alpha_j = 0.$$

Since $\left(\dfrac{\partial \varphi}{\partial y_1}\right)^2 + \left(\dfrac{\partial \varphi}{\partial y_2}\right)^2 > 0$, we can solve these linear equations.

As for the second condition, since the value of (α, β) is already determined, it gives a system of simultaneous algebraic equations for (γ, δ) alone, the form of the equations being given by

$$\frac{1}{2}\left(\frac{\partial^2 \Phi}{\partial u_j \partial u_k} - \frac{\partial^2 \Phi}{\partial v_j \partial v_k}\right) = \frac{\partial \varphi}{\partial y_1} \gamma_{jk} + \frac{\partial \varphi}{\partial y_2} \delta_{jk} + c_1 = 0,$$
$$\frac{1}{2}\left(\frac{\partial^2 \Phi}{\partial u_j \partial v_k} + \frac{\partial^2 \Phi}{\partial u_k \partial v_j}\right) = \frac{\partial \varphi}{\partial y_2} \gamma_{jk} - \frac{\partial \varphi}{\partial y_1} \delta_{jk} + c_2 = 0,$$

for all (j, k) with $(j, k = 1, 2, \ldots, n)$, where c_1 and c_2 are independent of (γ, δ). Since $\left(\dfrac{\partial \varphi}{\partial y_1}\right)^2 + \left(\dfrac{\partial \varphi}{\partial y_2}\right)^2 > 0$, we can solve this equation. We have thus found the desired characteristic surface σ. We formulate this result.

Suppose given a real continuous function $\varphi(x_1, x_2, \ldots, x_n)$ in a neighbourhood of a point (x^0) in the space (x) such that, if u_j, v_j are the real and imaginary parts of x_j $(j = 1, 2, \ldots, n)$, the function $\varphi(x) = \varphi(u, v)$ has continuous partial derivatives up to second order, we have $W(\varphi; \alpha, \beta) > 0$ and, further, at least one of the derivatives $\dfrac{\partial \varphi}{\partial u_j}$ $(j = 1, 2, \ldots, n)$ is non-zero at (x^0). We can then find a characteristic surface σ passing through (x^0) and remaining in the part of the space where $\varphi(x) > \varphi(x^0)$ except at (x^0) itself. Moreover, if one of the derivatives $\dfrac{\partial \varphi}{\partial u_n}$, $\dfrac{\partial \varphi}{\partial v_n}$ is not zero at (x^0), we can make σ take the form $x_n = f(x_1, x_2, \ldots, x_{n-1})$, f being a polynomial in $x_1, x_2, \ldots, x_{n-1}$ of degree at most 2.

17. Modification of Pseudoconvex Functions. We are now concerned with modifying pseudoconvex functions to make them satisfy the conditions of the above theorem.

Let us begin with the problem of raising the degree of continuity. This problem has already been solved for subharmonic functions by T. RADÓ[29] and F. RIESZ[30], whose methods can be applied to our case; we now explain this.

1°. Let $\mathfrak{D}$ be a domain over the space $(x_1, x_2, \ldots, x_n)$. Following the *book of Behnke-Thullen*, we consider the *boundary distance* (Randdistanz) as follows. Let P be an arbitrary point of $\mathfrak{D}$ and (x^0) its coordinates. Consider a polycylinder γ with centre (x^0) and radius r' such that there is a domain γ^* satisfying the conditions $\gamma^* \sim \gamma$, $\gamma^* \subseteqq \mathfrak{D}$ and $P \in \gamma^*$; γ^* will be called a *polycylindrical neighbourhood* on $\mathfrak{D}$ with centre P and radius r'. Let r be the upper bound of these r'. We call r the boundary distance of P with respect to $\mathfrak{D}$. (We shall sometimes call it the *polycylindrical boundary distance* to avoid confusion.) We denote by $\mathfrak{D}^{(\rho)}$ the set of points of $\mathfrak{D}$ such that the boundary distance is larger than ρ.

Let now $\varphi(P)$ be a singlevalued real upper semicontinuous *bounded* function on $\mathfrak{D}$. (Just for the present memoir, it is sufficient to start with continuous functions.) Let P be an arbitrary point of $\mathfrak{D}$, let P' be a definite point in $\mathfrak{D}_1 = \mathfrak{D}^{(r)}$ where r is a fixed positive number (such that $\mathfrak{D}^{(r)} \supset 0$). We consider a polycylindrical neighbourhood γ with centre P' and radius r, and the arithmetic mean of $\varphi(P)$ on γ using the *Lebesgue integral*

$$\varphi_1(P') = \frac{1}{V} \int_\gamma \varphi(P)\, dP, \qquad V = (\pi r^2)^n.$$

The function $\varphi_1(P)$ is defined on $\mathfrak{D}_1$ and takes a well-defined finite value at every point of $\mathfrak{D}_1$. It is obviously *continuous*. We shall denote, following F. RIESZ, the operation of constructing $\varphi_1(P)$ from $\varphi(P)$ by

$$\varphi_1(P) = A_r[\varphi(P)].$$

We suppose, in what follows, that $\varphi(P)$ is *pseudoconvex* and bounded on $\mathfrak{D}$. We shall first show that $\varphi_1(P)$ is *also pseudoconvex on* $\mathfrak{D}_1$. Suppose in fact that $\mathfrak{D}$ is a univalent domain in the space (x); there is no loss of generality in this. Let (x^0) be an arbitrary point of $\mathfrak{D}_1$ and L be a characteristic line passing through (x^0) but otherwise arbitrary. We draw a circumference C on L, around (x^0) and sufficiently small so that the circle is completely interior to $\mathfrak{D}_1$, and consider the arithmetic mean of φ_1 on C using the *Riemann integral*

$$m = \frac{1}{2\pi\rho} \int_C \varphi_1(x)\, dx,$$

ρ being the radius of C. It is sufficient to see that $m \geqq \varphi_1(x^0)$.

Let C_0 be the circumference around the origin such that we can carry it onto C by translation, and let (X) be an arbitrary point of C_0. We can then express m in the form

$$m = \frac{1}{2\pi\rho} \int_{C_0} \varphi_1(x^0 + X)\, d(X),$$

29) Cited earlier.

30) 1930, cited earlier.

the integral being taken in the sense of LEBESGUE ((X) stands for an arbitrary point of C_0 and has coordinates (X)). Let γ_0 be the polycylinder in the space (x) with centre the origin and radius r and let (X′) be an arbitrary point of γ_0. Then

$$m=\frac{1}{2\pi\rho V}\int_{C_0} d(x)\int_{\gamma_0}\varphi(x^0+X+X')\,d(X')$$

in the sense of LEBESGUE. Since the function $\varphi(x^0+X+X')$ is upper semicontinuous in the space (X, X'), it is BOREL measurable. Moreover, it is bounded. Consequently we can interchange the order of integration by virtue of LEBESGUE; we therefore have

$$m=\frac{1}{2\pi\rho V}\int_{\gamma_0} d(X')\int_{C_0}\varphi(x^0+X+X')\,d(X)\geqq\frac{1}{V}\int_{\gamma_0}\varphi(x^0+X')\,d(X')=\varphi_1(x^0).$$

Next, if $r_1\geqq r_2\geqq 0$, *we have*

$$A_{r_1}(\varphi)\geqq A_{r_2}(\varphi)\geqq\varphi$$

on $\mathfrak{D}^{(r_1)}$; the proof is similar and immediate.

Let r_p $(p=1,2,\ldots)$ be a decreasing sequence of positive numbers having the limit 0. The sequence of functions $A_{r_p}(\varphi)$ is then also decreasing, and being bounded, it is convergent. Let $\Phi(P)$ be the limit; it exists on $\mathfrak{D}$. Since $A_r(\varphi)\geqq\varphi$, we have $\Phi(P)\geqq\varphi(P)$. On the other hand, since $\varphi(P)$ is upper semicontinuous, we have $\Phi(P)\leqq\varphi(P)$. Hence $\Phi(P)=\varphi(P)$. We have thus seen that the sequence of functions $A_{r_p}(\varphi)$ $(p=1,2,\ldots)$ decreases and converges to φ, each $A_r(\varphi)$ being a pseudoconvex function, defined and continuous on $\mathfrak{D}^{(r_p)}$.

2°. Let us consider the *unbounded* case. Let $\varphi(P)$ be a pseudoconvex function on a domain $\mathfrak{D}$ over the space (x) which is not bounded on regions completely interior to $\mathfrak{D}$. We then construct the sequence of functions

$$\psi_p(P)=\max[\varphi(P),-p]\qquad(p=1,2,\ldots).$$

It decreases and tends to $\varphi(P)$. Since any upper semicontinuous function on $\mathfrak{D}$ is bounded above on any region completely interior to $\mathfrak{D}$, each $\psi_p(P)$ is bounded on any such region. This is just the preceding case.

3°. Suppose again that $\varphi(P)$ is a real *continuous* function on $\mathfrak{D}$. If we denote the real and imaginary parts of x_i $(i=1,2,\ldots,n)$ by u_i, v_i respectively, it is obvious that the function $\varphi_1(P)=A_r(\varphi)$ is defined and continuous on $\mathfrak{D}^{(r)}$, and that it has *continuous partial derivatives of first order with respect to* (u,v). And if $\varphi(P)$ already has these properties on $\mathfrak{D}$, then $\varphi_1(P)$ obviously has *continuous partial derivatives up to second order* with respect to (u,v) on $\mathfrak{D}^{(r)}$. Since $\varphi(P)$ is continuous on $\mathfrak{D}$, $A_r(\varphi)$ converges to φ as $r\to 0$, *uniformly* on any region completely interior to $\mathfrak{D}$.

Iterating this operation, we have

$$\varphi_2(P)=A_r[\varphi_1(P)]=A_r^{(2)}[\varphi(P)];$$

$\varphi_2(P)$ is defined on $\mathfrak{D}_1^{(r)}=\mathfrak{D}^{(2r)}$. Similarly, we consider $\varphi_3(P)=A_r^{(3)}[\varphi(P)]$ on $\mathfrak{D}_3=\mathfrak{D}^{(3r)}$. If $\varphi(x)$ is a pseudoconvex function on $\mathfrak{D}$, $\varphi_1(P)$ is pseudoconvex on $\mathfrak{D}_1$, $\varphi_2(P)$ on $\mathfrak{D}_2$, and consequently, $\varphi_3(P)$ is pseudoconvex on $\mathfrak{D}_3$. We have thus obtained the following result.

Suppose given a pseudoconvex function $\varphi(P)$ in a domain $\mathfrak{D}$ over the space $(x_1, x_2, \ldots, x_n)$ and a sequence of positive numbers r_p $(p=1,2,\ldots)$ decreasing to zero. We can construct a sequence of functions $\varphi_p(P)$ such that each $\varphi_p(P)$ is a continuous pseudoconvex function having continuous partial derivatives up to second order with respect to the real and imaginary parts of x_i $(i=1,2,\ldots,n)$ on $\mathfrak{D}^{(r_p)}$ and such that the sequence decreases to $\varphi(P)$ on $\mathfrak{D}$.

As an immediate consequence of the remark in No. 16, it follows that

Pseudoconvexity of any function is preserved by any biunique pseudoconformal transformation.

18. A Second Modification. Let $\varphi(P)$ be a *pseudoconvex* function on a domain $\mathfrak{D}$ over the space $(x_1, x_2, \ldots, x_n)$. In view of the preceding result, we can suppose that $\varphi(P)$ is continuous and has continuous partial derivatives up to second order with respect to u_j, v_j $(j=1,2,\ldots,n)$, the u_j, v_j being, respectively, the real and imaginary parts of x_j. To make $\varphi(P)$ have the main property (of No. 16), it is sufficient to make it satisfy the following conditions:

$$W(\varphi;\alpha,\beta)>0, \qquad \sum\left[\left(\frac{\partial\varphi}{\partial u_j}\right)^2+\left(\frac{\partial\varphi}{\partial v_j}\right)^2\right]>0, \quad (j=1,2,\ldots,n).$$

Recall that the second condition did not appear in the case of two complex variables. (No. 13, Memoir VI.)

1°. Let us consider the first condition, $W(\varphi;\alpha,\beta)>0$. Let $\psi(P)$ be another function having the same properties as $\varphi(P)$; we have

$$W(\varphi+\psi;\alpha,\beta)=W(\varphi;\alpha,\beta)+W(\psi;\alpha,\beta).$$

Since φ is pseudoconvex, we have $W(\varphi;\alpha,\beta)\geqq 0$. Hence if $W(\psi;\alpha,\beta)>0$, we necessarily have $W(\varphi+\psi;\alpha,\beta)>0$. Now, for

$$\psi(P)=\sum(u_j^2+v_j^2) \qquad (j=1,2,\ldots,n),$$

we have

$$W(\psi;\alpha,\beta)=4\sum(\alpha_j^2+\beta_j^2)>0.$$

Taking, therefore, a sequence of positive numbers ε_p $(p=1,2,\ldots)$ which decreases and tends to zero, we set

$$\varphi_p(P)=\varphi(P)+\varepsilon_p\psi(P);$$

the sequence of functions $\varphi_p(P)$ is then decreasing and tends uniformly to $\varphi(P)$ on any domain completely interior to $\mathfrak{D}$; here each $\varphi_p(P)$ has continuous partial derivatives up to second order and satisfies $W(\varphi_p;\alpha,\beta)>0$.

2°. There only remains the second condition. If we look at this condition, we find that we have to change the problem itself. Let us begin with a classification of pseudoconvex functions.

Let $\varphi(P)$ be a *pseudoconvex* function on a domain $\mathfrak{D}$ over the space $(x_1, x_2, \ldots, x_n)$. We shall say that $\varphi(P)$ has *property* (P_0) on $\mathfrak{D}$ if it is continuous, has continuous partial derivatives up to second order with respect to (u, v) in

$\mathfrak{D}$, and if, furthermore, it satisfies on $\mathfrak{D}$ the conditions

$$W(\varphi;\alpha,\beta)>0, \quad \sum\left[\left(\frac{\partial\varphi}{\partial u_j}\right)^2+\left(\frac{\partial\varphi}{\partial v_j}\right)^2\right]>0, \quad (j=1,2,\ldots,n).$$

We shall say that $\varphi(P)$ has *property* (P_1) on $\mathfrak{D}$ if, for any point P_0 of $\mathfrak{D}$, there is a polycylindrical neighbourhood γ in $\mathfrak{D}$ with centre P_0 such that $\varphi(P)$ is given, in γ, as the supremum of a finite number of pseudoconvex functions having the property (P_0).

Any pseudoconvex function $\varphi(P)$ *having the property* (P_1) *in* $\mathfrak{D}$ *possesses the main property on* $\mathfrak{D}$, more precisely, given a point P_0 of $\mathfrak{D}$ where $\varphi(P_0)=\alpha$, we can always find a characteristic surface on $\mathfrak{D}$ passing through P_0 and remaining in the portion $\varphi(P)>\alpha$ in the neighbourhood of P_0, except at P_0 itself. This is evident.

Problem. Given a continuous pseudoconvex function $\varphi(P)$ on a domain $\mathfrak{D}$ over the space $(x_1,x_2,\ldots,x_n)$, let $\mathfrak{D}_0$ be a domain such that $\mathfrak{D}_0\Subset\mathfrak{D}$ and let ε be a positive number; find a pseudoconvex function $\Phi(P)$ on $\mathfrak{D}_0$, having the property (P_1) and such that $|\varphi(P)-\Phi(P)|<\varepsilon$ on $\mathfrak{D}_0$.

We evidently have the following principle concerning this problem: Let a, b be two real numbers such that $a<b$, let $\mathfrak{D}_1$ be the part of $\mathfrak{D}$ over $u_1<b$ and $\mathfrak{D}_2$ the part of $\mathfrak{D}$ over $u_1>a$. If the problem is solvable for $\mathfrak{D}_1$ and for $\mathfrak{D}_2$, it is solvable for $\mathfrak{D}$.

It is therefore sufficient to solve the problem *locally*.

3°. Let us consider a univalent domain $\mathfrak{R}$ in the space $(x_1,x_2,\ldots,x_n)$ having the form $a_i<u_i<b_i$, $c_i<v_i<d_i$ $(i=1,2,\ldots,n)$ (where a_i, b_i, c_i, d_i are real numbers). We shall call any domain of this form a *rectangular domain.* Let $\varphi(x)$ be a real continuous function on the closed rectangular domain $\bar{\mathfrak{R}}$, let ε be a positive number. Then, thanks to WEIERSTRASS, we know that we can find a polynomial $\Phi(u,v)=\Phi(x)$ of the (u,v), having real coefficients, such that $|\varphi(x)-\Phi(x)|<\varepsilon$ on $\bar{\mathfrak{R}}$. We shall make some remarks in order to be able to apply this theorem.

Let $\varphi(x)$ be a real continuous function on a univalent domain $\mathfrak{D}$ in the space (x). We consider $\mathfrak{D}_1=\mathfrak{D}^{(r)}$, and

$$\varphi_1(x)=A_r[\varphi(x)]=\frac{1}{V}\int_\gamma\varphi(x')\,d\sigma \qquad (V=(\pi r^2)^n),$$

(where γ is the polycylinder with centre (x) and radius r, (x') represents an arbitrary point of γ, and $d\sigma$ is the volume element on γ). Suppose that $|\varphi(x)|\leqq M$ on $\mathfrak{D}$ (M being a positive number). Then, we have

$$|\varphi_1(x)|\leqq M$$

on $\mathfrak{D}_1$. To estimate $\dfrac{\partial\varphi_1}{\partial u_1}$, we consider two circles of radius r in the x_1-plane so that their centres are situated on a horizontal line and are at distance h, which is sufficiently small. We then have two narrow crescents, the sum of whose areas is smaller than $4rh$. We therefore have

$$\left|\frac{\partial \varphi_1}{\partial u_1}\right| \leqq \frac{4}{\pi r} M$$

on $\mathfrak{D}_1$. The same is true for the u_j, v_i ($j=2,3,\ldots,n$; $i=1,2,\ldots,n$). Let us estimate $\varphi_2 = A_r(\varphi_1)$ on $\mathfrak{D}_2 = \mathfrak{D}^{(2r)}$. Let ξ, η be any two of the u_i, v_i ($i=1,2,\ldots,n$); since $\dfrac{\partial^2 \varphi_2}{\partial \xi\, \partial \eta} = \dfrac{\partial}{\partial \eta} A_r\left(\dfrac{\partial \varphi_1}{\partial \xi}\right)$, we have, by the above inequality,

$$\left|\frac{\partial^2 \varphi_2}{\partial \xi\, \partial \eta}\right| \leqq \left(\frac{4}{\pi r}\right)^2 M$$

on $\mathfrak{D}_2$.

Let $\Phi(u,v) = \Phi(x)$ be a polynomial in (u,v) of degree at most ν and having real coefficients. Let γ_0 be a polycylinder with centre at the origin and radius r, let (x') be an arbitrary point of γ_0. $\varphi_1(x)$ can then be expressed in the form

$$\varphi_1(x) = \frac{1}{V} \int_{\gamma_0} \varphi(x+x')\, d\sigma.$$

As a special case, $\Phi_1(x) (= A_r[\Phi(x)])$ has the same properties as $\Phi(x)$, and consequently, the same is true of $\Phi_2(x)$.

Let $\varphi(x)$ be a real continuous function having continuous partial derivatives up to second order with respect to (u,v). We consider

$$W(\varphi;\alpha,\beta) = \sum_i \sum_j \left[\left(\frac{\partial^2 \varphi}{\partial u_i\, \partial u_j} + \frac{\partial^2 \varphi}{\partial v_i\, \partial v_j}\right)(\alpha_i \alpha_j + \beta_i \beta_j) + \left(\frac{\partial^2 \varphi}{\partial u_i\, \partial v_j} - \frac{\partial^2 \varphi}{\partial u_j\, \partial v_i}\right)(\alpha_i \beta_j - \alpha_j \beta_i)\right]$$

$(i,j=1,2,\ldots,n)$. Let us consider in the space (α,β) a hypersphere centred at the origin, of radius 1; let (α',β') be an arbitrary point on the boundary, and consider

$$w(\varphi(x)) = \min W(\varphi(x);\alpha',\beta').$$

In order that $W(\varphi(x);\alpha,\beta) > 0$, it is necessary and sufficient that $w(\varphi(x)) > 0$. Let $\varphi(x)$ be a given function having the property indicated on a univalent domain $\mathfrak{D}$; in order that $W(\varphi;\alpha,\beta) > 0$ on $\mathfrak{D}$, it is necessary and sufficient that the lower bound of $w(\varphi)$ be positive on any domain $\mathfrak{D}_0$ such that $\mathfrak{D}_0 \Subset \mathfrak{D}$. If $w(\varphi)$ has the lower bound w_0 on $\mathfrak{D}$, we obviously have

$$w(\varphi_1) \geqq w_0$$

on $\mathfrak{D}_1$.

Let now $\mathfrak{R}$ be a rectangular domain in the space (x) and let $\varphi(x)$ be a continuous pseudoconvex function having continuous partial derivatives with respect to (u,v) up to second order, and such that $W(\varphi;\alpha,\beta) > 0$ in a neighbourhood of the closed domain $\bar{\mathfrak{R}}$. For any positive number ε, we can find a polynomial $\Phi(x) = \Phi(u,v)$ in (u,v) with real coefficients in such a way that

$$|\Phi(x) - \varphi(x)| < \varepsilon$$

in a neighbourhood of $\bar{\mathfrak{R}}$. Consider $\Phi_2(x)=A_r^{(2)}[\Phi(x)]$ and $\varphi_2(x)=A_r^{(2)}[\varphi(x)]$, and let us choose ε sufficiently small, and a suitable r. In view of what we have seen, by comparing Φ_2 and φ with φ_2, we easily find that the polynomial $\Phi_2(x)$ has the same properties as $\varphi(x)$.

We suppose therefore that the function $\varphi(x)$ given above is a polynomial in (u,v) with real coefficients. There is just one condition to look at, viz

$$\sum\left[\left(\frac{\partial\varphi}{\partial u_i}\right)^2+\left(\frac{\partial\varphi}{\partial v_i}\right)^2\right]>0 \qquad (i=1,2,\dots,n).$$

Now, under these circumstances, we easily find, given a positive number ε, that we can find a polynomial $\psi(x)$ in (u,v) having the same properties as $\varphi(x)$, such that $|\varphi(x)-\psi(x)|<\varepsilon$ in a neighbourhood of $\bar{\mathfrak{R}}$, and moreover, concerning the last condition indicated above, the set of points where all derivatives of first order of $\psi(x)$ vanish simultaneously, are *isolated*[31].

4°. As we remarked above, we saw in Memoir VI that this circumstance creates no obstacle in the case $n=2$. It is this fact that we shall utilize. Let us recall the precise form of the theorem.

Let $\varphi(x_1,x_2)$ be a real continuous function having continuous partial derivatives up to second order with respect to u_i, v_i $(i=1,2)$ such that $W(\varphi;\alpha,\beta)>0$ in a neighbourhood of a point (x_1^0,x_2^0) in the space (x_1,x_2). If we have $\dfrac{\partial\varphi}{\partial u_1}=\dfrac{\partial\varphi}{\partial v_1}=\dfrac{\partial\varphi}{\partial u_2}=\dfrac{\partial\varphi}{\partial v_2}=0$ at (x^0), we can find a characteristic plane passing through (x^0) and remaining in the portion $\varphi(x)>\varphi(x^0)$ except at (x^0) itself in the neighbourhood of this point.

Let (x^0) be a fixed point in the space $(x_1,x_2,\dots,x_n)$, let $\varphi(x)$ be a polynomial in (u,v) with real coefficients such that $W(\varphi;\alpha,\beta)>0$ in a neighbourhood of (x^0), and such that at least one of the partial derivatives of first order is non-zero at every point of a neighbourhood of (x^0) except at this point itself, where they are all zero simultaneously. Let us suppose that (x^0) is the origin to simplify the notation. We set

$$\Phi(x_1,x_2)=\varphi(x_1,x_2,0,\dots,0).$$

We have $W(\Phi;\alpha,\beta)>0$ because $w(\Phi)\geqq w(\varphi)>0$ in a neighbourhood of the origin. We can therefore find a characteristic plane of the form $ax_1=x_2$ in the space (x_1,x_2) remaining in $\Phi(x_1,x_2)>\Phi(0,0)$ in a neighbourhood of the origin, except at the origin. We may suppose, without loss of generality, that $a=0$.

In the space $(x_1,x_2,\dots,x_n)$, we consider a polycylinder γ centred at the origin and radius r so small that all the above circumstances are valid in the polycylinder with centre at the origin and radius $2r$. Taking a positive number ρ smaller than r, let us set

$$\varphi_1=\varphi(x_1+\rho,x_2,\dots,x_n), \qquad \psi=\max(\varphi,\varphi_1).$$

Since, for $\varphi(x_1,0,\dots,0)=\chi(x_1)$, we have $\chi(x_1)>\chi(0)$ for $0<|x_1|<2r$, it is clear that ψ is a pseudoconvex function having property (P_1) in the polycylinder γ. Moreover, $|\psi-\varphi|$ tends to zero with ρ. *The problem posed in* (2°) *is thus solved.*

[31]) It is sufficient to consider a modification of the form $\psi=\varphi+L$, L being a suitable linear homogeneous polynomial in (u,v).

C. A Boundary Problem

19. A Boundary Problem. The aim of the present chapter is to solve the following problem.

Boundary Problem. *Given a pseudoconvex domain $\mathfrak{D}$ over the space $(x_1, x_2, \ldots, x_n)$, find a pseudoconvex function $\Phi(P)$ on $\mathfrak{D}$ having the following properties:* 1°. *$\Phi(P)$ possesses property* (P_1). 2°. *For any real number α, the set $\mathfrak{D}_\alpha$ of points of $\mathfrak{D}$ such that $\Phi(P) < \alpha$ satisfies the condition $\mathfrak{D}_\alpha \Subset \mathfrak{D}$.*

The significance of this problem will be understood if one recalls the results expounded in "III. A Complementary Problem" in Memoir VI.

Now, in the present case, since $\mathfrak{D}$ is not univalent, *we have to search for a new kind of pseudoconvex function on $\mathfrak{D}$*, apart from the HARTOGS radii, *in order to make $\mathfrak{D}_\alpha$ always compact on $\mathfrak{D}$*. (For the same reason, one sees that the *theorem of Cartan-Thullen* is not sufficient to treat the given problems on multivalent domains of holomorphy.) More precisely, the following problem appears:

Auxiliary Problem. Let $\mathfrak{D}$ be a domain over the space (x) and $\mathfrak{D}_0$ a subdomain whose polycylindrical boundary distance with respect to $\mathfrak{D}$ is not zero. Find a pseudoconvex function $\lambda_0(P)$ on $\mathfrak{D}_0$ such that for any real number α, the set Δ_α of points of $\mathfrak{D}_0$ such that $\lambda_0(P) < \alpha$ satisfies the condition $\Delta_\alpha \Subset \mathfrak{D}$.

20. Normal Families of Continuous Curves. We shall apply, for this purpose, Montel's concept[32)] to continuous curves[33)]. This is not indispensable, but is useful in handling the auxiliary problem.

Given a family $(\mathfrak{F})$ of *continuous curves* in the space $(x_1, x_2, \ldots, x_n)$, we shall say that $(\mathfrak{F})$ is a *normal family* if, from any infinite sequence C_j $(j = 1, 2, \ldots)$ in $(\mathfrak{F})$, we can extract an infinite subsequence C_{p_j} $(j = 1, 2, \ldots)$ in such a way that for each curve C_{p_j}, we can choose an equation of the form

$$x_i = \varphi_i^{(j)}(t), \qquad 0 \leqq t \leqq 1, \qquad (i = 1, 2, \ldots, n)$$

representing C_{p_j} (the $\varphi_i^{(j)}(t)$ being continuous functions) such that the sequences $\varphi_i^{(j)}(t)$ $(j = 1, 2, \ldots)$ are uniformly convergent on $0 \leqq t \leqq 1$.

For now, we content ourselves with remarking that one has the following *criterion:*

Given a family $(\mathfrak{F})$ of rectifiable curves in the space (x), if the total lengths of the curves in $(\mathfrak{F})$ form a bounded set, then $(\mathfrak{F})$ is a normal family.

Let, in fact, C be an arbitrary curve in $(\mathfrak{F})$ and let l be its total length. Let s be the length of the arc of C from the initial point to a variable point (x) on C. We take $t = s/l$ and set

$$x_i = \varphi_i(t), \qquad 0 \leqq t \leqq 1 \qquad (i = 1, 2, \ldots, n).$$

---

32) See: P. MONTEL, Lecons sur les familles normales.

33) We have already applied this concept to characteristic surfaces. See No. 7.

For any two points t, t' of the closed interval $[0, 1]$, we have

$$|\varphi_i(t)-\varphi_i(t')| \leqq l\,|t-t'|.$$

Since the set $\{l\}$ is bounded, for each i in $1, 2, \ldots, n$, the family $\{\varphi_i(t)\}$ is equicontinuous on $[0, 1]$. Hence, as is well known, this family is normal on $[0, 1]$, and $(\mathfrak{F})$ therefore forms a normal family because of the definition.

21. Solution of the Auxiliary Problem. In the auxiliary problem, let us for a moment ignore the condition of being pseudoconvex. We then find diverse functions of the kind wanted right at hand; of these we shall choose the following:

Let 0 be a fixed point of $\mathfrak{D}_0$ and P an arbitrary point of $\mathfrak{D}_0$. Let $d(P_1, P_2)$ be the distance between P_1 and P_2 on the domain $\mathfrak{D}_0$ (more precisely, the lower bound of the lengths of curves on $\mathfrak{D}_0$ joining P_1 and P_2) and let us set

$$\lambda(P)=d(0, P).$$

$\lambda(P)$ is a real continuous function on $\mathfrak{D}_0$. Let α be an arbitrary real number, and Δ_α the set of points of $\mathfrak{D}_0$ satisfying $\lambda(P)<\alpha$. Since the polycylindrical boundary distance of $\mathfrak{D}_0$ with respect to $\mathfrak{D}$ is not zero, it is clear that $\Delta_\alpha \Subset \mathfrak{D}$ because of the *criterion* we just saw. $\lambda(P)$ is thus one of the functions with the required property.

How can we make of $\lambda(P)$ a pseudoconvex function? Let us first remark that $\lambda(P)$ has the following property:

$$(1) \qquad |\lambda(P_1)-\lambda(P_2)| \leqq d(P_1, P_2).$$

We shall associate to this the method of taking arithmetic means (explained in No. 17).

Consider $\lambda_1(P)=A_r[\lambda(P)]$ on $\mathfrak{D}_1=\mathfrak{D}_0^{(r)}$. We denote any one of the u_i, v_i $(i=1, 2, \ldots, n)$ by ξ. By (1), we have

$$\left|\frac{\partial \lambda_1}{\partial \xi}\right| \leqq \frac{4}{\pi}$$

(see 3°, No. 18). We next consider $\lambda_2(P)=A_r[\lambda_1(P)]$ on $\mathfrak{D}_2=\mathfrak{D}_0^{(2r)}$. Let η be any one of the u_i, v_i. Since $\left|\frac{\partial \varphi_1}{\partial \xi}\right| \leqq \frac{4}{\pi r} \cdot M$ (where M is the upper bound of φ on $\mathfrak{D}$; 3°, No. 18), we have, by the above inequality,

$$\left|\frac{\partial^2 \lambda_2}{\partial \xi\, \partial \eta}\right| \leqq \left[\left(\frac{4}{\pi}\right)^2 \cdot \frac{1}{r} =\right] K.$$

Because of (1), we also have $|\lambda_2(P)-\lambda(P)| \leqq 2\sqrt{n}\, r$. Let (α', β') be an arbitrary point on the boundary of the hypersphere of centre the origin and radius 1 in the space (α, β). We then have

$$w(\lambda_2)=\min W(\lambda_2; \alpha', \beta') > \left[-8n^2 K = -\left(\frac{8n}{\pi}\right)^2 \cdot \frac{2}{r} =\right] -K_1.$$

Since for $\mu(x)=\sum(u_i^2+v_i^2)$ $(i=1, 2, \ldots, n)$ we have

$$W(\mu; \alpha', \beta')=4,$$

we take a positive number M such that $4M>K_1$ and construct

$$\lambda_0(P)=\lambda_2(P)+M\,\mu(x).$$

$\lambda_0(P)$ is then defined on $\mathfrak{D}_2$ and satisfies $W(\lambda_0; \alpha, \beta)>0$. It is obvious that $\lambda_0(P)$ preserves the (compactness) property in question.

The function $\lambda_0(P)$ so obtained is only defined on $\mathfrak{D}_2$. To obtain a function of this nature for $\mathfrak{D}_0$, we have only to start with a domain bigger than $\mathfrak{D}_0$. *The auxiliary problem is thus solved.*

22. Solution of the Boundary Problem; a Second Lemma. We consider the following problem to complete *the problem of No. 18.*

Problem. Let $\mathfrak{D}$ be a pseudoconvex[34)] domain over the space (x), let $\varphi(P)$ be a continuous pseudoconvex function on $\mathfrak{D}$ having the following *property* (α) (on $\mathfrak{D}$); for any real number α, the set of points of $\mathfrak{D}$ given by $\varphi(P)<\alpha$ is completely interior to $\mathfrak{D}$. Modify the function $\varphi(P)$ in such a way that it has property (P_1), preserving the other properties on $\mathfrak{D}$.

To solve this problem, let a_p $(p=1, 2, \ldots)$ be a sequence of positive numbers such that $a_p<a_{p+1}$, and having the limit $+\infty$. Let us denote the set of points of $\mathfrak{D}$ defined by $\varphi(P)<a_p$ by $\mathfrak{D}_p$; we then have $\mathfrak{D}_p \Subset \mathfrak{D}$. Since the problem of No. 18 is solvable, given a positive number ε, we can find a pseudoconvex function $\Phi_p(P)$ on $\mathfrak{D}_p$, having property (P_1) and such that

$$|\varphi(P)-\Phi_p(P)|<\varepsilon.$$

We choose ε so small that, if $\mathfrak{D}'_{p,q}$ is the set of points of $\mathfrak{D}_p$ defined by $\Phi_p(P)<a_q$, we have, for $q=2, 3, \ldots, p-1$,

$$\mathfrak{D}_{q-1} \Subset \mathfrak{D}'_{p,q} \Subset \mathfrak{D}_{q+1}.$$

We construct a sequence $\Phi'_p(P)$ successively from the sequence of functions $\Phi_p(P)$ $(p=1, 2, \ldots)$ as follows:

$$\Phi'_1=\Phi_6.$$

The function Φ'_1 is defined on $\mathfrak{D}_6$, is pseudoconvex, possesses the property (P_1), and

$$\Phi'_1>a_4 \quad \text{on} \quad (\mathfrak{D}_6-\mathfrak{D}_5).$$

To construct Φ'_2, we construct first Ψ_2 as follows:

$$\Psi_2=c_2(\Phi_7-a_3),$$

c_2 being a positive constant. Since $\mathfrak{D}_2 \Subset \mathfrak{D}'_{7,3} \Subset \mathfrak{D}_4$, we can take c_2 so large that

$$\Psi_2<\Phi'_1 \quad \text{on} \quad \mathfrak{D}_1, \qquad \Psi_2>\Phi'_1 \quad \text{on} \quad (\mathfrak{D}_6-\mathfrak{D}_5), \qquad \Psi_2>a_5 \quad \text{on} \quad (\mathfrak{D}_7-\mathfrak{D}_6).$$

---

34) Any domain possessing such a function $\varphi(P)$ is necessarily pseudoconvex.

Let us set

$$\Phi'_2 = \max(\Phi'_1, \Psi_2) \quad \text{on} \quad \mathfrak{D}_6,$$
$$= \Psi_2 \quad \text{on} \quad (\mathfrak{D}_7 - \mathfrak{D}_6).$$

Proceeding thus, we obtain a sequence of functions Φ'_p ($p=1, 2, \ldots$) such that each function Φ'_p has the following property: Φ'_p is a pseudoconvex function having property (P_1) on $\mathfrak{D}_{p+5}$ and satisfying the conditions $\Phi'_p = \Phi'_{p-1}$ on $\mathfrak{D}_{p-1}$ (if $p>1$), $\Phi'_p > a_{p+3}$ on $(\mathfrak{D}_{p+5} - \mathfrak{D}_{p+4})$. This sequence converges on $\mathfrak{D}$. Let $\Phi(P)$ be the limit. Φ is a pseudoconvex function having property (P_1) on $\mathfrak{D}$. It also has property (α) on $\mathfrak{D}$ because

$$\Phi > a_{p+3} \quad \text{on} \quad (\mathfrak{D} - \mathfrak{D}_{p+4}) \quad (p=1, 2, \ldots).$$

This problem having thus been solved, we have only to solve the following problem to solve the boundary problem.

Problem. Given a pseudoconvex domain $\mathfrak{D}$ over the space (x), find a continuous pseudoconvex function on $\mathfrak{D}$ having property (α).

This is the complete version of the auxiliary problem. We shall solve it using that problem and the function $d(P)$ formed with the help of HARTOGS radii.

We shall leave out the case when $\mathfrak{D}$ coincides with the space (x) because then $\mu(x) = \sum |x_i|^2$ ($i=1, 2, \ldots, n$) gives a solution. Let us consider the euclidean boundary distance $d(P)$ on $\mathfrak{D}$; $-\log d(P)$ is a continuous pseudoconvex function; we shall denote this function by $\delta(P)$ to simplify the notation. Let a_p ($p=1, 2, \ldots$) be a sequence of positive numbers such that $a_p < a_{p+1}$ which tends to $+\infty$. Consider the set $\mathfrak{D}_p$ of points of $\mathfrak{D}$ given by $\delta(P) < a_p$. The auxiliary problem being solvable, we can construct on $\mathfrak{D}_p$ a continuous pseudoconvex function $\varphi_p(P)$ such that the set of points of $\mathfrak{D}_p$ given by $\varphi_p(P) < \alpha$ is completely interior to $\mathfrak{D}$, α being an arbitrary real number. We choose a_1 sufficiently large so that $\mathfrak{D}_1 \ni 0$.

Let 0 be a fixed point of $\mathfrak{D}_1$ and let $d(P_1, P_2)$ be the distance between the points P_1, P_2 on $\mathfrak{D}$. We set $\lambda(P) = d(0, P)$. Let l_p ($p=1, 2, \ldots$) be a sequence of positive numbers, strictly increasing and tending to $+\infty$; we shall determine them recursively in what follows. Denote by Δ_p the set of points of $\mathfrak{D}_p$ defined by $\lambda(P) < l_p$; then $\Delta_p \Subset \Delta_{p+1}$ and the limit of Δ_p ($p=1, 2, \ldots$) is $\mathfrak{D}$.

From the sequence φ_p ($p=1, 2, \ldots$) we form, recursively, a sequence of functions $\Phi_p(P)$ in the following way. First, $\Phi_1 = \varphi_2$; Φ_1 is a continuous pseudoconvex function which is defined on $\mathfrak{D}_2$.

To construct Φ_2, we consider the set Δ'_1 of points of $\mathfrak{D}$ defined by

$$\delta < \frac{a_1 + a_2}{2}, \qquad \varphi_3 < \alpha_1,$$

where α_1 is a positive number, which we shall determine. Having chosen l_1 at random, let us choose α_1 so large that $\Delta'_1 \Supset \Delta_1$. *Since* $\Delta'_1 \Subset \mathfrak{D}_2$, we next choose l_2 so large that $\Delta_2 \Supset \Delta'_1$. With this configuration, we consider a function of the form

$$\Psi_2(P) = c_2 \max\left[\delta(P) - \frac{a_1 + a_2}{2}, \varphi_3(P) - \alpha_1\right],$$

where we choose the positive number c_2 sufficiently large. $\Psi_2(P)$ then has the following properties:

$$\Psi_2<\Phi_1 \quad \text{on} \quad \Delta_1, \qquad \Psi_2>a_2 \quad \text{on} \quad (\mathfrak{D}_3-\Delta_2), \qquad \Psi_2>\Phi_1$$

in a neighbourhood of the boundary of Δ_2, *in* Δ_2. We set

$$\begin{aligned}\Phi_2(P)&=\max(\Phi_1, \Psi_2) \quad \text{on} \quad \Delta_2,\\ &=\Psi_2 \qquad\qquad\quad \text{on} \quad (\mathfrak{D}_3-\Delta_2).\end{aligned}$$

Then $\Phi_2(P)$ is a continuous pseudoconvex function defined on $\mathfrak{D}_3$ such that $\Phi_2=\Phi_1$ on Δ_1 and such that $\Phi_2>a_2$ on $(\mathfrak{D}_3-\Delta_2)$. And so on.

We then have a sequence of continuous pseudoconvex functions $\Phi_p(P)$ ($p=1, 2, \ldots$) such that Φ_p is defined in $\mathfrak{D}_{p+1}$ and satisfies the conditions $\Phi_p=\Phi_{p-1}$ on Δ_{p-1}, $\Phi_p>a_p$ on $(\mathfrak{D}_{p+1}-\Delta_q)$ where $p\geqq q>1$, so long as $p>1$. This sequence converges on $\mathfrak{D}$; let $\Phi(P)$ be the limit. $\Phi(P)$ is pseudoconvex and continuous on $\mathfrak{D}$ and has property (α) on $\mathfrak{D}$ because $\Phi(P)>a_p$ on $\mathfrak{D}-\Delta_p$ ($p=2, 3, \ldots$).

We thus have the desired result:

Lemma II. *Given a pseudoconvex domain over the space* (x) *without points at infinity or interior points of ramification, we can always find a pseudoconvex function* $\Phi(P)$ *on* $\mathfrak{D}$ *having the following properties:* 1°. *$\Phi(P)$ possesses property* (P_1)[35]. 2°. *For any real number* α, *the set of points of* $\mathfrak{D}$ *given by* $\Phi(P)<\alpha$ *remains completely interior to* $\mathfrak{D}$.

23. Subsequent Problems. We shall inspect rapidly the nature of the boundary problem when one allows either points at infinity (in a suitable sense) or non-transcendental points of ramification as interior points.

In the first case, we cannot use the auxiliary function $\mu(x)=\sum|x_i|^2$ ($i=1, 2, \ldots, n$); this gives rise to a strange problem.

In the second case, the HARTOGS radii no longer play a role; this presents a difficulty which seems to me to be truly great.

As for details concerning the form of these difficulties, we shall content ourselves with presenting them in later memoirs.

Chapter III. The Main Problems

24. General Notions. Let us consider a domain $\mathfrak{D}$ over the space $(x_1, x_2, \ldots, x_n)$ (without points at infinity or interior points of ramification); let P be an arbitrary point of $\mathfrak{D}$ and (x) its coordinates. A function $f(P)$ of P is called *holomorphic* on $\mathfrak{D}$ if it is single valued on $\mathfrak{D}$ and, further, is a holomorphic function of (x) at every point of $\mathfrak{D}$. Let $f(P)$ be a holomorphic function (of P) on $\mathfrak{D}$. We say that $\mathfrak{D}$ is the domain of existence (or domain of holomorphy) of $f(P)$ if this function always has different TAYLOR expansions at any two different points of $\mathfrak{D}$ and in addition, if this function cannot be

[35]) For the property (P_1), see No. 18, also No. 16.

continued analytically across any boundary point of $\mathfrak{D}$. $\mathfrak{D}$ is called a *domain of holomorphy* if there exists at least one function for which $\mathfrak{D}$ is the domain of existence.

A domain of holomorphy is pseudoconvex by HARTOGS' theorem (1906). Here the notion of being pseudoconvex is a kind of *local convexity* (one can define it locally). Now, in 1926, JULIA proposed the following global problem: Is every domain of normality a domain of holomorphy? And in 1931, in order to handle JULIA's problem [36], H. CARTAN introduced, for the first time, a kind of *global convexity* (which cannot, à priori, be expressed locally) [37]; his idea is essentially as follows:

Let $\mathfrak{D}$ be a domain over the space (x) (not necessarily connected), and let $(\mathfrak{F})$ be a family of *holomorphic* functions on $\mathfrak{D}$. We say that $\mathfrak{D}$ is *convex with respect to* $(\mathfrak{F})$, or simply $(\mathfrak{F})$*-convex*, if it satisfies the following conditions:

1°. The family $(\mathfrak{F})$ contains at least one function $f_0(P)$ having different TAYLOR expansions at any pair of distinct points on $\mathfrak{D}$.

2°. For any domain Δ (connected or not) such that $\Delta \Subset \mathfrak{D}$, there exist two domains Δ', Δ'' (connected or not) such that $\Delta \subsetneqq \Delta' \Subset \Delta'' \subsetneqq \mathfrak{D}$ and such that to any point P_0 of $\Delta'' - \Delta'$ corresponds a function $f(P)$ of $(\mathfrak{F})$ for which $|f(P_0)| > \max |f(\Delta)|$ (where the term on the right denotes the upper bound of $|f(P)|$ on Δ).

As a special case, if $\mathfrak{D}$ is convex with respect to the family of holomorphic functions on $\mathfrak{D}$, we say, with BEHNKE, that $\mathfrak{D}$ is *holomorph-convex* (regulär-konvex). Concerning this notion, we shall see later that the first condition is necessarily a consequence of the second.

More generally, if $\mathfrak{D}$ is convex with respect to the family consisting of functions holomorphic on a domain $\mathfrak{D}'$ such that $\mathfrak{D}' \supsetneqq \mathfrak{D}$, we shall say that $\mathfrak{D}$ is *holomorph-convex with respect to* $\mathfrak{D}'$.

In 1932, H. CARTAN and THULLEN found that any univalent domain of holomorphy without points at infinity is holomorph-convex [38]. (This remains true for the case of a bounded number of sheets.) This is a truly fundamental fact for univalent domains, and gives the reason to distinguish domains of holomorphy from other kinds of pseudoconvex domains, domains of meromorphy, domains of normality, … . We have, in fact, solved the COUSIN problems and the problem of expansions for univalent (finite) domains of holomorphy with the help of this theorem.

However, in the present case, the (really simple) argument of CARTAN-THULLEN is no longer valid as we have already remarked. Consequently, one has to take the notion of holomorph-convex domains as the basic notion. This leads, first, to the following problem:

Is every pseudoconvex domain over the space (x) (without points at infinity or interior points of ramification) holomorph-convex?

[36]) As we said in the introduction to Memoir VI, JULIA's problem has been solved for univalent domains without points at infinity; the last step in this solution is due to BEHNKE and to STEIN. See: H. BEHNKE-K. STEIN, Konvergente Folgen von Regularitätsbereichen und die Meromorphiekonvexität, 1938 (Math. Annalen).

[37]) H. CARTAN, Sur les domaines d'existence des fonctions de plusieurs variables complexes, 1931 (Bull. de la Soc. math. de France).

[38]) H. CARTAN-P. THULLEN, Regularitäts- und Konvergenzbereiche (Math. Annalen).

We shall, from now on, call this the *inverse Hartogs problem.*

What have been (for us) the other main problems (since Memoir I), *the Cousin problems* and the *problem of expansions* remained open for a long time except on univalent cylindrical domains [39] as we said above. As for the *Cauchy theorems*, H. POINCARÉ generalized the first, but the second remained without progress.

Then, in 1935, WEIL generalized the second CAUCHY theorem [40] following the result of POINCARÉ, which, as an immediate consequence, makes the main problems solvable on rationally convex domains as we have said above (with references). And we also applied the WEIL integral to the inverse HARTOGS problem in Memoir VI.

In the work of WEIL, and already in the notion of $(\mathfrak{F})$-convexity one meets closed sets of the following form:

Let $\mathfrak{D}$ be a domain (not necessarily connected) over the space (x), without points at infinity and without transcendental ramification points in the interior. Let P be an arbitrary point of $\mathfrak{D}$ and let Δ be a closed set of points of $\mathfrak{D}$ such that $\mathfrak{D} \Supset \Delta$. We call Δ an *analytic polyhedron* if there exists a function $f_0(P)$ holomorphic on Δ and having different TAYLOR expansions at any distinct pair of regular points of Δ and if, furthermore, Δ can be expressed in the form

$$f_j(P) \in A_j \qquad (j=1, 2, \ldots, \nu),$$

where the $f_j(P)$ are holomorphic functions on $\mathfrak{D}$, and the A_j are (connected) closed domains, univalent and bounded in the plane (or, sometimes, more generally, simply bounded closed sets) [41].

It is for these analytic polyhedra that we established Theorem II in Memoir I, Theorem I in Memoir II and the fundamental lemma im Memoir VIII. Let us look at the form that these principles take in the present case.

Suppose that, in the above definition, $\mathfrak{D}$ does not contain any interior point of ramification. Since $f_0(P)$ always has different TAYLOR expansions at different points of $\mathfrak{D}$, we can find a finite system of functions $\varphi_k(P)$ $(k=1, 2, \ldots, p)$ $(p \leqq n+1$ if we want) such that the ordered systems of values of $(\varphi_1, \varphi_2, \ldots, \varphi_p)$ are different at any pair of distinct points of $\mathfrak{D}$ having the same coordinates. We can find a new representation of the polyhedron Δ in the following form:

$$\text{(1)} \qquad x_i \in A_i, \quad f_j(P) \in B_j \quad (i=1, 2, \ldots, n; j=1, 2, \ldots, \nu),$$

where the A_i, B_j are (connected) closed univalent bounded domains in the plane and the $f_j(P)$ are holomorphic functions on $\mathfrak{D}$ whose totality contains the functions $\varphi_k(P)$ $(k=1, 2, \ldots, p)$.

Let (x) be the coordinates of P. For every point P_0 of $\mathfrak{D}$, consider the element of the characteristic variety

$$y_j = f_j(x) \qquad (j=1, 2, \ldots, \nu)$$

---

39) P. COUSIN, 1985 (Acta Mathematica).

40) On this subject we also consulted the work of St. BERGMANN, and we shall probably have occasion to speak of them later.

41) In Memoir VIII, we followed H. CARTAN in imposing on A_j the condition of being simply connected, but we shall omit this from now on since it does not seem to me to be natural.

in the space $(x_1, \ldots, x_n, y_1, \ldots, y_\nu)$, where the $f_j(x)$ stand for the determinations in the neighbourhood of P_0. When P_0 describes $\mathfrak{D}$, the above element of variety traces a characteristic variety which we denote by Σ' and represent by the equations $y_j = f_j(P)$ $(P \in \mathfrak{D},\ j = 1, 2, \ldots, \nu)$. Because of the properties of the system (f), there is a *one-to-one correspondence* between the points P of $\mathfrak{D}$ and the points M of Σ' having coordinates $[x, f(P)]$.

Let us consider the closed cylindrical domain (A, B) in the space (x, y), and let us denote the part of Σ' in (A, B) by Σ. The point P of Δ and the point M: $[x, f(P)]$ of Σ correspond to each other biuniquely; moreover, *boundary points correspond to boundary points.* Let P be any point of Δ having the coordinates (x), and Q a point of Δ, different from P and having the same coordinates (x), but otherwise arbitrary. Let

$$\alpha(P) = \max |f_j(P) - f_j(Q)| \qquad (j = 1, 2, \ldots, \nu),$$

and let α_0 be the lower bound of $\alpha(P)$ (when P runs over Δ); we have $\alpha_0 > 0$. We take a positive number ρ such that $3\rho < \alpha_0$ and consider the set of points V in the space (x, y) such that to any point (x', y') of V, there corresponds at least one point P' of Δ having coordinates (x') and satisfying $|f_j(P') - y'_j| < \rho$ $(j = 1, 2, \ldots, \nu)$. To the point (x', y') of V, we make correspond the value $\varphi(P')$; the function thus defined is single valued since to (x') corresponds only one P'; it is holomorphic. We denote it, using the same letter φ, by $\varphi(x, y)$.

In the present case, the fundamental lemma reduces, therefore, to the following:

Lemma I. *With the above geometric configuration, let A_i^0, B_j^0 $(i = 1, 2, \ldots, n;\ j = 1, 2, \ldots, \nu)$ be closed sets such that $A_i^0 \Subset$ interior of A_i, $B_j^0 \Subset$ interior of B_j. There then exists a positive constant K having the following property: Given a holomorphic function $\varphi(P)$ in a neighbourhood of the analytic polyhedron Δ on $\mathfrak{D}$, having N as an upper bound for $|\varphi|$ on Δ, we can always find a holomorphic function $\Phi(x, y)$ in a neighbourhood of (A^0, B^0) such that $\Phi = \varphi$ on Σ and $|\Phi| < KN$ on (A^0, B^0).*

The present chapter consists of three parts, A, B, C. In part A, we shall treat the first Cousin problem and the problem of expansions on holomorph-convex domains because this is necessary in order to handle the inverse Hartogs problem. And in part B, we shall handle this problem.

As we have explained in Memoir VII, in the course of the present researches, we have met some problems which are arithmetical in nature, namely: problem (C_2) in Theorem II of Memoir I, problem (E) in Theorem I of Memoir II and problem (C_1) in the *Weil condition* explained in Memoir V. We solved them for polycylindrical domains in Memoir VII using the notion of *holomorphic ideal with indeterminate domains*[42] and *a theorem of H. Cartan*[43].

In part C, we shall deal with the second Cousin problem and with these three problems. We shall find a counterexample to problem (E). We shall see

[42]) In the theory of analytic continuation in a single complex variable, one has already considered function elements of the form (f, δ).

[43]) H. Cartan, Sur les matrices holomorphes de n variables complexes, 1940 (Journal de Mathématiques).

that the others are solvable (affirmatively) on pseudoconvex domains (without points at infinity and without interior ramification points).

A. Expansion of Holomorphic Functions; the First Cousin Problem

25. Expansion of Holomorphic Functions. Using the idea of Memoir I and Lemma I of No. 24, we obtain the following theorem.

Theorem I. *Let $\mathfrak{D}, \mathfrak{D}'$ be two domains over the space (x), without points at infinity or interior ramification points such that $\mathfrak{D} \subseteqq \mathfrak{D}'$; $\mathfrak{D}$ is not necessarily connected. If $\mathfrak{D}$ is holomorph-convex with respect to $\mathfrak{D}'$, every function holomorphic on $\mathfrak{D}$ can be expanded in a series of functions holomorphic on $\mathfrak{D}'$ converging uniformly on any region completely interior to $\mathfrak{D}$.*

In fact, let $f(P)$ be a function holomorphic on $\mathfrak{D}$. Since $\mathfrak{D}$ is holomorph-convex with respect to $\mathfrak{D}'$, it can be expressed as the limit of an increasing sequence of analytic polyhedra Δ_p of the form

$$|x_i| \leqq r_i, \quad |\varphi_j(P)| \leqq 1, \quad P \in \mathfrak{R} \quad (i=1, 2, \ldots, n;\ j=1, 2, \ldots, \nu),$$

where $\mathfrak{R}$ stands for a domain (connected or not) such that $\Delta_p \Subset \mathfrak{R} \subseteqq \mathfrak{D}$, the r_i are positive constants and the $\varphi_j(P)$ are holomorphic functions on $\mathfrak{D}'$ such that the ordered systems of values of $(\varphi_1, \varphi_2, \ldots, \varphi_\nu)$ are different at any pair of distinct points in a neighbourhood of Δ_p which have the same coordinates.

Let us consider the closed polycylinder C,

$$|x_i| \leqq r_i, \quad |y_j| \leqq 1 \quad (i=1, 2, \ldots, n;\ j=1, 2, \ldots, \nu)$$

in the space $(x_1, \ldots, x_n, y_1, \ldots, y_\nu)$ and the characteristic variety Σ given by $y_j = \varphi_j(P)$ $(j=1, 2, \ldots, \nu)$, $P \in \mathfrak{R}$ (the coordinates of P being (x)). By *Lemma* I, we can find a holomorphic function $F(x, y)$ in a neighbourhood of C such that $F = f$ on Σ. If we expand $F(x, y)$ in a TAYLOR series about the origin and substitute $y_j = \varphi_j(P)$ $(j=1, 2, \ldots, \nu)$, we obtain an expansion of $f(P)$ in a neighbourhood of Δ_p consisting of terms holomorphic on $\mathfrak{D}'$ and converging uniformly in a neighbourhood of Δ_p. The theorem follows immediately from this.

26. The First Cousin Problem. The same applies to the first COUSIN problem as we shall verify. We start by verifying the following lemma.

Let us consider the analytic polyhedron Δ over the space (x) as in Lemma I, and a hyperplane L meeting Δ. L partitions Δ into two closed sets Δ_1, Δ_2; let Δ_0 be the part of Δ on L. With this geometric configuration given a holomorphic function $\varphi(P)$ in a neighbourhood of Δ_0, we can find two holomorphic functions $\varphi_1(P), \varphi_2(P)$ in the neighbourhood of Δ_1 and of Δ_2 respectively such that we have $\varphi_1(P) - \varphi_2(P) = \varphi(P)$ identically.

In fact, suppose that L is defined by $u=0$, where, to simplify the notation, we assume that u is the real part of x_1; let Δ_1 be the part of Δ over $u \leqq 0$; Δ_2 is then that over $u \geqq 0$. Δ is given in the form $x_i \in A_i$, $f_j(P) \in B_j$ $(i=1,2,\ldots,n;\ j=1,2,\ldots,\nu)$ (with the same notation as indicated in Lemma I). Let us draw a circle of the form $|x_1|<R$ in the x_1-plane which contains A_1. Let L be the vertical diameter. By *Lemma* I, we have then a function $\Phi(x,y)$ holomorphic in a neighbourhood of the intersection $(x_1 \in L) \cap (A, B)$ such that $\Phi = \varphi$ on Σ' $(y_j = f_j(P),\ j=1,2,\ldots,\nu)$. Consider *the Cousin integral*

$$\frac{1}{2\pi i}\int_L \frac{\Phi(t, x_2, \ldots, x_n, (y))}{t - x_1}\, dt$$

where i is the imaginary unit and the integral is taken form bottom to top. Thanks to COUSIN, we know well the properties of this integral (as well as the role it plays). Substituting $y_j = f_j(P)$ $(j=1,2,\ldots,\nu)$, we obviously obtain the required pair of functions.

Let $\mathfrak{D}$ be a domain over the finite portion of the space (x), without interior ramification points. We can give ourselves a COUSIN problem on $\mathfrak{D}$ just as in the previous cases; we shall not repeat this.

Suppose that $\mathfrak{D}$ is *holomorph-convex*. $\mathfrak{D}$ is then the limit of an increasing sequence of analytic polyhedra Δ_p of the form $|x_i| \leqq r_i$, $|f_j(P)| \leqq 1$, $P \in \mathfrak{R}$ $(i=1,2,\ldots,n;\ j=1,2,\ldots,\nu)$, where $(\Delta \Subset \mathfrak{R} \subseteqq \mathfrak{D}$, the r_i are positive constants and) the $f_j(P)$ are holomorphic functions on $\mathfrak{D}$ having the properties indicated in Lemma I. Let $(\mathfrak{p})$ be the given poles on $\mathfrak{D}$. By the *lemma* which we have just established, we can find a meromorphic function $G_p(P)$ in a neighbourhood of Δ_p having the poles $(\mathfrak{p})$.

Let $H_p(P) = G_{p+1}(P) - G_p(P)$; H_p is a function holomorphic in a neighbourhood of Δ_p. Let ε_p be an arbitrary positive number. By *Theorem* I, we obtain a function $K_p(P)$ holomorphic on $\mathfrak{D}$ such that $|K_p - H_p| < \varepsilon_p$ on Δ_p. From this, we obtain by the usual argument

Theorem II. *The first Cousin problem is always solvable on a holomorph-convex domain over the space* (x) *without points at infinity or interior points of ramification.*

B. The Inverse Hartogs Problem. – The Point of Departure

27. An Auxiliary Problem. We shall deal with the inverse HARTOGS problem using Lemmas I, II and Theorems I, II. Following the idea of Memoir VI, we shall start by looking for something which plays the role of the COUSIN integral.

Let $\mathfrak{D}$ be a *bounded* domain over the space (x) (without interior ramification points). Let u be the real part of x_1, let a_1, a_2 be real numbers such that $a_1 < 0 < a_2$. We denote the part of $\mathfrak{D}$ over $u < a_2$ by $\mathfrak{D}_1$, that over $u > a_1$ by $\mathfrak{D}_2$, and the part over $a_1 < u < a_2$ by $\mathfrak{D}_3$. We suppose that $\mathfrak{D}_1$ and $\mathfrak{D}_2$ are non-empty and are *holomorphy-convex*. The same is then true of $\mathfrak{D}_3$.

Let $f_j(P)$ $(j=1,2,\ldots,\nu)$ be holomorphic functions on $\mathfrak{D}_3$. Consider the following subset E of $\mathfrak{D}$: Every point which does not belong to $\mathfrak{D}_3$ belongs to E. As for a point P of $\mathfrak{D}_3$, it belongs to E if and only if it satisfies simultaneously the conditions

$$|f_j(P)|<1 \quad (j=1,2,\ldots,\nu).$$

We suppose that E has connected components extending over both the portion $u<a_1$ and the portion $a_2<u$ at once; one of these components will be denoted by Δ. We suppose, relative to this Δ, *that the functions* (f) *satisfy the following conditions* (*which we call conditions* (C)):

1°. There exists a positive nubmer δ_1 such that if $\mathfrak{A}$ is the part of Δ over $|u|<\delta_1$, we have

$$\mathfrak{A}\Subset\mathfrak{D}.$$

2°. There exists a positive number δ_2 and a positive number ε_0 smaller than 1 such that, if p is any one of $1,2,\ldots,\nu$, the set of points of $\mathfrak{D}_3$ satisfying

$$|f_p(P)|\geqq 1-\varepsilon_0$$

remains outside $|u-a_1|<\delta_2$ and $|u-a_2|<\delta_2$.

3°. The ordered system of values of $(f_1, f_2, \ldots, f_\nu)$ do not coincide for any pair of disticnt points of $\mathfrak{A}$ which have the same coordinates.

Because of the second condition, Δ is a domain. We take a real number ρ_0 such that $1-\varepsilon_0<\rho_0<1$ and consider the subset Δ_0 of Δ given as follows: Every point of Δ not belonging to $\mathfrak{D}_3$ belongs to Δ_0; for any point P of Δ lying in $\mathfrak{D}_3$, it belongs to Δ_0 if and only if it satisfies simultaneously the conditions $|f_j(P)|<\rho_0$ $(j=1,2,\ldots,\nu)$. By the second condition, Δ_0 is an open set in $\mathfrak{D}$. We denote the part of Δ_0 over $u<0$ by Δ_0' and that over $u>0$ by Δ_0''. Under these conditions, we pose the following *auxiliary problem:*

Given a function $\varphi(P)$ holomorphic on $\mathfrak{A}$, find a function $\varphi_1(P)$ holomorphic on Δ_0' and a function $\varphi_2(P)$ holomorphic on Δ_0'' so that they remain holomorphic on the part of Δ_0 over $u=0$ and we have $\varphi_1(P)-\varphi_2(P)=\varphi(P)$ identically.

28. The Case of Holomorph-Convex Domains. We shall first consider the problem for $\mathfrak{D}_3$ instead of $\mathfrak{D}$. We shall obtain a solution (φ_1, φ_2) by the method expounded in No. 26; we are concerned with finding a suitable expression for the solution.

We consider the characteristic variety Σ in the space $(x_1, x_2, \ldots, x_n, y_1, y_2, \ldots, y_\nu)$ given by $y_j=f_j(P)$ $(j=1,2,\ldots,\nu)$, P being a general point of $\mathfrak{A}$ having coordinates (x). $\mathfrak{D}$ being bounded, we take a positive number r_0 so large that $\mathfrak{D}$ is contained in the polycylinder in the space (x) with centre the origin and radius r_0; and we consider a polycylinder C, $|x_i|<r$, $|y_k|<\rho$ $(i=1,2,\ldots,n;\ k=1,2,\ldots,\nu)$ in the space (x,y), where $r>r_0$, $\rho_0<\rho<1$. Consider the subset $\mathfrak{A}'$ of $\mathfrak{A}$ given by

$$|u|<\delta, \quad |f_k(P)|<\rho \quad (k=1,2,\ldots,\nu),$$

where δ is a positive number $<\delta_1$ (we necessarily have $\delta_1<-a_1, a_2$). Since $\mathfrak{A}\Subset\mathfrak{D}$, $\mathfrak{A}'$ is an open set such that $\mathfrak{A}'\Subset\mathfrak{A}$.

Since $\varphi(P)$ is a holomorphic function on $\mathfrak{A}$, we can construct, by Lemma I, a holomorphic function $\Phi(x,y)$ on the intersection of C and $|u|<\delta$ so that $\Phi=\varphi$ on Σ. Let l be a segment of the imaginary axis in the x_1-plane which contains the origin and has its extremities in the annulus $r_0<|x_1|<r$.

Consider the COUSIN integral

$$\Psi(x,y)=\frac{1}{2\pi i}\int_l \frac{\Phi(x_1',x_2,\dots,x_n,(y))}{x_1'-x_1}\,dx_1',$$

where i is the imaginary unit and the integral is taken from bottom to top. If we substitute $y_k=f_k(P)$ in the expression for $\Psi(x,y)$, we have

$$\psi(P)=\frac{1}{2\pi i}\int_l \frac{\Phi(x_1',x_2,\dots,x_n,f_1(P),\dots,f_\nu(P))}{x_1'-x_1}\,dx_1'.$$

The holomorphic function on the intersection $\Delta_0'\cap\mathfrak{D}_3$ defined by $\psi(P)$ will be denoted by $\psi_1(P)$, that on $\Delta_0''\cap\mathfrak{D}_3$ by $\psi_2(P)$. These functions $\psi_1(P)$ and $\psi_2(P)$ remain holomorphic on the part of Δ_0 over $u=0$ and satisfy the required relation $\psi_1(P)-\psi_2(P)=\varphi(P)$.

We shall modify this expression for the solution (ψ_1,ψ_2). Let us describe the circle Γ_k in the y_k-plane $(k=1,2,\dots,\nu)$ having the origin as centre and radius ρ_0; by CAUCHY's theorem, we have

$$\Phi(x,y)=\frac{1}{(2\pi i)^\nu}\int_{\Gamma_1}\frac{dy_1'}{y_1'-y_1}\int_{\Gamma_2}\frac{dy_2'}{y_2'-y_2}\cdots\int_{\Gamma_\nu}\frac{\Phi(x,y')}{y_\nu'-y_\nu}\,dy_\nu'$$

(the integrations are in the direct sense on Γ_k) for $|x_j|<r$, $|y_k|<\rho_0$ $(j=1,2,\dots,n;$ $k=1,2,\dots,\nu)$. We shall, for the moment, abbreviate this to

$$\Phi(x,y)=\frac{1}{(2\pi i)^\nu}\int_{(\Gamma)}\frac{\Phi(x,y')}{\prod(y_k'-y_k)}(dy')\qquad (k=1,2,\dots,\nu).$$

Substituting this expression for $\Phi(x,y)$ in the expression for $\psi(P)$, we obtain, with a similar abbreviation,

$$\psi(P)=\int_{(l,\Gamma)}\chi(x_1',y',P)\,\Phi(x_1',x_2,\dots,x_n,y')(dx_1',dy') \tag{1}$$

where

$$\chi(x_1',y',P)=[(2\pi i)^{\nu+1}(x_1'-x_1)(y_1-f_1(P))\dots(y_\nu-f_\nu(P))]^{-1}.$$

This is the expression we want for the solution of the problem on $\mathfrak{D}_3$.

29. The General Case. In the expression (1), the cylindrical set of points $(l,(\bar\Gamma))$[44] in the space (x_1',y') is the decreasing limit of a sequence of holomorph-convex domains which contain the set; let V be a domain of this sequence. We construct a meromorphic function $\chi_1(x_1',y',P)$ on $(V,\mathfrak{D}_1)$ which has the same poles as $\chi(x_1',y',P)$ in $(V,\mathfrak{D}_3)$, but is otherwise holomorphic; this is possible if V is chosen sufficiently close to $(l,(\bar\Gamma))$ because this is a first COUSIN problem in view of the second of the conditions (C) and because $(V,\mathfrak{D}_1)$ is holomorph-convex (Theorem II).

Since the function $(\chi-\chi_1)$ is holomorphic on $(V,\mathfrak{D}_3)$ and $(V,\mathfrak{D}_3)$ is holomorph-convex with respect to $(V,\mathfrak{D}_1)$, we can expand it in a series of functions holomorphic on $(V,\mathfrak{D}_1)$ converging uniformly on any region completely interior

[44] $(l,(\bar\Gamma))$ stands for the set of points $x_1\in l$, $y_j\in(\bar\Gamma_j)$, $(j=1,2,\dots,\nu)$, $(\bar\Gamma_j)$ being the closed circle whose circumference is Γ_j.

to $(V, \mathfrak{D}_3)$ (Theorem I). For a given positive number ε, we can therefore construct a function $F_1(x_1', y', P)$ holomorphic on $(V, \mathfrak{D}_1)$ such that $|(\chi-\chi_1)-F_1|<\varepsilon$ on $(V, \mathfrak{A})$, where V now stands for a new neighbourhood of $(l, (\bar{\Gamma}))$ completely interior to the old one. Set

$$K_1(x_1', y', P)=\chi-\chi_1-F_1;$$

K_1 is a holomorphic function on $(V, \mathfrak{D}_3)$ such that $|K_1|<\varepsilon$ on $(V, \mathfrak{A})$.

We construct, entirely analogously, a function $K_2(x_1', y', P)$ corresponding to $\mathfrak{D}_2$. We modify the integral (1), using these functions, as follows:

$$(2)\quad \begin{cases} I_1(P)=\int\limits_{(l,\Gamma)} [\chi(x_1', y', P)-K_1(x_1', y', P)]\, \Phi(x_1', x_2, \ldots, x_n, y')(dx_1', dy') \\ I_2(P)=\int\limits_{(l,\Gamma)} [\chi(x_1', y', P)-K_2(x_1', y', P)]\, \Phi(x_1', x_2, \ldots, x_n, y')(dx_1', dy'). \end{cases}$$

Let us look at $I_1(P)$. Since $\chi-K_1=\chi_1+F_1$, the function $(\chi-K_1)$ is meromorphic on $\mathfrak{D}_1$ for any fixed point (x_1', y') on (l, Γ) and is holomorphic for $P\in\Delta_0'$. Hence $I_1(P)$ is a holomorphic function on Δ_0'. Similarly, $I_2(P)$ is holomorphic on Δ_0''.

The functions $I_1(P)$, $I_2(P)$ remain holomorphic at every point of Δ_0 over $u=0$ because the function $\psi(P)$ given by (1) has this property and K_1, K_2 are holomorphic. From the properties of $\psi(P)$, it follows that

$$(3)\quad I_1(P)-I_2(P)=$$
$$\varphi(P)-\int\limits_{(l,\Gamma)} [K_1(x_1', y', P)-K_2(x_1', y', P)]\, \Phi(x_1', x_2, \ldots, x_n, y')(dx_1', dy').$$

Let us set

$$K(x_1', y', P)=K_1(x_1', y', P)-K_2(x_1', y', P),$$

and consider the identity (3). $\varphi(P)$ is holomorphic on $\mathfrak{A}$, K is holomorphic on $(V, \mathfrak{D}_3)$ and $\Phi(x, y)$ is holomorphic on $C\cap(|u|<\delta)$. The right hand side is therefore a holomorphic function on $\mathfrak{A}$, and the same is therefore true of left hand side. Let us set

$$\varphi_0(P)=I_1(P)-I_2(P);$$

this is the first approximation to $\varphi(P)$. We have thus seen *that the given function $\varphi(P)$ and its first approximation $\varphi_0(P)$ are holomorphic on one and the same domain (connected or not), namely* $\mathfrak{A}$. This is the fundamental idea.

Let us estimate the difference $\varphi'=\varphi-\varphi_0$. We have

$$\varphi'=\int\limits_{(l,\Gamma)} K\cdot\Phi(dx_1', dy').$$

Suppose that φ is bounded on $\mathfrak{A}$; there is no loss of generality in this assumption. Let M be the upper bound of $|\varphi|$ on $\mathfrak{A}$. By *Lemma* I, we can choose the function $\Phi(x, y)$ holomorphic on $C\cap(|u|<\delta)$ such that $|\Phi|<N\cdot M$, N being a constant independent of φ. As for K, we have $K=K_1-K_2$, where $|K_1|<\varepsilon$, $|K_2|<\varepsilon$ on $(l, (\bar{\Gamma}), \mathfrak{A})$. We therefore have

$$|\varphi'(P)|<2\varepsilon N\cdot N_1\cdot M, \qquad N_1=2r(2\pi\rho_0)^{\nu},$$

on $\mathfrak{A}$. Let us choose ε so that $2\varepsilon N \cdot N_1 < \lambda$, λ being a positive number smaller than 1. We then have $|\varphi'(P)| < \lambda M$.

If we start again with $\varphi'(P)$ instead of $\varphi(P)$, we obtain similarly a first approximation to $\varphi'(P)$, say $\varphi'_0(P)$, which is holomorphic on $\mathfrak{A}$ and such that, if $\varphi'' = \varphi' - \varphi'_0$, then $|\varphi''| < \lambda^2 M$ on $\mathfrak{A}$. $\varphi'_0(P)$ is given as the difference $I'_2(P) - I'_2(P)$, where

$$I'_1(P) = \int_{(l,\Gamma)} [\chi(x'_1, y', P) - K_1(x'_1, y', P)]\, \Phi'(x'_1, x_2, \ldots, x_n, y)\,(dx'_1, dy'),$$

$$I'_2(P) = \int_{(l,\Gamma)} [\chi(x'_1, y', P) - K_2(x'_1, y', P)]\, \Phi'(x'_1, x_2, \ldots, x_n, y)\,(dx'_1, dy'),$$

where Φ' is a holomorphic function on $C \cap (|u| < \delta)$ such that $|\Phi'| < \lambda N \cdot M$. Repeating this process an infinity of times, we obtain the required solution. Thus:

The auxiliary problem (of No. 27) is always solvable.

Solution of the Problem

30. The Idea of Memoir II and Lemma II. Let us recall the idea in Memoir II. Let $\mathfrak{D}$ be a domain over the space (x), let E be a set of points on $\mathfrak{D}$. We shall say that E is *holomorph-convex (from the outside) with respect to* $\mathfrak{D}$ if $E \Subset \mathfrak{D}$ and the closure[45] of E in $\mathfrak{D}$ is the limit of a decreasing sequence of subdomains (not neccssarily connected) of $\mathfrak{D}$ which are holomorph-convex with respect to $\mathfrak{D}$. If Δ_1, Δ_2 are two subdomains (not necessarily connected) of $\mathfrak{D}$ which are holomorph-convex with respect to $\mathfrak{D}$, the intersection $\Delta_1 \cap \Delta_2$ has the same property. Let E be a set of points such that $E \Subset \mathfrak{D}$ and F_1, F_2 two sets of points of $\mathfrak{D}$ containing E which are closed, compact, and holomorph-convex from the outside with respect to $\mathfrak{D}$. Then, by the preceding remark, the same is true of $F_1 \cap F_2$. Let F_p $(p = 1, 2, \ldots)$ be a decreasing sequence of sets of points of $\mathfrak{D}$ with this property; the same is then true of the limit.

Let $\mathfrak{D}$ be a *holomorph-convex* domain over the space (x) and let E be a set of points such that $E \Subset \mathfrak{D}$. There then exists a set F of points of $\mathfrak{D}$ containing E which is closed, compact and holomorph-convex from the outside with respect to $\mathfrak{D}$. Let H be the intersection of all these F; F is clearly a set of points of $\mathfrak{D}$ containing E, closed and compact in $\mathfrak{D}$. We shall call H the (H)-*closure of E in* $\mathfrak{D}$. We claim that H is the intersection of a countable number of suitable sets F as above.

In fact, if we recall the reasoning in Theorem I, we find immediately that if $\mathfrak{D}_0$ is a domain (not necessarily connected) such that $\mathfrak{D}_0 \Subset \mathfrak{D}$, and holomorph-convex with respect to $\mathfrak{D}$, then any function $f(P)$ holomorphic on $\mathfrak{D}_0$ can be expressed as a series of polynomials in a finite number of well-determined functions holomorphic on $\mathfrak{D}$ and converging uniformly on a given domain completely interior to $\mathfrak{D}_0$. We can restrict the real and imaginary parts of the coefficients of these polynomials to rational numbers. The above result follows immediately from this.

---

[45]) The smallest closed set on $\mathfrak{D}$ containing E.

Because of this result and what precedes, we find easily that:

The (H)*-closure of E in $\mathfrak{D}$ is holomorph-convex from the outside with respect to $\mathfrak{D}$.*

The (H)-closure has the following property relative to characteristic surfaces.

Let $\mathfrak{D}$ be a holomorph-convex domain over the space (x), *let E be a set of points such that $E \Subset \mathfrak{D}$ and let H be the* (H)*-closure of E with respect to $\mathfrak{D}$.*

Let σ_t be a piece of a characteristic surface depending on the parameter t in the interval $0 \leqq t \leqq 1$, given in the form $f(P,t)=0$, $P \in V$ where V is a domain such that $V \Subset \mathfrak{D}$ and $f(P,t)$ is a holomorphic function of (P,t) in a neighbourhood of $(V,[0,1])$ which is not identically zero[46]*. Then, H and the family $\{\sigma_t\}$ cannot be related to each other in the following way:* 1°. *The distance on $\mathfrak{D}$ between H and the boundary of any σ_t is larger than a positive constant.* 2°. *The distance between E and any σ_t is larger than a positive constant. And* 3°. *σ_0 passes through a point of H and σ_1 is situated at a positive distance from H.*

Suppose, on the contrary, that the family $\{\sigma_t\}$ really exists. We want to construct a meromorphic function $G(P,t)$ in a neighbourhood of $(H,[0,1])$ having the poles $[f(P,t)]^{-1}$. By the first condition, this ia a first COUSIN problem. H is holomorph-convex from the outside with respect to $\mathfrak{D}$; consequently, by Theorem II, this is possible.

Let α be the last of the values of t such that σ_t passes through H when t describes the closed interval $[0,1]$ from 0 to 1; by the third condition α is certainly well-defined. Let M be one of the points of H through which σ_α passes. Since (M,α) is a pole (not a point of indeterminacy) of $G(P,t)$, the modulus of $G(P,t)$ increases indefinitely when t tends to α on the open interval $(\alpha,1)$; on the other hand, by the second condition, $G(P,t)$ remains holomorphic when P lies in a neighbourhood of E, and consequently, if we choose β sufficiently close to α on $(\alpha,1)$, we have $\max|G(E,\beta)|<|G(M,\beta)|$, where the term on the left means the upperbound of $|G(P,\beta)|$ on E.

Since $G(P,\beta)$ is holomorphic on a neighbourhood of H, we can, by Theorem I, expand $G(P,\beta)$ in a series of functions holomorphic on $\mathfrak{D}$ converging uniformly on H. We can therefore find a function $\Phi(P)$ holomorphic on $\mathfrak{D}$ and satisfying $\max|\Phi(E)|<|\Phi(M)|$; this contradicts the minimality of H. Thus, the family $\{\sigma_t\}$ cannot exist.

If we apply this principle to the configuration of Lemma II, we obtain the following:

Let $\mathfrak{D}$ be a pseudoconvex domain over the space (x) *and $\varphi(P)$ a pseudoconvex function having the following properties:* 1°. *$\varphi(P)$ has property* (P_1). 2°. *If $\mathfrak{D}_\alpha$ is the set of points of $\mathfrak{D}$ given by $\varphi(P)<\alpha$ (α being an arbitrary real number), then $\mathfrak{D}_\alpha \Subset \mathfrak{D}$. Under these conditions, if, for some real number β, $\mathfrak{D}_\beta$ is holomorph-convex, then for any $\alpha<\beta$, $\mathfrak{D}_\alpha$ is holomorph-convex with respect to $\mathfrak{D}_\beta$. More-*

[46]) With respect to P for any fixed t.

over, if $\mathfrak{D}$ itself is holomorph-convex, then every $\mathfrak{D}_\alpha$ is holomorph-convex with respect to $\mathfrak{D}$.[47)]

In fact, it suffices to verify the second part of this statement. Let $\mathfrak{D}$ be holomorph-convex. Let Δ_α be the set of points of $\mathfrak{D}$ given by $\varphi(P) \leqq \alpha$. Since $\Delta_\alpha \Subset \mathfrak{D}$, it has a (H)-closure in $\mathfrak{D}$ which we denote by H. Suppose that $H \neq \Delta_\alpha$. Let $\max \varphi(H) = \beta$; then $\beta > \alpha$. Let P_0 be a point of H where $\varphi = \beta$. If we recall how we proved that pseudoconvex functions having property (P_0) also have the main property (No. 16), we easily realise, in a neighbourhood of P_0, the configuration we have just shown to be absurd. Thus $H = \Delta_\alpha$. Since this holds for any α, $\mathfrak{D}_\alpha$ is holomorph-convex with respect to $\mathfrak{D}$.[48)]

31. Solution of the Problem. Let us return to the auxiliary problem. We remark that, for the Δ of this problem, the first COUSIN problem is always solvable on Δ.

In fact, since $\mathfrak{D}_3$ is holomorph-convex with respect to $\mathfrak{D}_1$ and the $f_j(P)$ ($j = 1, 2, \ldots, \nu$) are holomorphic functions on $\mathfrak{D}_3$, we can, by Theorem I, expand each $f_j(P)$ in a series of functions holomorphic on $\mathfrak{D}_1$, converging uniformly on any domain completely interior to $\mathfrak{D}_3$. The part of Δ over $u < a_2$ is therefore holomorph-convex, as is the part over $a_1 < u$. Suppose given a first COUSIN problem on Δ. By Theorem II, this problem is solvable on the part of Δ over $u < a_2$ and for the part over $a_1 < u$. Since the auxiliary problem is always solvable, the given first COUSIN problem on Δ is solvable.

Let us consider how we might construct the domain Δ_0. Suppose given a domain $\mathfrak{D}$ over the space (x) *such that the domains $\mathfrak{D}_1, \mathfrak{D}_2$ are holomorph-convex.* Since any holomorph-convex domain is pseudoconvex, $\mathfrak{D}_1$ and $\mathfrak{D}_2$, and consequently $\mathfrak{D}$, are pseudoconvex. We can therefore apply Lemma II to $\mathfrak{D}$, namely: there exists a pseudoconvex function $\varphi(P)$ on $\mathfrak{D}$ having property (P_1) and such that, for any real number α, the set $\mathfrak{D}_\alpha$ of points of $\mathfrak{D}$ given by $\varphi(P) < \alpha$ satisfies $\mathfrak{D}_\alpha \Subset \mathfrak{D}$.

Let us denote the part of $\mathfrak{D}_\alpha$ over $u < a_2$ by $\mathfrak{D}'_\alpha$, that over $a_1 < u$ by $\mathfrak{D}''_\alpha$. We then have:

$\mathfrak{D}'_\alpha$ is holomorph-convex with respect to $\mathfrak{D}_1$, similarly $\mathfrak{D}''_\alpha$ with respect to $\mathfrak{D}_2$.

In fact, we take real numbers α', a'_2 such that $\alpha' < \alpha$, $a'_2 < a_2$ and consider the set F of points of $\mathfrak{D}_1$ given by $\varphi(P) \leqq \alpha'$, $u \leqq a'_2$ (supposing that $F \supset 0$). Since $F \Subset \mathfrak{D}_1$ and $\mathfrak{D}_1$ is holomorph-convex, we can consider the (H)-closure H of F in $\mathfrak{D}_1$. By the reasoning presented in the preceding section, we recognize immediately that $F = H$. Hence $\mathfrak{D}'_\alpha$ is holomorph-convex with respect to $\mathfrak{D}_1$. Entirely analogously, $\mathfrak{D}''_\alpha$ is holomorph-convex with respect to $\mathfrak{D}_2$.

Since we have this property for $\mathfrak{D}_\alpha$ and $\mathfrak{D}_\alpha$ is bounded, we shall construct Δ_0 in $\mathfrak{D}_\alpha$. For this, it is a question of finding a system of holomorphic functions $f_j(P)$ ($j = 1, 2, \ldots, \nu$) in the intersection $\mathfrak{D}'_\alpha \cap \mathfrak{D}''_\alpha$ satisfying conditions (C) (No. 27). Let us look at these conditions. Let (S_1), (S_2) be two finite systems

47) The first condition on $\varphi(P)$ is not necessary (the proof is immediate).

48) Using the same ideas, one can verify Theorem I of Memoir II in the present case in a simple way, without appealing to Memoir VIII.

of holomorphic functions on $\mathfrak{D}'_\alpha \cap \mathfrak{D}''_\alpha$; suppose that (S_1) satisfies the first and second conditions and that (S_2) satisfies the third and second conditions; we find then that the union $(S_1) \cup (S_2)$ fulfills the three conditions. It is therefore sufficient to construct them separately.

Let us start by constructing the system (S_2). Since $\mathfrak{D}_1$ is holomorph-convex, we can find a finite system of holomorphic functions on $\mathfrak{D}_1$ satisfying the third condition on $\mathfrak{D}_\alpha$. To make (S) satisfy the second condition on $\mathfrak{D}_\alpha$, we have only to multiply all the functions of (S) by a sufficiently small number.

As for (S_1), let P_0 be an arbitrary point of the set $\varphi(P)=\alpha$, $u=0$. We can find a characteristic surface σ_0 in a neighbourhood of P_0, remaining in the portion $\varphi(P)>\alpha$ except at P_0 itself. Let $f(P)=0$ be the equation of σ_0, $f(P)$ being a holomorphic function. Let β be a real number bigger than α. We want to construct a meromorphic function $G(P)$ in the part of $\mathfrak{D}_\beta$ over $u<a_2$ having the poles $[f(P)]^{-1}$. Since this part of $\mathfrak{D}_\beta$ is holomorph-convex, this is possible if we take β sufficiently close to α because of Theorem II. We again take for a_2 a real number a little bit smaller than the old one; the function $G(P)$ is then holomorphic on $\mathfrak{D}'_\alpha$ meromorphic in a neighbourhood of $\mathfrak{D}'_\alpha$ and has the single pole P_0 on the boundary of the set $\mathfrak{D}'_\alpha$. Since P_0 is an arbitrary point of the set $\varphi=\alpha$, $u=0$, we can easily find a system (S_1) using the BOREL lemma. If we recall the reasoning just made, we find the following:

Let α, α' be two real numbers such that $\alpha'<\alpha$, but else arbitrary. We can construct a Δ_0 in $\mathfrak{D}_\alpha$ in such a way that $\Delta_0 \Supset \mathfrak{D}_{\alpha'}$.

This holds under the assumption that $\mathfrak{D}_1$ and $\mathfrak{D}_2$ are holomorph-convex. With the same hypothesis, we shall prove that:

$\mathfrak{D}_\alpha$ is holomorph-convex.

Let us first verify the first condition that there exists a function $f_0(P)$ holomorphic on $\mathfrak{D}_\alpha$ and having different TAYLOR expansions at any pair of distinct points of $\mathfrak{D}_\alpha$. Let P_1, P_2 be a pair of distinct points of $\mathfrak{D}_\alpha$ having the same coordinates, but otherwise arbitrary. We wish to construct a holomorphic function $f(P)$ on $\mathfrak{D}_\alpha$ such that $f(P_1) \neq f(P_2)$. We shall apply to this problem *the method of H. Cartan*[49]. Let (x^0) be the common coordinates of P_1 and P_2. The point (x^0) lies in one, at least, of the portions $u<a_2$, $a_1<u$, say in $u<a_2$. Since $\mathfrak{D}_1$ is holomorph-convex, we can find a holomorphic function $\psi(P)$ on $\mathfrak{D}_1$ such that $\psi(P_1) \neq \psi(P_2)$. By what we have seen above, we can find a meromorphic function $G(P)$ on $\mathfrak{D}_\alpha$ having the poles $\psi(P)(x_1-x_1^0)^{-1}$. Set $f(P)=(x_1-x_1^0)\, G(P)$. The function $f(P)$ is then holomorphic on $\mathfrak{D}_\alpha$ and takes the values $\psi(P)$ on $x_1-x_1^0=0$. It is therefore one of the functions desired. From this, we can easily construct the required function $f_0(P)$.

Let us pass to the second condition that, for each domain (connected or not) Δ with $\Delta \Subset \mathfrak{D}_\alpha$, we can find two domains (connected or not) Δ', Δ'' satisfying $\Delta \subsetneqq \Delta' \Subset \Delta'' \subseteqq \mathfrak{D}_\alpha$ such that for any point P_0 of $\Delta''-\Delta'$, there exists a holomorphic function $f(P)$ on $\mathfrak{D}_\alpha$ for which $|f(P_0)|>\max|f(\Delta)|$. Now, for any point P_0 of $\mathfrak{D}$ such that $\varphi(P_0)=\alpha$, we can find a function $G(P)$ holomorphic on $\mathfrak{D}_\alpha$, meromorphic in a neighbourhood of $\mathfrak{D}_\alpha$ having a single pole (which is not a

[49]) This is a way of applying the first COUSIN problem. See Note 8 on p. 4 of Memoir I (this translation).

point of indeterminacy) at the point P_0 on the set $\varphi=\alpha$. It follows immediately from this that $\mathfrak{D}_\alpha$ satisfies the second condition (where we can in fact take $\Delta'' = \mathfrak{D}_\alpha$).

Let next $\mathfrak{D}$ be a *pseudoconvex* domain over the space (x). We shall prove, with the configuration of Lemma II, the following:

(α) If every $\mathfrak{D}_\alpha$ is holomorph-convex, then so is $\mathfrak{D}$.

In fact, *by what we proved at the end of the preceding section*, we see that if $\alpha<\beta$, then $\mathfrak{D}_\alpha$ is holomorph-convex with respect to $\mathfrak{D}_\beta$. Consequently, by Theorem I, every function holomorphic on $\mathfrak{D}_\alpha$ can be expanded in a series of functions holomorphic on $\mathfrak{D}_\beta$, uniformly convergent on any domain completely interior to $\mathfrak{D}_\alpha$. Let α_p $(p=1,2,\ldots)$ be a sequence of positive numbers such that $\alpha_p<\alpha_{p+1}$ increasing indefinitely (where we choose α_1 so large that $\mathfrak{D}_{\alpha_1}\supset 0$). Let $f(P)$ be a holomorphic function on $\mathfrak{D}_{\alpha_2}$ and ε a given positive number. We want to construct a holomorphic function $F(P)$ on $\mathfrak{D}$ such that $|f(P)-F(P)|<\varepsilon$ on $\mathfrak{D}_{\alpha_1}$. Let ε_p $(p=1,2,\ldots)$ be a sequence of positive numbers such that $\sum\varepsilon_p$ converges to a number $<\varepsilon$. By the preceding result, we can find a holomorphic function $f_1(P)$ on $\mathfrak{D}_{\alpha_3}$ such that $|f(P)-f_1(P)|<\varepsilon_1$ on $\mathfrak{D}_{\alpha_1}$. We next construct a holomorphic function $f_2(P)$ on $\mathfrak{D}_{\alpha_4}$ such that $|f_1(P)-f_2(P)|<\varepsilon_2$ on $\mathfrak{D}_{\alpha_2}$ and so on. We then have a sequence $f_2(P), f_3(P), \ldots, f_p(P), \ldots$ which converges uniformly on any domain completely interior to $\mathfrak{D}$. Consequently, the limit $F(P)$ is a holomorphic function on $\mathfrak{D}$; and on $\mathfrak{D}_{\alpha_1}$, we have $|f(P)-F(P)|<\varepsilon$. This being the case, let us examine the first condition for $\mathfrak{D}$ to be holomorph-convex. Let P_1, P_2 be a pair of different points of $\mathfrak{D}$ having the same coordinates, but otherwise arbitrary. If we take α_1 sufficiently large, $\mathfrak{D}_{\alpha_1}$ contains these two points. Since $\mathfrak{D}_{\alpha_2}$ is holomorph-convex, we can find a holomorphic function $f(P)$ on $\mathfrak{D}_{\alpha_2}$ such that $f(P_1)\neq f(P_2)$. Choose ε so that $|f(P_1)-f(P_2)|\geqq 2\varepsilon$. Let F be the function indicated above which corresponds to these f, ε. $F(P)$ is holomorphic on $\mathfrak{D}$ and $F(P_1)\neq F(P_2)$. We can therefore construct a holomorphic function on $\mathfrak{D}$ having different Taylor expansions at any pair of distinct points of $\mathfrak{D}$.

Let us examine the second condition. Let Δ be a domain such that $\Delta\Subset\mathfrak{D}$. For sufficiently large α_1, we have $\Delta\Subset\mathfrak{D}_{\alpha_1}$. Since $\mathfrak{D}_{\alpha_1}$ is holomorph-convex with respect to $\mathfrak{D}_{\alpha_2}$, there exist two domains (not necessarily connected) Δ', Δ'' so that $\Delta\subseteqq\Delta'\Subset\Delta''\subseteqq\mathfrak{D}_{\alpha_1}$ and such that to any point P_0 of $\Delta''-\Delta'$, there correponds a holomorphic function $f(P)$ on $\mathfrak{D}_{\alpha_2}$ satisfying $|f(P_0)|>\max|f(\Delta)|$. We choose ε such that $|f(P_0)|-\max|f(\Delta)|>2\varepsilon$ and construct the function $F(P)$. $F(P)$ is then holomorphic on $\mathfrak{D}$ and $|F(P_0)|>\max|F(\Delta)|$. The two conditions being thus fulfilled, $\mathfrak{D}$ is holomorph-convex.

Combining these two intermediate results, we obtain:

(β) Let $\mathfrak{D}$ be a domain such that $\mathfrak{D}_1$ and $\mathfrak{D}_2$ are holomorph-convex. Then $\mathfrak{D}$ is also holomorph-convex.

Let now $\mathfrak{D}$ be a pseudoconvex domain over the space (x). Let P_0 be an arbitrary point on the boundary of the set $\mathfrak{D}_\alpha$. From the reasoning given in proving the theorem in No. 16 that every function having property (P_0) has also the main property, we deduce immediately that the intersection of $\mathfrak{D}_\alpha$ and a sufficiently small polycylindrical neighbourhood with centre P_0 is holomorph-convex. Consequently, by (β) above, $\mathfrak{D}_\alpha$ is holomorph-convex; hence, by (α), so is $\mathfrak{D}$. We formulate the result so obtained:

Theorem III. *Every pseudoconvex domain over the space (x) without points at infinity or interior points of ramification is holomorph-convex.*

From this, it follows at once that in the definition of *holomorph-convex domains*, or of *analytic polyhedra, the first condition is but a consequence of the second.*

32. Pseudoconvex Domains and Domains of Holomorphy

Corollary. *Every pseudoconvex domain over the space (x) without points at infinity or interior points of ramification is a domain of holomorphy.*

Let, in fact, $\mathfrak{D}$ be a pseudoconvex domain over the space (x). By Theorem III, $\mathfrak{D}$ is holomorph-convex. There therefore exists a holomorphic function $F_1(P)$ on $\mathfrak{D}$ having distinct TAYLOR expansions at any pair of distinct points of $\mathfrak{D}$. Suppose that there is a holomorphic function $F_2(P)$ on $\mathfrak{D}$ such that it cannot be analytically continued across any boundary point of $\mathfrak{D}$, and consider $F(P)=F_1(P)+c\,F_2(P)$, c being a constant. We find that if we choose c suitably, F conserves both the character of F_1 and that of F_2 (for this, it suffices to avoid a countable number of values). It is therefore sufficient to construct a function $F_2(P)$.

Since the domain $\mathfrak{D}$ is holomorph-convex, it is the limit of a sequence of analytic polyhedra Δ_p $(p=1,2,\ldots)$ such that $\Delta_p \Subset \Delta_{p+1}$. Each Δ_p is given in the form

$$|f_j(P)| \leqq 1, \quad P \in \mathfrak{R}_p \quad (j=1,2,\ldots,\nu),$$

where the $f_j(P)$ are holomorphic functions on $\mathfrak{D}$ (depending on p), $\mathfrak{R}_p$ is a subdomain (connected or not) of $\mathfrak{D}$ such that $\mathfrak{R}_p \Supset \Delta_p$ (and ν also depends on p).

As for boundary points of $\mathfrak{D}$, we can choose a sequence of points Q_p $(p=1,2,\ldots)$ of $\mathfrak{D}$ with the following properties: 1°. $Q_p \in (\mathfrak{R}_p - \Delta_p)$. 2°. Every boundary point of $\mathfrak{D}$ can be expressed by a suitable subsequence of Q_p $(p=1,2,\ldots)$.

Let ε_p $(p=1,2,\ldots)$ be a sequence of positive numbers such that the sum $\sum \varepsilon_p$ is convergent. To each Q_p corresponds one of the functions $f_j(P)$ such that, if we denote it by $\varphi_p(P)$, we have $\varphi_p(Q_p)=a_p$ where $|a_p|>1$ and $|\varphi_p(P)| \leqq 1$ on Δ_p. Let us set $\psi_p(P)=[\varphi_p(P)-a_p]^p$. Let σ_p be the characteristic surface on $\mathfrak{D}$ given by $\varphi_p(P)=a_p$; σ_p passes through Q_p. ψ_p is then a holomorphic function on $\mathfrak{D}$ having zeros of order p on σ_p; it does not have zeros on Δ_p.

By *Theorem* I, we can find a holomorphic function $\chi_p(P)$ on $\mathfrak{D}$ such that we have

$$\left|\frac{1}{\psi_p(P)} - \chi_p(P)\right| < \frac{\varepsilon_p}{|\psi_p(P)|}$$

on Δ_p. Let us set

$$F_2(P) = \prod [\psi_p(P) \cdot \chi_p(P)] \quad (p=1,2,\ldots).$$

This product converges absolutely and uniformly on any domain completely interior to $\mathfrak{D}$ since

$$|\psi_p(P)\,\chi_p(P) - 1| < \varepsilon_p \quad \text{on} \quad \Delta_p.$$

The limit $F_2(P)$ is therefore a holomorphic function on $\mathfrak{D}$ whose zeros are given as the union of the zeros of the factors $(\psi_p \cdot \chi_p)$. Hence $F_2(P)$ has zeros of order at least p on the characteristic surface σ_p which passes through Q_p (p is arbitrary). Hence we cannot continue $F_2(P)$ analytically across any boundary point of $\mathfrak{D}$. $\mathfrak{D}$ is therefore a domain of holomorphy.

C. The Other Problems

33. The Second Cousin Problem. The case considered in the present memoir is the last for which the second Cousin problem holds in its original form and generality. If we introduce points at infinity, it is à priori obvious that holomorphic solutions do not exist in general. If we permit non-transcendental interior points of ramification, then, as we remarked in the preceding memoir, zeros distributed on a characteristic surface are not even locally the set of zeros of a single holomorphic function in general. In the present case, the inverse Hartogs problem having been solved, it is sufficient to treat the problem for holomorph-convex domains.

We dealt with this problem in Memoir III on univalent domains without points at infinity; in a word, we extended the problem itself to the domain of continuous functions to acquire an auxiliary notion with the help of which we expressed our conclusion, namely: if there exists a non-analytic solution, analytic solutions also exist. It is this result which we shall extend.

Let us recall briefly the general notions in the present case. Let $\mathfrak{D}$ be a domain over the space $(x_1, x_2, \ldots, x_n)$ (without points at infinity or interior ramification points) and let P be an arbitrary point of $\mathfrak{D}$. Let $f_1(P), f_2(P)$ be two continuous functions on $\mathfrak{D}$;[50] we shall say that f_1 and f_2 are *equivalent* ($f_1 \sim f_2$) on $\mathfrak{D}$ if there exists a continuous function $\lambda(P)$, nowhere zero, such that we have $f_1(P) = \lambda(P) f_2(P)$ identically on $\mathfrak{D}$. If $f_1 \sim f_2$ on $\mathfrak{D}$ and the set of zeros has no interior point, $\lambda(P)$ is *unique*; otherwise $\lambda(P)$ may not be determined uniquely, which gives rise to strange phenomena. We shall *therefore exclude this case* in what follows (when a continuous function vanishes identically in the neighbourhood of a point).

Suppose given for any point P_0 of $\mathfrak{D}$ a polycylindrical neighbourhood γ around P_0 and a continuous function $f(P)$ defined on γ in such a way that for any pair of contiguous γ, the corresponding functions are equivalent on the intersection (equivalence condition). Find a continuous function $F(P)$ on $\mathfrak{D}$ such that we always have $F(P) \sim f(P)$ on γ. This is the *generalized second Cousin problem.* If the function $F(P)$ exists, we shall say that $F(P)$ has the given zeros $(\mathfrak{z})$ in $\mathfrak{D}$, where $(\mathfrak{z})$ is expressed by the system $\{(f, \gamma)\}$ satisfying the equivalence condition.

To distinguish the Cousin problem itself from the generalized problem, we use the terms *analytic zeros*, *analytic solutions*; sometimes, if our aim is the opposite of this, we use the terms *continuous zeros*, *non-analytic solutions*, ... If we are given analytic zeros $(\mathfrak{z})$ on a domain $\mathfrak{D}$, the two problems arise simultaneously.

[50]) It is always understood that continuous functions are single valued.

Given zeros $(\mathfrak{z})$ on a domain $\mathfrak{D}$, we shall say that $(\mathfrak{z})$ *can be swept out* if to each point P_0 of $\mathfrak{D}$, there exists a polycylindrical neighbourhood γ with centre P_0 and a continuous function $f(P,t)$ defined for $P\in\gamma$, $0\leqq t\leqq 1$ satisfying the following three conditions:

1°. $f(P,0)$ has the zeros $(\mathfrak{z})$, $f(P,1)$ does not vanish anywhere.

2°. $f(P,t)$ does not vanish identically on any portion having real dimension $(2n+1)$.

3°. Let (f_1,γ_1), (f_2,γ_2) be any two couples in the system $\{(f,\gamma)\}$ and let $\delta=\gamma_1\cap\gamma_2$; if δ is not empty, we have $f_1(P,t)\sim f_2(P,t)$ on $(P\in\delta, 0\leqq t\leqq 1)$.

Given zeros $(\mathfrak{z})$ *on a domain* $\mathfrak{D}$, *if there exists a continuous function* $F(P)$ *on* $\mathfrak{D}$ *having the zeros* $(\mathfrak{z})$, *then* $(\mathfrak{z})$ *can, necessarily, be swept out on* $\mathfrak{D}$.

To see this, we have only to consider

$$f(P,t)=(1-t)F(P)+t.$$

If the zeros $(\mathfrak{z})$ can be swept out, we can choose à system $\{(f(P,t),\gamma)\}$ so that we always have $f(P,1)=1$ identically; further, if the $(\mathfrak{z})$ are also analytic, we can choose the system so that, in addition, $f(P,0)$ is always holomorphic.

Relative to this notion, let $\lambda(P)$ be a continuous function *nowhere zero* on a domain $\mathfrak{D}$. We say that $\lambda(P)$ satisfies condition (α) on $\mathfrak{D}$ if there exists a continuous function $\lambda(P,t)$ without zeros on $(\mathfrak{D},[0,1])$ such that $\lambda(P,0)=\lambda(P)$, $\lambda(P,1)=1$ for any P in $\mathfrak{D}$. We have the following lemma: *In order that every determination of* $\log\lambda(P)$ *be single valued on* $\mathfrak{D}$, *it is necessary and sufficient that* $\lambda(P)$ *satisfy condition* (α) *on* $\mathfrak{D}$. As for the proof, since there is nothing to modify, we shall not repeat it.

We are extending the conclusion of Memoir III to the present case. Now, for this, the most favourable method to use is that of *Memoir* II. For this reason, let us extend *Theorem* I of Memoir II.

Let us consider a closed polycylinder C: $|x_i|\leqq r_i$ $(i=1,2,\dots,n)$ *in the space* (x) *and a characteristic variety* Σ *defined in a neighbourhood of* C. *Let* Σ_0 *be the part of* Σ *on* C. *Then* Σ_0 *is holomorph-convex from the outside with respect to the space* (x) *(in other words, with respect to polynomials).*

In fact, let us consider the geometric ideal $(\mathfrak{I})$ with indeterminate domain attached to Σ, which is defined in a certain neighbourhood of C. By *the theorem of H. Cartan*, $(\mathfrak{I})$ has local pseudobases (see Memoir VIII). Consequently, by *Theorem* III *of Memoir* VII, $(\mathfrak{I})$ possesses a pseudobasis on a neighbourhood of C. The above lemma follows from this.

Theorem IV. *Given analytic zeros* $(\mathfrak{z})$ *on a pseudoconvex domain over the space* (x) *(without points at infinity or interior ramification points), if there exists a non-analytic solution, there also exists an analytic solution.*

In fact, the zeros $(\mathfrak{z})$ can be swept out, so that there exists a system $\{(f(P,t),\gamma)\}$ satisfying the conditions indicated relative to $(\mathfrak{z})$; we suppose also that each $f(P,0)$ is holomorphic and that each $f(P,1)$ is the constant 1.

The domain $\mathfrak{D}$ can be expressed as the limit of a sequence of analytic polyhedra Δ_p $(p=1,2,\ldots)$ such that $\Delta_p \Subset \Delta_{p+1}$. Each Δ_p is represented, in the usual sense, in the form

$$|x_i| \leqq r_i, \quad |g_j(P)| \leqq 1, \quad P \in \mathfrak{R} \quad (i=1,\ldots,n; j=1,\ldots,\nu),$$

($g_j(P)$ being holomorphic on $\mathfrak{D}$). Let us consider the configuration of Lemma I relative to this polyhedron Δ_p; we have C, Σ, V in the space $(x_1,\ldots,x_n,y_1,\ldots,y_\nu)$ (No. 24). We have seen that if we are given a holomorphic function $\varphi(P)$ on Δ_p, we can regard it as being defined on V. By the same method, we can regard the zeros $(\mathfrak{z})$ (which can be swept out) as well as the corresponding system $\{(f(P,t),\gamma)\}$ as being given on V; here V is holomorph-convex with respect to the space (x,y) because of the above lemma if it is suitably chosen sufficiently close to Σ. Hence, by what we have seen in Memoir III (No. 8), we obtain a special solution, having the property indicated there. Substituting $y_j=g_j(P)$ $(j=1,2,\ldots,\nu)$ in this solution, we obtain a special solution on Δ_{p-1} $(p>1)$.

Thus, we have a function $F_p(P,t)$ on Δ_p; it is continuous on $[P\in\Delta_p, 0\leqq t\leqq 1]$ and such that $F_p(P,0)$ is a holomorphic function on Δ_p and $F_p(P,1)=1$ identically; furthermore $F_p(P,t)\sim f(P,t)$ on $[P\in(\gamma), 0\leqq t\leqq 1]$. The theorem results from this; the argument being word for word the same as in Memoir III (No. 9), we omit it.

Applying the same method to the generalized problem, we obtain the following criterion more easily than the above.

In order that the generalized second Cousin problem be solvable, it is necessary and sufficient that the given zeros be such that they can be swept out.

34. The Problems (C_1), (C_2), (E); Incomplete Solutions. We shall follow the plan of Memoir VII and apply *the method of Memoir* I to problems (C_1), (C_2) and (E).

Let us begin with *the generalized problem* (C_1) as presented in Chapter I.

(α) Suppose given a functional equation

$$(1) \qquad \Phi_i = \sum A_j F_{ij} \quad (i=1,2,\ldots,p; j=1,2,\ldots,q)$$

in the neighbourhood of an analytic polyhedron Δ, where the Φ_i, F_{ij} stand for given holomorphic functions and the A_j are unknown functions. If this equation has a local solution everywhere in a neighbourhood of Δ, it has a (global) solution on a neighbourhood of Δ.

Suppose, in fact, that Δ is given, in the usual sense, in the form $|x_i|\leqq r_i$, $|g_j(P)|\leqq 1$, $P\in\mathfrak{R}$ $(i=1,2,\ldots,n; j=1,2,\ldots,\nu)$. In the space (x,y), consider the closed polycylinder C, $|x_i|\leqq r_i$, $|y_j|\leqq 1$ and the characteristic variety Σ given by $y_j=f_j(P)$, $P\in\mathfrak{R}$. With this configuration, we can, by Lemma I, find holomorphic functions $G_{kl}(x,y)$, $\Psi_k(x,y)$ $(k=1,2,\ldots,p; l=1,2,\ldots,q)$ in a neighbourhood of C such that we have $G_{kl}=F_{kl}$, $\Psi_k=\Phi_k$ on Σ. Let $(\mathfrak{I}_0)$ be the geometric ideal (with indeterminate domains) attached to Σ and defined in a neighbourhood of C, and let $(H_1,H_2,\ldots,H_\lambda)$ be a pseudobasis of $(\mathfrak{I}_0)$ on a neighbourhood of C.

Consider the functional equation

$$(2) \qquad \Psi_i = \sum A_j G_{ij} + \sum B_{ik} H_k \quad (i=1,\dots,p; j=1,\dots,q; k=1,\dots,\lambda)$$

in a neighbourhood of C; the A_j, B_{ik} are the unknown functions. Since the equation (1) has a local solution at every point of a neighbourhood of Δ, the same is true of the equation (2) in a neighbourhood of C. Hence, by what we have seen in No. 4, we can find holomorphic functions A_j, B_{ik} in a neighbourhood of C satisfying the equation (2). Substituting $y_l = g_l(P)$ $(l=1,2,\dots,\nu)$, the desired result follows from this.

Let us consider *problem* (C_2):

(β) Let Δ be an analytic polyhedron over the space (x), let $F_1(P)$, $F_2(P),\dots,F_\mu(P)$ be a system of functions holomorphic on a neighbourhood of Δ. Suppose that for any point P_0 in a neighbourhood of Δ, we are given a polycylindrical neighbourhood γ with centre P_0 and a holomorphic function $\varphi(P)$ on γ in such a way that for any pair (γ_1,γ_2) of these polycylindrical neighbourhoods, the corresponding functions φ_1,φ_2 satisfy $\varphi_1 \equiv \varphi_2 \bmod (F)$ at every point of the intersection $\gamma_1 \cap \gamma_2$. We can then find a holomorphic function $\Phi(P)$ such that $\Phi(P) \equiv \varphi(P)$ at any point of γ, for an arbitrary γ.

In fact, with the same geometric configuration as above, let $G_i(x,y)$ $(i=1,2,\dots,\mu)$ be functions holomorphic in a neighbourhood of C such that $G_i = F_i$ on Σ. We consider a problem (C_2) in a neighbourhood of C in the following way: To each point (x^0,y^0) in a neighbourhood of C we attach a function $\psi(x,y)$; if (x^0,y^0) is a point of Σ, we take for ψ a function which takes the value φ on Σ; if not, we take ψ at random. Under these conditions, find a function $\Psi(x,y)$, holomorphic in a neighbourhood of C such that $\Psi \equiv \psi \bmod (G,H)$ at every point (x^0,y^0), (H) being the same pseudobasis of $(\mathfrak{I}_0)$ as above. This is obviously a problem (C_2), hence $\Psi(x,y)$ exists by Theorem II of VII. If we set $\Phi(P) = \Psi(x, g(P))$, $\Phi(P)$ is the required function.

Next, let $\mathfrak{D}$ be a domain over the space (x). We consider a pair (f,δ) where δ is a subdomain of $\mathfrak{D}$ (not necessarily connected) and f is a holomorphic function defined on δ. With these (f,δ), we can define, exactly as in Memoir VII, *(holomorphic) ideals (with indeterminate domains) on* $\mathfrak{D}$, (finite) *pseudobases*, ... ; we shall not repeat this.

Concerning *problem* (E), we have

Theorem V. *Suppose given, on a pseudoconvex domain* $\mathfrak{D}$ *over the space* (x) *without points at infinity or interior points of ramification, a holomorphic ideal* $(\mathfrak{I})$ *with indeterminate domains which has a finite pseudobasis at every point of* $\mathfrak{D}$. *We can find a finite pseudobasis of* $(\mathfrak{I})$ *on any given domain* $\mathfrak{D}_0$ *such that* $\mathfrak{D}_0 \Subset \mathfrak{D}$.

In fact, let us consider the usual configuration of Lemma I with the same notation as in the propositions (α), (β); we suppose that $\mathfrak{D}_0 = \Delta$ (possible since $\mathfrak{D}$ is pseudoconvex).

Let V be a neighbourhood of C, sufficiently close to C. Let (φ,δ') be an ordered couple, where δ' is a univalent domain (connected or not) in the space (x,y) and φ is a holomorphic function defined on δ'. We consider the set (J) of the (φ, δ') such that one has $\delta' \subseteq V$ and, if $[x, g(P)]$ is an arbitrary point of Σ in δ', the function $\varphi[x, g(P)] = f(P)$ belongs locally to $(\mathfrak{I})$. Clearly (J) is an ideal without lacunary points in V.

Let P_0 be an arbitrary point in the neighbourhood of Δ and $[f_1(P), f_2(P), \ldots, f_\mu(P)]$ a pseudobasis of $(\mathfrak{I})$ at P_0. Let (x^0) be the coordinates of P_0 and let M_0 be the point on Σ having the coordinates $[x^0, g(P^0)]$. Let $\varphi_i(x, y)$ $(i = 1, 2, \ldots, \mu)$ be a holomorphic function at M_0 such that $\varphi_i = f_i$ on Σ. Then $(\varphi_1, \ldots, \varphi_\mu, H_1, \ldots, H_\lambda)$ clearly is a pseudobasis of (J) at M_0 ((H) is the pseudobasis indicated above). At any point of V which does not belong to Σ, (J) has (1) as pseudobasis. Hence, by Theorem III of VII, (J) has a (finite) pseudobasis on a neighbourhood of C. The theorem follows from this on substituting $y_j = g_j(P)$ $(j = 1, 2, \ldots, \nu)$.

35. Complete Solution of the Congruence Problems. We are concerned with solving these problems on pseudoconvex domains themselves. This is possible for *problems* $(C_1), (C_2)$, as we shall now show.

Theorem VI. *Suppose given on a pseudoconvex domain $\mathfrak{D}$ over the space (x) without points at infinity or interior points of ramification, a functional equation*

$$\Phi_i = \sum A_j F_{ij} \quad (i = 1, 2, \ldots, \mu; j = 1, 2, \ldots, \mu'), \tag{1}$$

where Φ_i, F_{ij} represent given holomorphic functions, and the A_j stand for unknown functions. If this equation has a local solution everywhere on $\mathfrak{D}$, it has a solution on $\mathfrak{D}$.

In fact, $\mathfrak{D}$ can be expressed as the limit of a sequence of analytic polyhedra Δ_p $(p = 1, 2, \ldots)$ such that $\Delta_p \Subset \Delta_{p+1}$, each Δ_p being defined by functions holomorphic on $\mathfrak{D}$. By the intermediate result (α), there exists a solution $(A_{p1}, A_{p2}, \ldots, A_{p\mu'})$ on a neighbourhood of Δ_p. We shall construct another solution (A'_p) from this as follows:

For (A'_1), we take $A'_{1i} = A_{1i}$ $(i = 1, 2, \ldots, \mu')$.

To construct (A'_2), set $B_i = A_{2i} - A'_{1i}$ $(i = 1, 2, \ldots, \mu')$. The system (B) satisfies the linear functional equation

$$B_1 F_{i1} + B_2 F_{i2} + \ldots + B_{\mu'} F_{i\mu'} = 0 \quad (i = 1, 2, \ldots, \mu) \tag{2}$$

on a neighbourhood of Δ_1.

Let $(\mathfrak{I}_1)$ be the set (B_1, δ) such that, to each (B_1, δ) there corresponds a solution of (2) on δ of the form $(B_1, B_2, \ldots, B_{\mu'})$. $(\mathfrak{I}_1)$ is an ideal and, by Theorem IV of VII has a pseudobasis at every point of a neighbourhood of Δ_1. Consequently, by Theorem V above, $(\mathfrak{I}_1)$ has a pseudobasis on a neighbourhood of Δ_1. It follows immediately from this, since equation (2) is arbitrary, that there exists a formula for the solutions of (2) on a neighbourhood of Δ_1, which we can put in the form $B_i = \sum C_j \Pi_{ij}$ $(i = 1, \ldots, \mu'; j = 1, \ldots, \lambda)$.

Consider again $B_i = A_{2i} - A'_{1i}$ $(i = 1, 2, \ldots, \mu')$. Because of the result (α), we can find a system (C) of holomorphic functions in a neighbourhood of Δ_1 corresponding to the system of functions (B) such that we have $B_i = \sum C_j \Pi_{ij}$ identically. Let ε_p $(p = 1, 2, \ldots)$ be a sequence of positive numbers such that the sum

$\sum \varepsilon_p$ is convergent. By Theorem I, we can find holomorphic functions $C'_j(P)$ on $\mathfrak{D}$ such that we have

$$|B_i - \sum C'_j \Pi_{ij}| < \varepsilon_1$$

on Δ_1. Let us set

$$A'_{2i} = A_{2i} - (\sum C'_j \Pi_{ij});$$

the A'_{2i} are holomorphic on a neighbourhood of Δ_2, the system (A'_2) satisfies equation (1) and $|A'_{2i} - A'_{1i}| < \varepsilon_1$ on Δ_1.

We construct (A'_3) similarly, and so on. The sequence of functions A'_{pi} ($p = 1, 2, \ldots$) converges uniformly on any domain completely interior to $\mathfrak{D}$. Let A'_i be the limit. The A'_i are holomorphic and satisfy the equation (1).

Theorem VII. *Let $\mathfrak{D}$ be a pseudoconvex domain over the space (x) without points at infinity or interior ramification points. Let $F_1(P), F_2(P), \ldots, F_\mu(P)$ be a system of functions holomorphic on $\mathfrak{D}$. Suppose that we are given, for every point P_0 of $\mathfrak{D}$, a polycylindrical neighbourhood γ with centre P_0 and a holomorphic function $\varphi(P)$ defined on γ such that for any pair (γ_1, γ_2) of these polycylindrical neighbourhoods, the corresponding functions φ_1, φ_2 satisfy $\varphi_1 \equiv \varphi_2 \bmod (F)$ at every point of the intersection $\gamma_1 \cap \gamma_2$. We can then find a holomorphic function $\Phi(P)$ on $\mathfrak{D}$ such that we have $\Phi(P) \equiv \varphi(P) \bmod (F)$ for every γ.*

In fact, continuing to regard $\mathfrak{D}$ as the limit of the sequence Δ_p ($p = 1, 2, \ldots$), we have, by the intermediate result (β), a solution $\Phi_p(P)$ on a certain neighbourhood of Δ_p (because, by Theorem VI, the following two terms mean the same thing: $\Phi \equiv \varphi$ at every point of γ; and: $\Phi \equiv \varphi$ on γ). Let ε_p ($p = 1, 2, \ldots$) be a sequence of positive numbers such that the sum $\sum \varepsilon_p$ is convergent. We shall, recursively, construct a new sequence of solutions Φ'_p ($p = 1, 2, \ldots$). First, let $\Phi'_1(P) = \Phi_1(P)$. For $p = 2$, let $\Psi_1(P) = \Phi_2(P) - \Phi'_1(P)$. Ψ_1 is a holomorphic function on a neighbourhood of Δ_1 such that $\Psi_1 = A_{11} F_1 + A_{12} F_2 + \ldots + A_{1\mu} F_\mu$, the A_{1i} ($i = 1, 2, \ldots, \mu$) being holomorphic on a neighbourhood of Δ_1 (Theorem VI). By Theorem I, we can find holomorphic functions $A'_{1i}(P)$ ($i = 1, 2, \ldots, \mu$) on $\mathfrak{D}$ such that, on Δ_1, we have

$$|\sum A'_{1i} F_i - \sum A_{1i} F_i| < \varepsilon_1.$$

We set

$$\Phi'_2(P) = \Phi_2 - \sum A'_{1i} F_i;$$

Φ'_2 is then also a solution on a neighbourhood of Δ_2, and we have $|\Phi'_2 - \Phi_2| < \varepsilon_1$ on Δ_1.

Similarly for $p > 2$; we obtain a sequence Φ'_p ($p = 1, 2, \ldots$) such that each $\Phi'_p(P)$ is a solution on a neighbourhood of Δ_p, and $|\Phi'_{p+1} - \Phi'_p| < \varepsilon_p$ on Δ_p. This sequence converges uniformly on any domain completely interior to $\mathfrak{D}$. The limit $\Phi(P)$ is therefore a holomorphic function on $\mathfrak{D}$ and it satisfies $\Phi(p) \equiv \varphi(P) \bmod (F)$ identically on each γ.

36. Problem (E) and a Counterexample. As for problem (E), it is not necessarily solvable on pseudoconvex domains (over the space (x), without points at infinity or interior ramification points); an example effectively showing this can already be found among *geometric ideals with indeterminate domains.*

Example. We take a positive integer ν such that $\nu \geqq 3$ and consider the following four functions in the space (x_1, x_2, y_1, y_2):

$$F_1 = y_1^\nu - x_1^{\nu-1}, \quad F_2 = y_2^\nu - x_2^\nu x_1, \quad F_3 = y_1 y_2 - x_1 x_2, \quad F_4 = x_1^{\nu-2} + x_2^\nu;$$

we denote the characteristic variety defined by $F_1 = F_2 = F_3 = F_4 = 0$ by Σ and the *geometric ideal* with indeterminate domain attached to Σ by $(\mathfrak{J}_0)$; this is defined on the space (x_1, x_2, y_1, y_2). We shall prove that *the number of elements in any pseudobasis of the ideal* $(\mathfrak{J}_0)$ *at the origin is* $\geqq \nu - 1$.

Suppose that the ideal $(\mathfrak{J}_0)$ has a pseudobasis $[\Phi_1(x, y), \Phi_2(x, y), \ldots, \Phi_\rho(x, y)]$ at the origin, where $\rho = \nu - 2$. Consider the RIEMANN surface (of complex dimension 2) of the function $t = x_1^{1/\nu}$ over the space (x_1, x_2); we denote it by $\mathfrak{R}$. Let P be an arbitrary point of $\mathfrak{R}$ and (x_1, x_2) the coordinates of P, and let t be the value of $x_1^{1/\nu}$ at P. Let S be the characteristic variety defined by $F_1 = F_2 = F_3 = 0$, and let M be an arbitrary point of S; what are the coordinates of M?

Since $F_1 = 0$, we have, first, $y_1 = x_1^{(\nu-1)/\nu} = t^{\nu-1}$. From $F_2 = 0$, we next obtain $y_2 = x_2 x_1^{1/\nu} = \varepsilon x_2 t$, where ε is a number such that $\varepsilon^\nu = 1$, but is otherwise arbitrary. Now, since $F_3 = 0$, we must have $x_1 x_2 = y_1 y_2 = \varepsilon x_2 t^\nu = \varepsilon x_1 x_2$, that is, $\varepsilon = 1$. Hence M is given by $(x_1, x_2, t^{\nu-1}, x_2 t)$. Let us make the point $P(x_1, x_2, t)$ of $\mathfrak{R}$ and the above point M correspond to each other; this correspondenc is biunique.

Let (J_0) be the image on $\mathfrak{R}$ of the restriction of the ideal $(\mathfrak{J}_0)$ to S; more precisely, let $(\mathfrak{J}_0) = \{(f, \delta)\}$, let δ' be the image on $\mathfrak{R}$ of the intersection of δ with S, let $f(x_1, x_2, t^{\nu-1}, x_2 t) = \varphi(t, x_2) = \varphi(P)$, and let $(J_0) = \{(\varphi, \delta')\}$. What is the nature of the set (J_0)?

Let δ' be a subdomain (not necessarily connected) of $\mathfrak{R}$, and let $\varphi(P)$ be a holomorphic function defined on δ', which has the property (H)[51] with respect to S, but is otherwise arbitrary. Let us consider the set of ordered pairs (φ, δ') and denote it by $(\mathfrak{D}_S)$. Suppose that a subset $(\mathfrak{J})$ of $(\mathfrak{D}_S)$ satisfies the following two conditions: 1°. If $(\varphi, \delta') \in (\mathfrak{J})$, $(\alpha, \delta'') \in (\mathfrak{D}_S)$, we necessarily have $\alpha \varphi \in (\mathfrak{J})$ on $\delta' \cap \delta''$. 2°. If $(\varphi_1, \delta_1) \in (\mathfrak{J})$, $(\varphi_2, \delta_2) \in (\mathfrak{J})$, we necessarily have $\varphi_1 + \varphi_2 \in (\mathfrak{J})$ on $\delta_1 \cap \delta_2$. We shall then call $(\mathfrak{J})$ an *ideal in* $(\mathfrak{D}_S)$.

The set (J_0) introduced above is then an ideal in $(\mathfrak{D}_S)$. Moreover, since $F_4 = x_1^{\nu-2} + x_2^\nu = 0$ does not have multiple factors, (J_0) is a geometric ideal attached to $F_4 = 0$ and defined on the whole RIEMANN surface $\mathfrak{R}$ (the definition of these terms is the same as before).

The point of $\mathfrak{R}$ over the origin is unique; we denote it by P_0. A pseudobasis of (J_0) at P_0 is given by

$$\Phi_i(x_1, x_2, t^{\nu-1}, x_2 t) = \varphi_i(t, x_2) = \varphi_i(P) \qquad (i = 1, 2, \ldots, \rho).$$

Let us consider the following holomorphic functions on $\mathfrak{R}$:

$$f_i(t, x_2) = F_4 t^{i-1} = (x_1^{\nu-2} + x_2^\nu) t^{i-1} \qquad (i = 1, 2, \ldots, \rho + 1).$$

---

[51]) "φ has property (H) with respect to S" means that "φ is locally the image on $\mathfrak{R}$ of the restriction to S of a function holomorphic in the space (x_1, x_2, y_1, y_2)".

Each f_i has property (H) with respect to S; in fact, we have on S

$$t^{\alpha}(x_1^{\nu-2}+x_2^{\nu})=y_1^{\nu-\alpha}x_1^{\alpha-1}+y_2^{\alpha}x_2^{\nu-\alpha} \qquad (\alpha=1,2,\ldots,\rho).$$

Moreover, the function $f_i(t,x_2)$ vanish identically on $F_4=0$; they therefore belong to (J_0) on $\mathfrak{R}$. Consequently, we have, in the neighbourhood of P_0,

$$f_i=c_{i1}\varphi_1+c_{i2}\varphi_2+\ldots+c_{i\rho}\varphi_\rho \qquad (i=1,2,\ldots,\rho+1),$$

where the c_{ij} $(j=1,2,\ldots,\rho)$ represent holomorphic functions of (t,x_2) in the neighbourhood of $(0,0)$ which have property (H) with respect to S.

Consider a holomorphic function $f(P)$ in the neighbourhood of P_0 having property (H) with respect to S; say $f(P)=f(t,x_2)$. There then exists a holomorphic function $F(x_1,x_2,y_1,y_2)$ in the neighbourhood of the origin of the space (x_1,x_2,y_1,y_2) such that we have, identically, $F(x_1,x_2,t^{\nu-1},x_2t)=f(t,x_2)$. If we expand $f(t,x_2)$ in a TAYLOR series about the origin and set $x_2=0$, we therefore have, necessarily,

$$f(t,0)=a_0+a_{\nu-1}t^{\nu-1}+a_\nu t^\nu+\ldots,$$

$(a_0,a_{\nu-1},a_\nu,\ldots$ being constants).

The functions $\varphi_i(t,x_2)$ must vanish on $F_4=0$. Since $F_4=0$ does not have multiple factors, they must be of the form

$$\varphi_i(t,x_2)=(x_1^{\nu-2}+x_2^{\nu})\,h_i(t,x_2),$$

the $h_i(t,x_2)$ being holomorphic. We therefore have, in the neighbourhood of $t=0$,

$$\begin{aligned}\varphi_i(t,0)&=x_1^{\nu-2}(A_{i1}+A_{i2}t+A_{i3}t^2+\ldots+A_{i,\rho+1}t^\rho)+Bx_1^{\nu-2}t^{\rho+1}+\ldots\\&=\psi_i(t)+Bx_1^{\nu-2}t^{\rho+1}+\ldots\end{aligned}$$

(where $A_{ij}(j=1,2,\ldots,\rho+1)$ and B are constants).

Let us set

$$c_{ij}(t,0)=\gamma_{ij}+\delta_{ij,\nu-1}t^{\nu-1}+\delta_{ij\nu}t^\nu+\ldots \qquad (i=1,\ldots,\rho+1;j=1,\ldots,\rho).$$

Since $\nu-1=\rho+1$ and

$$f_i(t,0)=x_1^{\nu-2}t^{i-1}=c_{i1}(t,0)\,\varphi_1(t,0)+\ldots+c_{i\rho}(t,0)\,\varphi_\rho(t,0)$$

$(i=1,2,\ldots,\rho+1)$, we necessarily have

$$x_1^{\nu-2}t^{i-1}=\gamma_{i1}\psi_1(t)+\gamma_{i2}\psi_2(t)+\ldots+\gamma_{i\rho}\psi_\rho(t).$$

Comparing coefficients of $x_1^{\nu-2}t^{j-1}$, we have

$$\gamma_{i1}A_{1j}+\gamma_{i2}A_{2j}+\ldots+\gamma_{i\rho}A_{\rho j}=\varepsilon_{ij} \qquad (i=1,\ldots,\rho+1;j=1,\ldots,\rho+1),$$

where

$$\varepsilon_{ij}=1 \quad \text{if } i=j, \quad \varepsilon_{ij}=0 \quad \text{if } i\neq j.$$

Introducing independent complex variables $z_1,z_2,\ldots,z_\rho$, we consider

$$L_k(z)=A_{k1}z_1+A_{k2}z_2+\ldots+A_{k\rho}z_\rho \qquad (k=1,2,\ldots,\rho);$$

we then have

$$\gamma_{\rho+1,1}L_1+\gamma_{\rho+1,2}L_2+\ldots+\gamma_{\rho+1,\rho}L_\rho=0$$

identically, where one at least of the $\gamma_{\rho+1,k}$ $(k=1,2,\ldots,\rho)$ is not zero because $\gamma_{\rho+1,1}A_{1,\rho+1}+\ldots+\gamma_{\rho+1,\rho}A_{\rho,\rho+1}=1$; thus $L_1, L_2, \ldots, L_\rho$ are linearly dependent. On the other hand, since

$$\gamma_{i1}L_1+\gamma_{i2}L_2+\ldots+\gamma_{i\rho}L_\rho=z_i \quad (i=1,2,\ldots,\rho),$$

$L_1, L_2, \ldots, L_\rho$ are necessarily linearly independent. These conclusions are contradictory. Hence, the number of elements in any pseudobasis of $(\mathfrak{I}_0)$ at the origin must be $\geqq \nu-1$.

Let us now consider, in the space $(x_1, x_2, y_1, y_2, y_3)$ the characteristic variety T_ν given by

$$y_3=b_\nu, \quad F_i(x_1,x_2,y_1,y_2)=0 \quad (i=1,2,3,4),$$

where ν is a positive integer with $\nu\geqq 3$, b_ν is a constant, and $F_i(x_1,x_2,y_1,y_2)$ stand for the functions we have just considered. For the geometric ideal attached to T_ν and defined in a domain containing the point $(0,\ldots,0,b_\nu)$, the number of elements in a local pseudobasis at this point is always $\geqq \nu-1$.

Let us consider the polycylinder C, $|x_i|<1$, $|y_j|<1$ $(i=1,2; j=1,2,3)$ in the space (x_1,x_2,y_1,y_2,y_3), let us take the sequence $b_\nu=1-2^{-\nu}$ $(\nu=3,4,\ldots)$ and consider the characteristic variety $T=T_3+T_4+\ldots$ in C. Let $(\mathfrak{I}_1)$ be the *geometric ideal* with indeterminate domains attached to T. Because of what we have just seen, it is clear that $(\mathfrak{I}_1)$ *does not posses any finite pseudobasis on C.*

Commentaire de H. Cartan

Ce Mémoire IX représente le couronnement de l'œuvre de Kiyoshi Oka. Il rassemble les solutions des problèmes étudiés dans tous les Mémoires antérieurs (sauf le Mémoire VIII), et ceci dans un cadre plus général: en effet Oka traite ici le cas des “domaines étalés” dans $\mathbb{C}^n$, tandis que dans les Mémoires I à VII il s'était borné au cas des domaines univalents. Cela l'oblige à développer de nouvelles méthodes. En outre la théorie des domaines pseudo-convexes est traitée ici dans le cas de n variables complexes, et non plus seulement dans le cas de deux variables comme dans le Mémoire VI. Les nouvelles méthodes n'utilisent plus l'intégrale de Weil, mais seulement l'intégrale simple de Cauchy.

Il y a trois chapitres. Le chapitre I donne des précisions quantitatives au sujet des résultats des Mémoires VII et VIII. Il s'agit de majorations que Oka obtient en reprenant toutes les démonstrations. On pourrait obtenir toutes ces majorations beaucoup plus simplement; en effet, dans chacun des cas étudiés, il est question d'une application linéaire continue f d'un certain espace de Banach E dans un autre espace de Banach E', et on a prouvé chaque fois que f est surjective. Or un théorème classique de Banach permet alors d'affirmer que l'application f est *ouverte*, ce qui fournit aussitôt les majorations cherchées. Ces majorations seront, bien entendu, utilisées dans la suite de ce Mémoire.

Le chapitre II est consacré aux “domaines étalés”, à la notion de pseudo-convexité pour ces domaines, et à l'existence, dans chaque domaine pseudo-convexe, de fonctions plurisous-harmoniques ayant des propriétés convenables.

Comme pour le Mémoire VI, nous adoptons dans ce commentaire la terminologie de "fonction plurisous-harmonique" (en abrégé, fonction *psh*) au lieu de "fonction pseudo-convexe" (terminologie utilisée par Oka), et nous réservons le qualificatif de "pseudo-convexe" pour les domaines étalés dans $\mathbb{C}^n$.

Il faut d'abord préciser les notions et définitions relatives aux "domaines étalés" dans $\mathbb{C}^n$ (notions et définitions introduites dans le mémoire de Cartan-Thullen de 1932). Dans la terminologie d'aujourd'hui, un domaine étalé dans $\mathbb{C}^n$ est un espace topologique séparé et connesce D muni d'une application continue $p: D \to \mathbb{C}^n$ qui est *étale* (i.e.: pour tout point $x \in D$ il existe un ouvert U de D contenant x et un ouvert V de $\mathbb{C}^n$ contenant $p(x)$ tels que f induise un homéomorphisme de U sur V). Alors D se trouve muni d'une structure de variété analytique de dimension n. Si on a deux domaines étalés $p: D \to \mathbb{C}^n$ et $p_1: D_1 \to \mathbb{C}^n$, et si on s'est donné des points-base $x_0 \in D$ et $x_1 \in D_1$ tels que $p(x_0) = p_1(x_1)$, on dit que D_1 *est étalé sur* D (et on écrit $D_1 \subset D$) s'il existe une application continue $\varphi: D_1 \to D$ telle que $\varphi(x_1) = x_0$ et $p_1 = p \circ \varphi$; une telle φ est alors unique, et c'est une application étale de D_1 dans D. On définit alors facilement l' "intersection" de deux domaines étalés à point-base.

Rappelons que (D, p) est un *domaine d'holomorphie* s'il jouit de la propriété suivante: chaque fois que l'on a un domaine étalé $p_1: D_1 \to \mathbb{C}^n$, deux points $x_0 \in D$ et $x_1 \in D_1$ tels que $p(x_0) = p_1(x_1)$, une fonction f holomorphe dans D et une fonction f_1 holomorphe dans D_1 telles que $f \circ p^{-1}$ et $f_1 \circ p_1^{-1}$ coïncident au voisinage du point $p(x_0) \in \mathbb{C}^n$, alors D_1 est étalé dans D. C'est cette caractérisation des domaines d'holomorphie qui permet de prouver pour les domaines d'holomorphie ce que nous appellerons la "condition de Hartogs" (cf. notre commentaire du Mémoire VI):

$H(D)$: Soit $\lambda: K \times B \to \mathbb{C}^n$ une application holomorphe étale d'un voisinage de $K \times B$ dans $\mathbb{C}^n$ (où K désigne le disque-unité compact $|z| \leqq 1$ et B une boule ouverte de centre b_0 dans $\mathbb{C}^{n-1}$). Si la restriction de λ à un voisinage U de la réunion $(\partial K \times B) \cup (K \times \{b_0\})$ se factorise en $U \to D \xrightarrow{p} \mathbb{C}^n$, alors cette factorisation se prolonge en $K \times B \to D \xrightarrow{p} \mathbb{C}^n$.

Cette condition $H(D)$ est vérifiée par tout domaine d'holomorphie $p: D \to \mathbb{C}^n$. L'intersection de deux domaines étalés qui y satisfont y satisfait aussi. Un des buts du présent Mémoire est de prouver que, réciproquement, tout domaine étalé qui satisfait à $H(D)$ est un domaine d'holomorphie.

Mais on doit d'abord prouver que la condition $H(D)$ a un caractère *local* vis-à-vis de la *frontière* de D. Pour cela, il faut d'abord définir avec précision ce qu'on entend par point-frontière de D, et par voisinage d'un tel point-frontière. Soit donné $y \in \mathbb{C}^n$; on va définir ce que c'est qu'un point-frontière $\bar{x}$ de (D, p) au-dessus de y: pour chaque voisinage ouvert V de y dans $\mathbb{C}^n$, considérons l'ensemble des composantes connexes de $p^{-1}(V)$ (qui sont des ouverts de X), et soit $C(V)$ le sous-ensemble des composantes connexes U de $p^{-1}(V)$ telles que $y \notin p(U)$; si V' est un voisinage ouvert de y tel que $V' \subset V$, tout ouvert de X appartenant à $C(V')$ est contenu dans un (unique) ouvert appartenant à $C(V)$. Un *point-frontière* $\bar{x}$ au-dessus de y est défini par le choix, pour tout V, d'un ouvert $U_V \in C(V)$ de façon que si $V' \subset V$, on ait $U_{V'} \subset U_V$. Ces ouverts U_V forment un système fondamental de "voisinages" du point-frontière $\bar{x}$ (ce sont des voisinages formés uniquement de points de D). On écrit $\bar{p}(\bar{x}) = y$, ce qui définit l'application $\bar{p}$ de la frontière de D dans $\mathbb{C}^n$. Bien entendu, les U_V sont étalés dans $\mathbb{C}^n$ (par la restriction à U_V de l'application p: $D \to \mathbb{C}^n$). Cela dit, la *condition locale de Hartogs* pour le domaine étalé D s'exprime comme suit: pour tout point-frontière $\bar{x}$ de D, les domaines étalés U_V ("voisinages" de $\bar{x}$) satisfont à la condition $H(U_V)$ dès que V est un polycylindre de centre $\bar{p}(\bar{x})$ et de rayon assez petit. Les D qui satisfont à cette condition locale sont appelés par Oka "pseudo-convexes (C)". Il reste alors à prouver que cette condition locale (qui est une conséquence de la condition globale $H(D)$) est en réalité équivalente à $H(D)$. Pour prouver cette équivalence, Oka introduit deux autres conditions (A) et (B) pour un domaine étalé; et il prouve que la pseudo-convexité (C) entraîne (A), que (A) entraîne (B), et que (B) entraîne la condition de Hartogs globale. Désormais, on appellera "pseudo-convexe" tout domaine étalé qui satisfait à l'une de ces conditions équivalentes.

L'un des principaux problèmes qui restent à résoudre est le suivant: étant donné un D pseudo-convexe, montrer qu'il existe dans D une fonction *psh* φ qui jouit de la "propriété (P_1)" (i.e.: φ est localement la borne supérieure d'un nombre fini de fonctions de classe C^2, strictement *psh*, et dont la différentielle ne s'annule pas) et de la "propriété (α)" (i.e.: pour tout $\alpha \in \mathbb{R}$, l'ensemble D_α des $x \in D$ tels que $\varphi(x) < \alpha$ est relativement compact dans D). Le lemme I du n° 22 affirme qu'il en est bien ainsi. Mais pour le prouver il aura fallu surmonter des difficultés qui ne se présentaient pas dans le cas des domaines univalents. Certes, on utilise, comme dans le Mémoire VI, le fait que $-\log d_D(x)$ est une fonction continue *psh* dans D (en désignant par $d_D(x)$ la distance du point $x \in D$ à la frontière de D; on peut prendre cette distance dans des sens divers). Mais si D n'est pas univalent l'ensemble des $x \in D$ tels que $d_D(x) > r$ (avec $r > 0$) n'est pas relativement compact en général, et ceci nécessite la résolution d'un problème auxiliaire (fin du n° 19): si D_0 est un ouvert d'un domaine étalé D dont tous les points sont à une distance de la frontière de D au moins égale à r ($r > 0$), on construit une fonction λ, continue et *psh dans* D_0, telle que, pour tout $\alpha \in \mathbb{R}$, l'ensemble des $x \in D_0$ tels que $\lambda(x) < \alpha$ soit relativement compact *dans* D.

La démonstration du lemme I achève le chapitre II. Au chapitre III, on prouve les théorèmes fondamentaux du Mémoire, et notamment le théorème III qui entraîne l'équivalence des deux assertions: "D est pseudo-convexe" et "D est un domaine d'holomorphie". (On sait déjà que la seconde entraîne la première.) Pour y parvenir la route sera longue et sinueuse. Oka fait intervenir (ce qui est bien naturel) la notion de "domaine D holomorphiquement convexe" (c'est une notion valable pour toute variété analytique complexe), notion introduite par Cartan-Thullen qui ont prouvé en 1932 que tout domaine holomorphiquement convexe est un domaine d'holomorphie; ils ont aussi prouvé la réciproque, mais seulement dans le cas où le domaine étalé p: $D \to \mathbb{C}^n$ n'a qu'un nombre fini de "feuillets" (i.e.: le cardinal de $p^{-1}(y)$ est borné quand y parcour $\mathbb{C}^n$). Dans ce cas, en effet, l'ensemble des $x \in D$ tels que $d_D(x) > r$ (avec $r > 0$) est relativement compact. Ici Oka prouve la réciproque sans aucune restriction. Il lui suffit, pour cela, de démontrer son théorème III: tout domaine pseudo-convexe est holomorphiquement convexe.

Voici quelques étapes de cette démonstration. On sait déjà que si D est pseudo-convexe, il existe dans D une fonction *psh* φ jouissant des propriétés

(P_1) et (α); on prouve alors (fin du n° 31) que si D est en outre holomorphiquement convexe, tout D_α (défini par $\varphi(x)<\alpha$) est holomorphiquement convexe, et même holomorphiquement convexe relativement à D (c'est-à-dire en utilisant seulement des fonctions holomorphes dans D). Il s'ensuit que toute fonction holomorphe dans D_α est limite, uniformément sur tout compact, de fonctions holomorphes dans D (il s'agit là d'un théorème connu, que OKA rappelle sous l'énoncé du théorème I (n° 25)). La suite des raisonnements utilise le fait que le premier problème de COUSIN est résoluble dans tout D holomorphiquement convexe (ce que OKA prouve en utilisant les polyèdres analytiques contenus dans D puis en passant à la limite: c'est l'objet du théorème II). Ensuite OKA prouve ce que nous avons appelé un "lemme de recollement" dans notre commentaire du Mémoire VI: si D est un domaine étalé dans $\mathbb{C}^n$ (coordonnées $x_1,\ldots,x_n$), et si D_1 (défini comme l'ensemble des $x\in D$ dont l'image dans $\mathbb{C}^n$ satisfait à $\mathrm{Re}(x_1)>-1$) et D_2 (défini comme l'ensemble des points de D dont l'image dans $\mathbb{C}^n$ satisfait à $\mathrm{Re}(x_1)<+1$) sont holomorphiquement convexes, alors D est holomorphiquement convexe. C'est là l'énoncé (β) du n° 31.

La preuve de ce lemme de recollement est très technique. On observe d'abord que D_1 et D_2 sont des domaines d'holomorphie (plus exactement, leurs composantes connexes sont des domaines d'holomorphie), donc ils sont pseudo-convexes. Or la pseudo-convexité a un caractère local à la frontière; il s'ensuit que D lui-même est pseudo-convexe, ce qui permet (grâce au lemme I du chapitre II) de construire dans D une fonction $psh\,\varphi$ ayant les propriétés (P_1) et (α). On définit D_α par $\varphi(x)<\alpha$. Soient $D'_\alpha=D_\alpha\cap D_1$, $D''_\alpha=D_\alpha\cap D_2$. Comme D_1 est holomorphiquement convexe, un résultat antérieur permet de conclure que D'_α est holomorphiquement convexe relativement à D_1, et D''_α est holomorphiquement convexe par rapport à D_2. Une série de constructions subtiles, qui utilisent notamment la solution du premier problème de COUSIN pour les domaines holomorphiquement convexes, amène finalement à prouver que D_α est holomorphiquement convexe. Alors, si $\alpha<\beta$, D_α est holomorphiquement convexe relativement à D_β; par un passage à la limite, on en déduit que D est holomorphiquement convexe. Ceci achève de prouver le lemme de recollement.

On peut enfin démontrer le théorème III: car si D est pseudoconvexe, on construit comme ci-dessus les D_α à l'aide d'une fonction $psh\,\varphi$ dans D. Mais parce que φ jouit de la propriété (P_1), on constate que chaque D_α possède la propriété suivante: par chaque point frontière P de D_α passe un germe d'hypersurface analytique qui ne rencontre l'adhérence $\bar{D}_\alpha$ qu'au point P. On en déduit que l'intersection de D_α et d'un polycylindre convenable de centre P est holomorphiquement convexe. Vu la compacité de $\bar{D}_\alpha$ et le lemme de recollement, on conclut que D_α est holomorphiquement convexe. D'après un résultat précédent, si $\alpha<\beta$, (D_α, D_β) est une paire de RUNGE. Un passage à la limite permet enfin de conclure que D est holomorphiquement convexe.

OKA a ainsi atteint le but recherché. Arrivé à ce point, le lecteur reste confondu d'admiration devant une telle virtuosité.

Grâce au théorème III, on sait désormais que tout domaine d'holomorphie est une variété de STEIN. Les résultats globaux du Séminaire H. CARTAN 1951/52 peuvent donc s'appliquer aux domaines d'holomorphie. Les exposés de ce séminaire ayant été rédigés en octobre 1952, il semble que Oka n'en ait pas eu connaissance lorsqu'il écrivit ce Mémoire IX, terminé en octobre 1953. Aussi éprouve-t-il le besoin, bien naturel, d'étendre les résultats globaux de son Mémoire VII (où ils étaient énoncés pour les polycylindres compacts) à tous les domaines d'holomorphie. C'est l'objet de la dernière partie du chapitre III. Oka y montre aussi sur un exemple que, en général, les sections d'un faisceau cohérent sur D ne sont pas combinaisons linéaires (à coefficients holomorphes dans D) d'un nombre *fini* d'entre elles.

X. A New Method of Generating Pseudoconvex Domains

Une mode nouvelle engendrant les domaines pseudoconvexes,

Japanese Journal of Mathematics **32** (1962), p. 1–12

Introduction. In a series of earlier Memoirs[1], we treated a group of problems which were posed precisely in the beautiful book of H. BEHNKE-P. THULLEN. But, as we remarked in the preceding memoir, there still remain unsolved problems, some of which seem to us to be among the most difficult. It is our intention to treat these problems successively.

I shall not go into technical detail in this introduction. Rather, I should like to refer to the feeling for the seasons, which has been special to japanese people since time immemorial, to explain what I feel in completing the present Memoir.

There is a tendency towards abstraction in the progress of the mathematical sciences of today.

Even in the field of our own research, theorems have become more and more general, and some of them have even gone outside the space of complex variables. I felt that this was winter. I have waited for a long time for the return of spring and have wanted to make some studies which would make this felt. The present memoir is but the first result. (See the theorem in No. 6.)

1. For simplicity, we shall restrict ourselves to the space of two complex variables x, y. Let us consider a univalent domain D without points at infinity and a characteristic surface Σ in D. Let M_0 be a point of Σ. Suppose that, in the neighbourhood of M_0, we can represent the point $M(x, y)$ of Σ by

$$x=\varphi(t), \qquad y=\psi(t),$$

where $\varphi(t)$ and $\psi(t)$ are holomorphic functions of t for $|t|<1$. We are concerned with the area of Σ in the neighbourhood of M.

Let us express the real and imaginary parts of x, y, t as follows:

$$x=u_1+iu_2, \qquad y=u_3+iu_4, \qquad t=t_1+it_2,$$

i being the imaginary unit. In the cartesian space (u_1, u_2, u_3, u_4) of real dimension 4, the area of the element of surface on Σ is given by

$$d\sigma=\sqrt{EG-F^2}\,dt_1\,dt_2,$$

$$E=\sum\left(\frac{\partial u_j}{\partial t_1}\right)^2, \quad G=\sum\left(\frac{\partial u_j}{\partial t_2}\right)^2, \quad F=\sum\frac{\partial u_j}{\partial t_1}\frac{\partial u_j}{\partial t_2},$$

[1] See: Sur les fonctions analytiques de plusieurs variables, Iwanami, Tokyo, 1961. (On analytic functions of several variables, I–IX.)

where $j=1,2,3,4$. In view of the CAUCHY-RIEMANN condition, this reduces to

$$d\sigma=\left|\frac{dx}{dt}\right|^2 dt_1\,dt_2+\left|\frac{dy}{dt}\right|^2 dt_1\,dt_2.$$

We see thus that the *area of the characteristic surface Σ is expressed as the sum of the areas of its projections on the coordinate planes.*

2. Let us consider again the characteristic surface Σ introduces above, and a dicylinder $[(C),(C')]$ in the interior of the domain D (that is to say, $D \Supset [(C),(C')]$), and we shall find a representation of Σ in the dicylinder following COUSIN. The projection of the portion of Σ in the dicylinder onto the x-

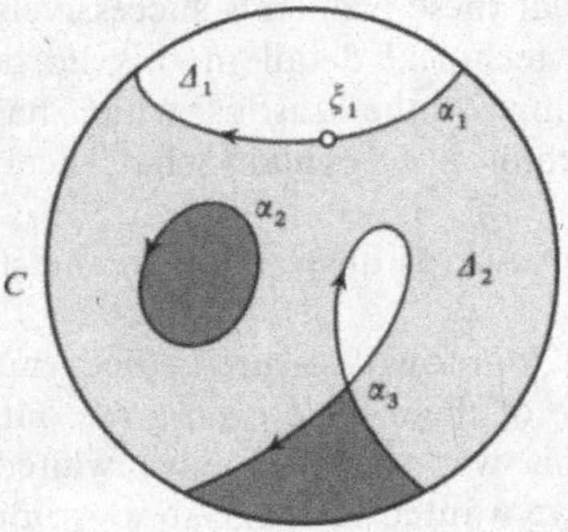

plane consists of pieces of RIEMANN surfaces (as in the Figure) which we denote by $\mathfrak{R}_x$. Let us suppose, for simplicity, that the characteristic surface Σ does not contain any plane of the form $x=$const. or $y=$const., and that the circles (C), (C') are given by

$$|x|<R, \qquad |y|<R'$$

repectively.

Let us look at the RIEMANN surface $\mathfrak{R}_x$ (not necessarily connected). Let x', x'' be two boundary points of $\mathfrak{R}_x$ such that

$$|x'|<R, \qquad |x''|<R,$$

and let us suppose, to facilitate our explanations, that they cannot have the same coordinate except for a finite number of points.

Let us look at the figure, from which we can easily find the general case. The circle (C) is partitioned into 5 portions $\Delta_1, \Delta_2, \ldots$, by 3 curves $\alpha_1, \alpha_2, \alpha_3$ which, together with the circumference C constitute the projection on the x-plane of the boundary of $\mathfrak{R}_x$. The number of sheets of $\mathfrak{R}_x$ is 0 over Δ_1, 1 over Δ_2, which one recognises from the direction of the curve α_1.

We are going to decompose the curves $\alpha_1, \alpha_2, \alpha_3$ into pieces δ_i $(i=1,2,\ldots,t)$. Let α_j^* $(j=1,2,3)$ be the primitive curve of α_j (that is, α_j is the projection on the x-plane of α_j^*), and let β_j be the projection of α_j^* on the y-plane; we are regarding β_j to be on the RIEMANN surface (obtained by continuing $\mathfrak{R}_y$) and is situated over the circumference C'. Let ξ_0 be the starting point of α_1, let η_0 be the corresponding point of β_1 and having the coordinate η_0 (this notation is general). The positive sense of β_j is that of the projection, C', and that of α_j is the direction corresponding to this. Let $q+1$ be the number of sheets of the RIEMANN surface over the point η_0. If we let y trace out the curve β_1 starting at η_0, y will pass successively through q points (over η_0) which we denote by $\eta_1, \eta_2, \ldots, \eta_q$; let $\xi_1, \xi_2, \ldots, \xi_q$ be the corresponding points of α_1. The arc α_1 is then divided into $q+1$ arcs (ξ_0, ξ_1), (ξ_1, ξ_2), ..., (ξ_{q-1}, ξ_q) and what remains; we denote these by $\delta_1, \delta_2, \ldots, \delta_{q+1}$. As for α_2, since it is a closed curve, there is just one difference: we take first a point η_0 over the point R'. Just as above, we obtain points of division $\xi_{q+1}, \xi_{q+2}, \ldots, \xi_{q+r}$ on α_2.[2] We mark points of division $\xi_{q+r+1}, \xi_{q+r+2}, \ldots, \xi_{q+r+s}$ on α_3. We thus obtain $\delta_1, \delta_2, \ldots, \delta_t$, where $t = q+r+s+2$.

To every x in Δ_j, there correspond j' points in (C'):

$$y_{j,1}, y_{j,2}, \ldots, y_{j,\nu}, \ldots, y_{j,j'}$$

such that the points $(x, y_{j,\nu})$ lie on Σ. If we regard x as variable in the neighbourhood of the initial position, these $y_{j,\nu}$ represent functions of x, holomorphic on the RIEMANN surface. To the domain Δ_j, we make correspond the holomorphic function $\varphi_j(x, y)$ as follows:

$$\begin{aligned} &\varphi_j(x,y) = (y - y_{j,1})(y - y_{j,2}) \ldots (y - y_{j,j'}) && \text{if } j' \neq 0, \\ &\varphi_j(x,y) = 1 && \text{if } j' = 0. \end{aligned}$$

Let x be an arbitrary point on the curve α_j considered on the RIEMANN surface (obtained by continuation of $\mathfrak{R}_x$); to x, there corresponds a single point y_j on β_j. This determines a function $y_j(x)$ attached to α_j. In the situation of the figure, we have

$$\varphi_2 = (y - y_1)\varphi_1. \tag{1}$$

There is a similar relation for every pair of contiguous domains.

Let δ_j be one of the arcs $(j = 1, 2, \ldots, t)$ introduced above, and let α_μ the curve of which δ_j forms a part. Let a_j be the initial point of δ_j, b_j the final point. We consider the following COUSIN integral:

$$I_j(x,y) = \frac{1}{2\pi i} \int_{\delta_j} \frac{\operatorname{Log}[y - y_j(t)]}{t - x} dt, \tag{2}$$

where i is the imaginary unit and the integral is taken in the positive sense on δ_j. We must specify the determination of $\operatorname{Log}[y - y_j(t)]$; for this, it is sufficient to determine $\operatorname{Log}[-y_j(a_j)]$ as follows:

$$\operatorname{Log}[-y_j(a_j)] = \log R' + i\theta_0, \qquad -2\pi < \theta_0 \leqq 0,$$

$\log R'$ being real.

We know well the properties of the COUSIN integral in the general case. However, we have a special case. Let us look therefore at the points of division. It is obvious that α_μ can be arbitrary, open or closed. Let us look

[2]) Here r is the number of points on $|y| = R'$.

therefore at the point ξ_1 on the curve α_1. We are concerned with $I_1(x,y)+I_2(x,y)$. Now since

$$\text{(3)} \qquad \operatorname{Log}[y-y_2(x)]-\operatorname{Log}[y-y_1(x)]=-2\pi i,$$
$$I_2(x,y)+I_1(x,y)-\operatorname{Log}(\xi_1-x)$$

is a holomorphic function in the neighbourhood of $x\in\xi_1$.

For each domain of the form $[\Delta_n,(C')]$, $n=1,2,\dots$, we now consider the function

$$F(x,y)=\varphi_n\cdot e^{-\Sigma I_j}\prod(\xi_k-x),$$

where the summation $\sum$ extends over all the arcs δ_j, and the product $\prod$ is taken over all the points of division. By what we have seen above, we find that $F(x,y)$ is holomorphic on a neighbourhood of the dicylinder $[(C),(C')]$ having the property that $F(x,y)/\varphi_n(x,y)$ is holomorphic and non-zero in the neighbourhood of each Δ_n.

Let us consider the general case. One finds without essential difficulties that we can construct the function $F(x,y)$ by the same method and that this function has the same property.

We also remark that while we have chosen a dicylinder among all dicylindrical domains (A,A'), this is just for simplicity; the essential condition is that *A' be simply connected.*

3. We now assert the following: Consider circles (C) and (Γ) in the x-plane, (Γ) being interior to (C) and concentric circles (C'), (C'') in the y-plane, (C'') being interior to (C'). Let Σ be a characteristic surface defined in a neighbourhood of the dicylinder $[(C),(C')]$.

With this geometric configuration, suppose that the area of Σ is restricted as follows:

1°. *If (C'',C') is the annulus in the y-plane between C' and C'', the area of Σ in the dicylindrical domain $[(C),(C'',C')]$ is smaller than Ω.*

2°. *The same holds for the area of Σ in the dicylinder $[(\Gamma),(C')]$.*

Let C'_1 be a circumference concentric with C' and C'' and situated between them.

We can then find a holomorphic function $F(x,y)$ defining the characteristic surface Σ on $[(C),(C'_1)]$ and having the following bounds from the two sides:

1) *Let (C_1) be a given circle concentric with (C) and smaller than (C); we have, on $[(C_1),(C'_1)]$,*

$$\text{(1)} \qquad \log|F(x,y)|<A\cdot\Omega.$$

2) *For an arbitrary dicylinder $[(\gamma),(\gamma')]$ of given radii (ρ,ρ') in the interior of*

$$G=[(\Gamma),(C'_1)]\cup[(C),(C'',C'_1)],$$

we have

$$\text{(2)} \qquad m(\log|F(x,y)|)>-B\cdot\Omega,$$

where m is the arithmetic mean on $[(\gamma),(\gamma')]$ and A and B are positive constants.

In fact, suppose that the circles (C), (C') are given by $|x|<R$, $|y|<R'$ respectively, where $R \leqq \frac{1}{2}$, $R' \leqq \frac{1}{2}$, which is permitted. Suppose, in addition, that the characteristic surface Σ does not contain any characteristic plane of the form $x=\text{const.}$ or $y=\text{const.}$; this is also permissible.

Let $\mathfrak{R}'_x$, $\mathfrak{R}'_y$ be the RIEMANN surfaces over the x-, y-planes respectively corresponding to the portion of Σ in $[(C),(C'',C')]$. If we denote their areas by A_x, A_y, we have

$$(3) \qquad A_y = \int_{\mathfrak{R}'_y} d\sigma, \qquad A_x = \int_{\mathfrak{R}'_y} \left|\frac{dx}{dy}\right|^2 d\sigma,$$

where $d\sigma$ is the area element on the RIEMANN surface $\mathfrak{R}'_y$ and $x(y)$ is the function defined by Σ. Now, since

$$A_x + A_y < \Omega, \qquad 1 + X^2 \geqq 2X$$

for any positive number X, we have

$$(4) \qquad \int_{\mathfrak{R}'_y} \left(1 + \left|\frac{dx}{dy}\right|\right) d\sigma < \frac{3}{2}\Omega.$$

Let us draw on Σ the following (one or several) curves:

$$|x| \leqq R, \qquad |y| = R'_2, \qquad R'_1 < R'_2 < R',$$

R'_1 being the radius of C'_1. Let us denote by α_j, β_j the projections of these curves on the x- and y-planes respectively and their lengths by s_j, s'_j. It then follows from the above inequality that there exists a radius R'_2 in the interval (R'_1, R') satisfying

$$(5) \qquad \sum (s_j + s'_j) < \frac{3\Omega}{2(R' - R'_1)} = K_1 \cdot \Omega.$$

Let (C'_2) be the circle $|y| < R'_2$.

We apply the method of the preceding section to the dicylinder $[(C),(C'_2)]$ thus obtained to construct the function $F(x,y)$ associated to Σ. We continue to use the same notations.

We look first at the *upper bound*. We set

$$\log|F| = u = u_1 + u_2 + u_3,$$

where

$$u_1 = \log|\varphi|, \quad u_2 = \Sigma \log|\xi_k - x|, \quad u_3 = \mathfrak{R}[-\Sigma I_j],$$

where the logarithms are real and the operation $\mathfrak{R}$ means that we take the real part. Since $R \leqq \frac{1}{2}$, $R' \leqq \frac{1}{2}$, we find at once that

$$u_1 \leqq 0, \qquad u_2 \leqq 0.$$

Let R_1 be the radius of (C_1). Let $R - R_1 = \delta$, $R' - R'_2 = \delta'$. Let (ξ, η) be an arbitrary point of the dicylinder $[(C_1), (C'_2)]$. We describe a circle (δ): $|x - \xi| < \delta$ in (C). Since $u(x,\eta)$ is subharmonic it follows that

$$u(\xi,\eta) \leqq m_{(\delta)}\, u(x,\eta) \leqq m_{(\delta)} u_3(x,\eta),$$

where $m_{(\delta)}$ stands for the arithmetic mean on the circle (δ).

Let us recall the form of the integrals I (see formula (2) in No. 2); it is the imaginary part θ of $\mathrm{Log}[y-y_j(t)]$ which is now in question.

1°. The initial value θ_0 is defined to be in $-2\pi<\theta_0\leqq 0$.

2°. When y changes from the origin to its present value, if t remains fixed at its initial value t_0, the corresponding variation of θ does not surpass $\pi/2$ in absolute value.

3°. If y is fixed and t traces δ_j positively, the argument θ increases, but the variation is at most 2π.

Thus, we have

$$|\theta|<\tfrac{5}{2}\pi.$$

Consequently, if we consider the function

$$v(x)=\frac{L}{2\pi}\sum\int_{\delta}\frac{1}{|t-x|}\,ds$$

where

$$L^2=(-\log\delta')^2+\left(\frac{5\pi}{2}\right)^2,\qquad L>0,\qquad \delta=\delta_j,$$

we have

$$|u_3(\xi,\eta)|\leqq v(x).$$

Now, we see immediately that we have

$$m_{(\delta)}\frac{1}{|x-t|}\leqq m_{(\delta)}\frac{1}{|x-\xi|}=\frac{2}{\delta};$$

we have, therefore,

$$m_{(\delta)}\,v(x)=\frac{L}{2\pi}\left\{\sum\int_{\delta}\left[m_{(\delta)}\frac{1}{|t-x|}\right]ds\right\}\leqq\frac{L}{\pi\delta}[\Sigma s],$$

$$\therefore\ m_{(\delta)}\,v(x)\leqq\frac{LK_1}{\pi\delta}\Omega=A\cdot\Omega.$$

We therefore have

$$\log|F(\xi,\eta)|<A\cdot\Omega,$$

for

$$|\xi|<R_1,\qquad |\eta|<R_2'.$$

We are concerned next with the *lower bound*. We consider a dicylinder $[(\gamma),(\gamma')]$ in G, where (γ) and (γ') are given by $|x-\xi|<\rho$ and $|y-\eta|<\rho'$ respectively. We are going to apply the well-known theorem concerning the mean of a logarithmic potential on a circumference to the different summands in $u=u_1+u_2+u_3$.

1°. For the function $u_1(x,y)$ given by

$$u_1(x,y)=\sum_{\nu=1}^{n'}\log|y-y_{n,\nu}(x)|,$$

where (x,y) is supposed to lie in $[\Delta_n,(C_1')]$, we first remark that

$$\pi\rho^2\,m_{(\delta)}n'<\Omega.$$

Now, let $m_{(\gamma')}$ be the mean with respect to (γ'). We have

$$m_{(\gamma')} u_1(x, y) \geqq \frac{2n'}{\rho'} \int_0^{\rho'} \log r \cdot r dr = n'(\log \rho' - \tfrac{1}{2}),$$

$$\therefore\ m u_1(x, y) \geqq \frac{\Omega}{\pi \rho^2} (\log \rho' - \tfrac{1}{2}) = -B_1 \cdot \Omega.$$

2°. The function $u_2 = \sum \log |\xi_k - x|$ is independent of y. If we count the number λ of points of division, we find that

$$2\pi R_1 \cdot \lambda < \Sigma s' < K_1 \Omega, \therefore\ m u_2 > \frac{K_1 \Omega}{2\pi R_1} (\log \rho - \tfrac{1}{2}) = -B_2 \Omega.$$

3°. As for $u_3 = \Re[-\Sigma I]$, let us set

$$w(x, y) = \frac{1}{2\pi} \sum \int_{\delta_j} \frac{\log |y - y_j(t)| - 5\pi/2}{|t - x|} ds.$$

We then have

$$u_3(x, y) > w(x, y).$$

Now, in this case, all the masses are distributed on the cirumference C_2' exterior to (γ'); we therefore have

$$m_{(\gamma')} w(x, y) = \frac{1}{2\pi} \sum \int_{\delta_j} \frac{\log d - 5\pi/2}{|t - x|} ds$$

where $d = R_2' - |\eta|$. From this, it follows that

$$m w(x, y) > \frac{K_1 \Omega}{\pi \rho} (\log \rho' - 5\pi/2) = -B_3 \Omega.$$

We thus achieve the required inequality

$$m(\log |F(x, y)|) > -B \cdot \Omega,$$

$B > 0$, for any $[(\gamma), (\gamma')]$ in G. Q.E.D.

4. We are concerned now with a certain converse of the preceding proposition. We shall prove the following:

Let $F(x, y)$ be a holomorphic function on the polycylinder $[(C), (C')]$ and having the following bounds:

1°. $\log |F| \leqq M$,

2°. $m(\log |F|) \geqq -K_1 \cdot M$

on $[(C), (C')]$, K_1, M being positive constants. Then, for any given polycylinder $[(C_1), (C_1')]$ in the interior of $[(C), (C')]$, the area Ω_1 of the characteristic surface Σ given by $F(x, y) = 0$ is restricted by the inequality

$$\Omega_1 \leqq K_2 \cdot M,$$

K_2 being a positive constant independent of M and of F.

1°. To verify this, we equip ourselves with the following auxiliary proposition (which one could also reduce to the Jensen formula).

Let (C), (C_1) be the circles $|z|<R$, $|z|<R_1$ respectively, where $R_1<R$. Let $f(z)$ be a holomorphic function on (C), bounded in such a way that $|f(z)|\leqq M$, and having at least n zeros on the closed circle $(\bar{C}_1)$; we then necessarily have

$$|f|\leqq M\lambda^n$$

on $(\bar{C}_1)$, λ being a positive constant independent of M and of f such that $0<\lambda<1$.

To check this, we may suppose that $R=1$. We denote R_1 by θ.

Suppose first that $f(z)$ is zero at the point a on $(\bar{C}_1)$. If $a=0$, we have, by SCHWARZ's lemma,

$$|f(z)|\leqq M\cdot\theta$$

on $(\bar{C}_1)$. Suppose, therefore, that $a\neq 0$. We can suppose that a is real and positive. Consider the well-known transformation

$$z'=\frac{1}{a}\frac{z-a}{b-z} \qquad (ab=1),$$

which leaves the circle (C) invariant. Let

$$\lambda=\max|z'| \quad \text{for} \quad |z|\leqq\theta.$$

We see that

$$0<\lambda<1, \qquad \theta<\lambda.$$

By virtue of SCHWARZ, we have

$$|f|\leqq M\lambda$$

on $(\bar{C}_1)$.

Suppose next that f is zero at a, b. We can suppose that $a\neq 0$ and that a is real and positive. Consider the function

$$\varphi=\frac{1}{z'}f.$$

By the preceding result, we have

$$|\varphi|\leqq M\lambda$$

on $(\bar{C}_1)$. We therefore have

$$|f|=|\varphi|\cdot|z'|\leqq M\cdot\lambda^2.$$

Continuing in this way, we arrive at the result indicated.

2°. Under the circumstances explained at the beginning of this section, suppose that Σ does not contain any piece of characteristic surface of the form $x=\text{const.}$ or $y=\text{const.}$ Suppose that (C), (C_1), (C'), (C'_1) are given by $|x|<R$, $|x|<R_1(<R)$, $|y|<R'$, $|y|<R'_1(<R')$ respectively and that $R\leqq\frac{1}{2}$, $R'\leqq\frac{1}{2}$. Let $\mathfrak{R}_x, \mathfrak{R}_y$ be the projections of Σ on the x- and y-planes respectively. Let Ω'_1, Ω''_1, Ω_1 be the areas of the portions of $\mathfrak{R}_x$, of $\mathfrak{R}_y$ and of Σ in the dicylinder $[(C_1), (C'_1)]$ respectively. We then have

$$\Omega_1=\Omega'_1+\Omega''_1.$$

We shall therefore estimate the area Ω'_1.

Let us use the same notation as in No. 2, except that we replace $[(C), (C')]$ by $[(C_1), (C_1')]$. We have

$$\sum \int_{\Delta_p} p'\, d\sigma = \Omega_1'.$$

Let x_0 be a point of Δ_p; we have p' roots of $F(x_0, y)=0$ in (C_1'). By the preliminary proposition, we have

$$|F(x_0, y)| \leqq e^M \cdot \lambda^{p'}$$

for $x_0 \in \Delta_p$, $|y| \leqq R_1'$. Hence

$$\log |F(x_0, y)| \leqq M - p' \log \frac{1}{\lambda},$$

$$\int_{(C_1)} \log |F(x, y)|\, d\sigma \leqq M \cdot \pi \cdot R_1^2 - \log \frac{1}{\lambda} \cdot \Omega_1',$$

$$m(\log |F(x, y)|) \leqq M - \frac{1}{\pi R_1^2} \log \frac{1}{\lambda} \cdot \Omega_1'.$$

Now $m(\log |F(x, y)|) \geqq -K_1 M$,

$$\therefore \Omega_1' \leqq \frac{\pi R_1^2 (K_1+1)}{\log \frac{1}{\lambda}} \cdot M = K_2' M.$$

3°. Let us consider the general case in which Σ can be expressed in the form

$$\Sigma = \Sigma_1 + \Sigma_2 + \Sigma_3,$$

where Σ_2 consists of pieces of characteristic planes of the form $x=$const., Σ_3 of pieces $y=$const., and Σ_1 is the rest.

Let us estimate the area Ω_2 of the portion of Σ_2 in the dicylinder $[(C_1), (C_1')]$. Let m' be the number of planes of the form $x=$const., multiplicity being taken into account; the area Ω_2 is then

$$\Omega_2 = m' \pi R_1'^2.$$

Now, by the preliminary proposition,

$$|F| \leqq e^M \lambda^{m'}$$

on $[(\bar{C}_1), (\bar{C}_1')]$. Hence

$$-K_1 M \leqq M - m' \log \frac{1}{\lambda}, \quad \therefore m' < \frac{(K_1+1)}{\log \frac{1}{\lambda}} \cdot M.$$

The area Ω_2 is therefore bounded in the way indicated. The same holds for Ω_3. For the area Ω_1, consider the function

$$\Phi(x, y) = F(x, y) / \prod (x - a_j) \prod (y - b_k),$$

where $x = a_j$ $(j=1, 2, \ldots, m')$ and $y = b_k$ $(k=1, 2, \ldots, n)$ represent roots of the equation $F(x, y)=0$, multiplicity being taken into account. We then have, by the maximum modulus principle,

$$\log|\Phi| < M + m' \log\frac{1}{R-R_1} + n\log\frac{1}{R'-R} = K'M,$$

K' being a positive constant.

We still have

$$m\log|\Phi| > -K_1 M.$$

By the preceding case, we therefore know that the area Ω_1' is also bounded in the way indicated above, and consequently, we have

$$\Omega \leqq K_2 M. \qquad \text{Q.E.D.}$$

5. Let us draw two circles (C) and (Γ_0) of the form $|x|<R$, $|x|<R_0$, $R>R_0$, in the x-plane, and two circles (C'), (C'') of the form $|y|<R'$, $|y|<R''$, $R'>R''$ in the y-plane. Let us consider a characteristic surface Σ in the neighbourhood of $[(C),(C')]$ (not necessarily irreducible). We consider the case when Σ does not contain any characteristic plane of the form $x=$const. or $y=$const. We suppose that the area of Σ in $[(C),(C'',C')]$ is smaller than Ω, as is the area of Σ in $[(\Gamma_0),(C')]$.

Let C_1' be a given circumference of the form $|y|=R_1'$, $R''<R_1'<R'$. As we saw in No. 3, we find another circumference C_2' of the form $|y|=R_2'$, $R_1'<R_2'<R'$ having the following property: Let $\mathfrak{R}_x$ be the projection on the x-plane of the portion of Σ in $[(C),(C_2')]$, and $\mathfrak{R}_y$ the corresponding projection on the y-plane. Let β_i, $i=1,2,\ldots,p$ be the curves forming part of the boundary of $\mathfrak{R}_y$ for which $|y|=R_2'$, and let α_i be the corresponding curves in the x-plane (multiplicity being taken into account). Let L be the total length of the curves α_i, and L' be that of the β_i. We then have:

$$L+L' \leqq K_1\Omega, \quad K_1 = \frac{3}{2(R'-R_1')}.$$

The circle (C) is then partitioned by the curves α_i into several portions Δ_n where $\mathfrak{R}_x$ has n sheets. Let x_0 be point of Δ_n; we write

$$n=\psi(x_0).$$

Consider a circle (Γ_1) of the form $|x|<R_1(R_0>R_1)$ and the annulus (Γ_1,Γ_0). The area of the part of $\mathfrak{R}_x$ in this annulus is given by

$$\int_{(\Gamma_1,\Gamma_0)} \psi(x)\,d\sigma,$$

which we can write

$$\int_{R_1}^{R_0} dr \int_0^{2\pi} \psi(x)\,r\,d\theta \qquad (x=re^{i\theta}).$$

Since this area is smaller than Ω, we can find a radius R_2 in $R_1<R_2<R_0$ such that

$$\int_0^{2\pi} \psi(x)R_2\,d\theta \frac{\Omega}{R_0-R_1} = K_2\cdot\Omega.$$

We denote this circumference $|x|=R_2$ by Γ_2.

The curves α_i are analytic, and its elements are regular except perhaps at finitely many points. Let us consider the following problem.

Problem. Let x_0 be an arbitrary point of the annulus (Γ_2, C). We define the function $\varphi(x_0)$ as follows: Consider the ray issuing from the origin and passing through x_0; let ξ_0 be the point of intersection of this ray and the circumference Γ_2. When x traces this ray from ξ_0 to x_0, x will meet the curves α_i several times; an intersection is positive if x crosses the curve from right to left and negative in the opposite case. Let q be the number of positive intersections and q' the number of negative intersections, where we count the intersection at the point ξ_0 if it is present, but not at the point x_0. Let us define

$$\varphi(x_0)=\psi(\xi_0)+q-q'.$$

Consider

$$A=\int_{(\Gamma_2,C)}\varphi(x)\,d\sigma,$$

($d\sigma$ being the area element in the annulus). What is the maximum of the quantity A if one draws the curves α_i at random, the only conditions being that that they fulfil the following:

1°. the curves are analytic and finite in number; its elements are regular except perhaps at a finite number of points;

2°. the total length is L;

3°. their extremities lie on C or on Γ_2.

The answer to this problem is obvious: the maximum of A occurs when all the curves lie on Γ_2 and have negative orientation with respect to Γ_2. The maximum, A_0, is given by

$$A_0=\int_{(\Gamma_0,C)}[\psi(\xi_0)+q]r\,dr\,d\theta\,(x_0=re^{i\theta})<(R^2-R_2^2)\left(\frac{K_2}{2R_0}+\frac{K_1}{2R_2}\right)\Omega=K_3\Omega.$$

We therefore know that the area of the projection on the x-plane of the part of Σ in $[(C),(C')]$ is smaller than $(K_3+1)\cdot\Omega$.

It is easy to see, using the linear transformation introduced earlier, that we can replace the special circle (Γ_0) by an arbitrarily given circle (Γ) in the interior of (C) without changing the conclusion. Hence, repeating the reasoning of No. 3, we see without difficulty that if *$F(x,y)$ is the function constructed in No. 3 for the dicylinder* $[(C),(C')]$, *we also have*

$$m\log|F(x,y)|>-B\Omega,$$

B being a positive constant independent of Σ.

In view of what we have seen in No. 4, we have therefore the following result.

Lemma. *Let (C) be a circle in the x-plane and (Γ) a circle interior to (C). Let (C'), (C'') be two concentric circles in the y-plane, of which (C'') is the interior circle. Let Σ be a characteristic surface in the dicylinder $[(C),(C')]$ (irreducible or not) such tht the area of the portion of Σ in $[(C),(C'',C')]$ is smaller than Ω, and* *that the same holds also for* $[(\Gamma), (C')]$. *The area of* Σ *in the interior of* $[(C), (C')]$ *is then finite and* $\leqq \cdot \Omega$, *K being a positive constant independent of* Σ.

6. Let us consider a univalent domain D in the space (x, y); we shall suppose that D has no points at infinity to simplify the language. We consider a sequence

$$\text{(S)} \qquad \Sigma_1, \Sigma_2, \ldots, \Sigma_n, \ldots$$

of characteristic surfaces in D and a sequence

$$\text{(T)} \qquad \nu_1, \nu_2, \ldots, \nu_n, \ldots$$

of positive numbers. The sequence (T) is arbitrary, but we shall regard it as being unbounded, in general.

Using these sequences (S) and (T), we distinguish between two kinds of points in D.

1°. If, for a point P_0, there exists a dicylinder (γ) with centre P_0 such that the sequence

$$\text{(1)} \qquad \frac{\omega_1}{\nu_1}, \frac{\omega_2}{\nu_2}, \ldots, \frac{\omega_n}{\nu_n}, \ldots,$$

where ω_n is the area of Σ_n in (γ), is bounded, we shall call P_0 a *point of the first kind.*

2°. If there exists no dicylinder (γ) having this property, we shall call P_0 a *point of the second kind.*

We shall see that the following result holds:

Theorem. *The set of points of the first kind is pseudoconvex in the domain D.*

In fact, it is sufficient to verify the continuity theorem (C). Let E be the set of points of the second kind in D. E is clearly closed in D. Let us consider a dicylinder $[(C), (C')]$ in D. Let (Γ) be a circle interior to (C) and let (C'') be a circle concentric with (C') and interior to it. Suppose that there are no points of E in $[(\Gamma), (C')]$, nor in $[(C), (C'', C')]$. It is sufficient to affirm that there are no points of E in $[(C), (C')]$.

Suppose, by contradiction, that there exists a point (ξ, η) of E in $[(C), (C')]$. Let us consider a dicylidner $[(C_1), (C_1')]$ concentric with $[(C), (C')]$ in $[(C), (C')]$ such that it contains (ξ, η), (C_1) contains (Γ) and the circumference C_1' is contained in the annulus (C'', C'). Let (Γ_0) be a circle concentric with (Γ) and smaller than (Γ). Since every point P of the closed set $[(\bar{\Gamma}_0), (\bar{C}_1')] \cup [(\bar{C}_1), C_1']$ is the centre of a dicylinder (γ) such that the sequence (1) is bounded, we can cover this closed set by a finite number of these dicylinders (γ) by the Borel lemma. Let Ω_n be the area of Σ_n in $[(\Gamma_0), (C_1')]$. The sequence Ω_n/ν_n $(n=1,2,\ldots)$ is therefore bounded. Let (C_3', C_2') be an annulus concentric with C_1', containing C_1' and sufficiently narrow. Let Ω_n' be the area of Σ_n in $[(C_1), (C_3', C_2')]$; the sequence Ω_n'/ν_n is also bounded. Let $[(C_2), (C_4')]$ be a dicylinder concentric with $[(C_1), (C_1')]$, smaller than $[(C_1), (C_1')]$ but continuing to contain the point (ξ, η).

By the lemma, if Ω_n'' is the area of Σ_n in $[(C_2),(C_4')]$, we have

$$\Omega_n'' \leqq K \max(\Omega_n, \Omega_n'),$$

K being a positive constant in dependent of n. The sequence Ω_n''/v_n is therefore bounded. This contradicts (ξ, η) being a point of the second kind.

Commentaire de H. Cartan

Dans ce Mémoire X, OKA aborde un problème particulier et se restreint au cas de 2 variables complexes x, y. Dans le dicylindre

$$(\Delta) \qquad |x|<1, \quad |y|<1$$

toute hypersurface analytique (ici, sous-ensemble analytique de dimension complexe un dans Δ) peut être définie par une équation $F(x,y)=0$, où F est holomorphe dans Δ. Ce résultat remonte à Cousin, et se précise comme suit: étant donné un *diviseur positif* Σ dans Δ (c'est-à-dire une famille localement finie de sous-ensembles analytiques de dimension un, dont chacun est affecté d'un entier positif), il existe une fonction holomorphe $F(x,y)$ dans Δ qui s'annule sur chaque composante de Σ avec l'ordre de multiplicité indiqué, et ne s'annule pas ailleurs. Evidemment, F détermine Σ, tandis que Σ ne détermine F qu'à un facteur près de la forme e^G, où G est holomorphe dans Δ. OKA étudie alors les relations existant entre d'une part une majoration de l'aire de Σ, d'autre part des majorations de $|F(x,y)|$ et de la moyenne de $\log|1/F(x,y)|$ dans Δ ou dans des dicylindres partiels.

En ce qui concerne l'*aire* de Σ, OKA observe d'abord qu'elle est égale à la somme des aires de ses projections sur le plan (x) et sur le plan (y); c'est un résultat classique qui s'étend à n variables. Le principal résultat nouveau concernant l'aire de Σ est énoncé sous forme de "lemme" à la fin du n° 5. Nous y reviendrons dans un instant.

Auparavant OKA établit diverses propositions concernant les relations entre la fonction F et l'aire de Σ. L'une d'elles, énoncée au début du n° 4, pourrait être précisée comme suit: si $F(x,y)$ est holomorphe et non identiquement nulle dans Δ, avec $|F(x,y)|<1$, et si on note $\Omega(r,F)$ l'aire de la portion du diviseur Σ (défini par F) contenue dans le dicylindre Δ_r ($|x|<r$, $|y|<r$, $0<r<1$), on a la majoration

$$\Omega(r,F) \leqq \frac{4\pi}{1-r^2} m_\Delta(\log|1/F(x,y)|),$$

où m_Δ désigne la moyenne prise dans le dicylindre Δ par rapport à la mesure de LEBESGUE. Ce résultat se démontre facilement à partir de la relation

$$m\left(\log\left|\frac{1-\bar{a}x}{x-a}\right|\right) = \frac{1-a\bar{a}}{2},$$

où m désigne cette fois la moyenne dans le disque $|x|<1$, et a un nombre complexe de module inférieur à un.

En sens inverse, OKA cherche des majorations pour la moyenne de $\log|1/F(x,y)|$ en fonction de l'aire du diviseur Σ_F défini par F. Plus précisément, un diviseur Σ d'aire finie étant donné, on cherche, parmi les F associées à Σ, une F particulière qui satisfasse à certaines majorations. Cela donne lieu à une analyse assez subtile d'OKA (voir les n°s 2 et 3).

Venons-en au contenu du "lemme" du n° 5. Il concerne la "marmite de HARTOGS", qui a joué un rôle essentiel dans l'étude de la "pseudo-convexité" des domaines (cf. Mémoires VI et IX). Rappelons que la "marmite" $M(r,r')$ du dicylindre Δ est la réunion des ouverts $|x|<r$, $|y|<1$ et $|x|<1$, $1-r'<|y|<1$ (avec $0<r<1$, $0<r'<1$). Le lemme en question affirme qu'il existe un nombre $K>0$ jouissant de la propriété suivante: si Σ est un diviseur dans Δ, et si Ω désigne l'aire (supposée finie) de la portion de Ω située dans la marmite $M(r,r')$, alors l'aire totale de Σ dans Δ est finie et majorée par $K\Omega$.

Le commentateur n'est pas certain que ce résultat soit vraiment démontré comme conséquence des énoncés antérieurs, car ceux-ci sont souvent ambigus, OKA ne précisant pas toujours si les dicylindres considérés sont compacts ou ouverts. Le commentateur a l'impression que la conclusion du "lemme" devrait être affaiblie comme suit: pour chaque dicylindre Δ_ρ de rayon ρ, il existe une constante K_ρ telle que l'aire de la portion de Σ contenue dans Δ_ρ soit majorée par $K_\rho\Omega$.

De toute façon, cet énoncé, même ainsi affaibli, permet de prouver le "théorème" du n° 6 sur lequel s'achève ce Mémoire X. Il s'agit d'un problème dont la solution conduit à un domaine pseudo-convexe (ceci explique le titre du Mémoire). Voici le problème: on se donne, dans un domaine $D\subset\mathbb{C}^2$, une suite infinie de diviseurs Σ_n et, pour chacun d'eux, un nombre réel v_n (au lieu d'une suite, on pourrait se donner une famille quelconque); on considère l'ouvert U formé des points de D qui sont centre d'un dicylindre γ tel que la suite des nombres

$$\frac{\text{aire de } (\Sigma_n\cap\gamma)}{v_n}$$

soit bornée. Alors U est "pseudo-convexe dans D" (ce qui veut dire que U est pseudo-convexe en chacun des points de la frontière de U situés dans D). Tel est le contenu du "théorème" du n° 6.

Note on Families of Multivalued Analytic Functions etc.

Note sur les familles de fonctions analytiques multiformes etc.

Journal of Science of the Hiroshima University **4** (1934), p. 93–98

1. The aim of the researches whose outline and principal results will be announced briefly in what follows[1] is to extend the series of remarkable results of Messrs. WEIERSTRASS, STIELTJES, VITALI and MONTEL in the classical theory of convergence of single valued analytic functions of one complex variable first to multivalued analytic functions, and then, beyond that. However, the second generalisation is possible only up to the theorem of Mr. STIELTJES[2].

I. Normal Families of Characteristic Surfaces

2. In treating an analytic function $f(x)$ of a complex variable x, in general multi-valued, it will be more convenient, and more general, to look at it in the following 4 dimensional space:

Let y be a second complex variable, and let $x=x_1+ix_2$, $y=y_1+iy_2$, $i=\sqrt{-1}$. With these x, y, we shall always consider a space with rectangular coordinates (x_1, x_2, y_1, y_2), and we shall represent them by (x, y) to abbreviate the notation.

In a given domain $\varDelta$ in the space (x, y), the set of points satisfying the equation $y=f(x)$ consists, if it exists, of one or several continuous surfaces.

Let S be any one of these surfaces; it will be called *analytically connected* if any two of its analytic elements, considered again in the space (x, y), can be analytically continued into each other along a suitable path in $\varDelta$ (which necessarily lies on the surface S).

This being said, we shall understand by the words "*a characteristic surface in $\varDelta$*" a surface S as above which is analytically connected; we have in mind "characteristic planes" of the form $x=$ const.

Let S be again a characteristic surface in a domain $\varDelta$. It it can be represented by an equation of the form $F(x, y)=0$ in a domain (δ) contained in $\varDelta$, $F(x, y)$ being a holomorphic function of the two complex variables x, y in (δ), we shall, for brevity, call the function $F(x, y)$ an *adjoint holomorphic function* of the characteristic surface S on the domain (δ).

We distinguish between two kinds of points on the characteristic surface S: A point of S will be of *the first kind* if we can find a sufficiently small hypersphere about the point on which there exists an adjoint holomorphic function. If not, the point will be of *the second kind*.

[1] Details will be published soon.

[2] I know of only one memoir devoted to a similar goal: G. JULIA, *Sur les familles de fonctions analytiques de plusieurs variables* (Nos. 72–80). Acta. 1926.

3. Definition of Normal Families. Let (F) be a family of characteristic surfaces without points of the second kind in a domain Δ in (x, y) space.

The family (F) will be normal at a point of the domain if we can find a hypersphere (w) about this point and so small that we can find adjoint holomorphic functions $F(x, y)$ in the hypersphere for the characteristics of the family in such a way that the family of functions $F(x, y)$ is normal[3] in (w), and, in addition, the constant zero is not among the limits of the family.

It should be remarked that if a point of Δ does not belong to the limit points of the family of surfaces, the family is normal at this point because of the definition.

The family (F) will be said to be *normal in the domain* Δ if it is normal at every point of the domain.

A point of the domain will be said to belong to *the (J)-points of the family if the family is not normal* at this point.

4. Let S be a characteristic surface without points of the second kind in a spatial domain Δ. Corresponding to any dicylinder in the interior of Δ, we can form, by the classical method of Mr. Cousin, a certain adjoint holomorphic function in such a way that one can *estimate, from both sides, bounds for the mean of the modulus of the function by means of a single quantity, the area of the surface S in the dicylinder*. The following theorems are obtained from this.

Theorem 1. *Let (F) be a family of characteristic surfaces without points of the second kind in a domain Δ in the space (x, y). For the family (F) to be normal in Δ, it is necessary and sufficient that the areas of the surfaces of the family in an arbitrary domain (δ) in the interior of Δ have an upper bound depending only on (F) and on (δ).*

Theorem 2. *The set of (J)-points of the family (F) belongs to the class (H)*, which will be defined in what follows.

II. Sets of Class (*H*)

5. Definition and Examples. *We say that a set E of points in (x, y)-space belongs to the class (H)[4] on a spatial domain Δ (after the name of the discoverer,* Mr. F. Hartogs[5]) *if it has the following properties:*

1°. *The set E is closed in the interior of Δ.*

2°. *Let O be a fixed point, given arbitrarily in the finite part of space. The distance of the point O from an arbitrary point of E never attains a relative maximum (even in the wider sense of non-strict maximum) in the finite part of E in the domain Δ.*

[3] In the sense of Mr. Julia, *loc. cit.*
[4] Or, simply, that E is an (H)-set.
[5] Math. Annalen 62; Acta. 32, etc.

3°. *The above properties are preserved by any biunique analytic transformation of* (x, y)*-space.*

More precisely, if a domain (ω) in Δ is subjected to a biunique analytic transformation, together with the part of E in (ω), then the transformed set still has properties (1°, 2°) on the new domain.

The importance of the above properties was noted for the first time by Mr. HARTOGS when he studied the set of singular points of an analytic function of two complex variables, and these sets give us, with a little modification, examples of sets of class (H). Other examples will be found in the *memoirs of* E.E. LEVI[6] and Mr. G. JULIA[7]. And characteristic surfaces without points of the second kind form the simplest example.

The theory of sets of class (H) was developed by these mathematicians and others; naturally, their results were not proved abstractly. However, one will find that almost all these results are established using just the three properties given in the definition; for example, the results of No. 25 to No. 44 of the memoir of Mr. JULIA cited above.

6. Given an arbitrary set E of points of (x, y)-space, to a point x_0 of the x-plane, there correspond points (x_0, y_0), (x_0, y_1), ... on the set E, which, as a set, can be empty, countable or non-countable; to this, there corresponds the set of points $y_0, y_1, \ldots$ in the y-plane, which we shall denote by $\mathfrak{E}(x_0)$. This is the *section of the set E* by the characteristic plane $x = x_0$.

Conversely, if we are given a correspondence $\mathfrak{E}(x)$, a variable set depending on x, we can determine the original set E; we shall represent this by $y = \mathfrak{E}(x)$ exactly as with characteristic surfaces.

As a special case, let E be a set of class (H) on a dicylindrical domain (x in D, y in D'). The corresponding section $\mathfrak{H}(x)$ behaves, in the domain D, in a way which is very analogous to analytic functions if we limit ourselves to looking at the set in the domain D'. It is by means of such sections $\mathfrak{H}(x)$ that I shall re-establish the theorem of STIELTJES in its general form. I shall first make precise two ways of passing to the limit.

7. 1°. **Passage to the limit in the space (x, y).** Let (F) be a family of (H)-sets on a spatial domain Δ, *whose limit is* E_0. This means that for any neighbourhood of an arbitrary point[8] of E_0, there exist infinitely many sets of the family having at least one point in the neighbourhood.

Theorem 3. *The limit* E_0 *of the family* (F) *in* (x, y)*-space also belongs to the class* (H).

2°. **Passage to the limit in the y-plane.** Let (Σ): $E_1, E_2, \ldots, E_\nu, \ldots$ be a sequence of (H)-sets on a cylindrical domain (D, D') whose spatial limit is E_0. Let $\mathfrak{H}_\nu(x)$ be the section corresponding to the set E_ν, $\nu = 0, 1, 2, \ldots$. On the other

---

6) Annali di Matematica 17, 18, series III.
7) Loc. cit.
8) Which can be at infinity.

hand, let us consider a correspondence $\mathfrak{K}_0(x)$ such that, for any neighbourhood of an arbitrary point of $\mathfrak{K}_0(x')$, x' being any fixed point in D, we can find an infinity of sets of the sequence $\mathfrak{H}_\nu(x')$ having at least one point in the neighbourhood[9].

Theorem 4. *On an arbitrarily given rectifiable Jordan curve, the set consisting of points ξ such that the boundary of the set $\mathfrak{H}_0(\xi)$ exists (in D') and is not contained in that of $\mathfrak{K}_0(\xi)$, is of measure zero.*

In view of the preceding results, it is sufficient to look at a single set of the class (H).

III. Generalisations of a Theorem of Mr. Hartogs

8. We shall first make some remarks on the *capacity* of sets of points in the plane with respect to logarithmic potentials, which will play a fundamental role in what follows. The theory of capacities, which notion was introduced by Mr. H. LEBESGUE, was developed later[10]. However, almost all research on the subject is devoted to three dimensions, and some words are necessary to give the basic notions in our case. I shall first give the definitions.

1°. *The capacity of an open set O contained in a circle of radius smaller than* $\frac{1}{2}$ is the upper bound of positive masses which one can distribute in the interior of O without the corresponding logarithmic potentials exceeding 1.

2°. *The outer capacity of a set E contained in the interior of a circle of radius smaller than* $\frac{1}{2}$ is the lower bound of the capacities of open sets containing E and contained in the same circle.

3°. *A bounded set E will be of capacity zero* if, for any circle of radius smaller than $\frac{1}{2}$, the outer capacity of the part of E in the circle is zero. If not, E will have *non-zero capacity*.

For *closed* and *bounded* sets, the family of sets of capacity zero, or non-zero, is as follows.

The family of sets of *capacity zero* contains all *countable* sets, and extends beyond that.

The family of sets of *non-zero capacity* contains the non-pointwise[11] sets, and extends beyond that.

9. In what follows, we shall always consider a set E of class (H) on a dicylindrical domain (x in D, y arbitrary), whose sections will be denoted by $\mathfrak{H}(x)$. We suppose, in addition, that the section $\mathfrak{H}(x)$ is *bounded* in the domain D.

Lemma. *The diameter $d(x)$ of $\mathfrak{H}(x)$ is a logarithmically subharmonic function on the domain D.*

---

9) Of course the set $\mathfrak{K}_0(x')$ contains all such points.

10) See: DE LA VALLIE POUSSIN, Note II, Annales de l'Institut Henri Poincaré, 1932.

11) In the sense of PAINLEVÉ.

That is to say, the function $\log d(x)$ is subharmonic in the sense of Mr. F. RIESZ[12]; however, in the present case, these functions can become "very infinite".

From this lemma, one obtains the following theorems, which may be regarded as generalisations of a well-known theorem of Mr. HARTOGS.

Theorem 5. *If, for every point x in a set of non-zero capacity in the domain D, the set* $\mathfrak{H}(x)$ *contains only a finite number of points of the y-plane (which can change with x), the same is true for every point of the domain D, and the points are such that the section* $\mathfrak{H}(x)$ *consists of a finite number of algebroid functions.*

Theorem 6. *If, for every point x in a set of non-zero capacity in the domain D, the set* $\mathfrak{H}(x)$ *contains only a countable set of points of the y-plane (which can change with x), the same is true for every point of the domain D; the precise form is as follows:*

1°. One can find a transfinite ordinal α of class at most II, independent of x, such that $\mathfrak{H}^{(\alpha)}(x)=0$, where $\mathfrak{H}^{(\alpha)}(x)$ represents the derived set of order α of the set of points $\mathfrak{H}(x)$, x being supposed fixed.

2°. One can entirely cover the set E by a countable number of characteristic surfaces S which are respectively defined in hyperspheres (ω) and which have no points of the second kind in (ω).

3°. One can choose the preceding family of characteristic surfaces in such a way that the analytic continuation of each characteristic S outside its hypersphere (ω) in the dicylindrical domain (x in D, y arbitrary) does not give rise to any new points outside E.

12) Acta 48.

On Pseudoconvex Domains

Sur les domaines pseudoconvexes

Proceedings of the Imperial Academy, Tokyo (1941), p. 7–10

At the start of recent progress in the theory of analytic functions of several variables, F. HARTOGS discovered that every *domain of holomorphy*[1] is a *pseudoconvex domain*. These two notions have become extremely important because of the development of the theory; however, the converse problem remains largely open even today. We shall treat this problem in the present Note, in which we shall place ourselves, for simplicity, in the space of 2 complex variables, but the conclusion applies, I think, to any number of variables.

1. In the space of 2 complex variables x, y, let us consider a finite univalent domain D. We shall call it *pseudoconvex* if the set E complementary to D satisfies *the continuity theorem* in the neighbourhood of an arbitrary finite point P of E and if, in addition, this is true after any biunique pseudoconformal transformation in the neighbourhood of P. Here, the first condition means the following: if we draw a sufficiently small hypersphere around P and take, arbitrarily, a point (a, b) in the hypersphere and a circumference of the form $x = a$, $|y-b|=r$, if (a, b) belongs to E without any point of the circumference doing so, one can find a positive number d in such a way that to any x' in $|x-a|<d$, there corresponds at least one y' in $|y-b|<r$ such that (x', y') belongs to E.

Then, we have the

Theorem. *In the space of* 2 *complex variables, every finite, univalent, pseudoconvex domain is a domain of holomorphy.*

The proof will be given in a later memoir. In what follows, we shall expound rapidly the essential part of the proof.

2. Starting with 2 domains of holomorphy which overlap each other, we shall construct a pseudoconvex domain satisfying rather complicated conditions. Consider a *univalent and bounded* domain D in the space (x, y) and three hyperplanes of the form $x_1 = a$, $x_1 = a_1$, $x_1 = a_2$, which we shall denote by L, L_1, L_2 respectively; here x_1 represents the real part of x and $a_2 < a < a_1$. Suppose that each of these hyperplanes crosses D and let us denote the parts of D to the left of L_1, to the right of L_2 and between L_1 and L_2 by D_1, D_2, and D_3 respectively. Let us suppose first that every connected component of the sets D_1, D_2 is a domain of holomorphy. Next, with functions $X_j(x, y)$ ($j = 1, 2, \ldots, \nu$) holomorphic on D_3:

[1] A domain is called a domain of holomorphy if it is that of at least one function.

1°. *Suppose that the set of points of D_3 satisfying $|X_j(x, y)| \leqq 1$ $(j=1, 2, \dots, \nu)$ does not have points in the neighbourhood of the intersection of the boundary of D with L.*

2°. *Suppose that for a sufficiently small positive number ε and for each j from $1, 2, \dots, \nu$, the set of points of D_3 satisfying $|X_j(x, y)| > 1-\varepsilon$ does not have points in the neighbourhood of either L_1 or of L_2.*

3°. The points of D which either do not belong to D_3 or satisfy $|X_j(x, y)| < 1$ $(j=1, 2, \dots, \nu)$ constitute an open set because of the preceding hypothesis. *Suppose that this set has a connected compound Δ extending from the left of L_2 to the right of L_1* [2].

4°. *Suppose that we have, identically,*

$$(X_j - X_j^0)R = (x-x_0)P_j + (y-y_0)Q_j \qquad (j=1, 2, \dots, \nu)$$

when $(x, y) \in D_3$, $(x_0, y_0) \in D_3$, where X_j^0 stands for $X_j(x_0, y_0)$, P_j, Q_j and R are holomorphic functions of the variables x, y, x_0, y_0 and R reduces to 1 for $x=x_0$, $y=y_0$. [3]

5°. *Suppose that for each X_j, the number of points in the interior of D_3 satisfying*

$$\frac{\partial X_j}{\partial y}=0, \qquad x_1=a, \qquad |X_j(x, y)|=1$$

is at most finite.

6°. We shall denote the analytic variety $x_1=a$, $|X_j(x, y)|=1$ defined in D_3 by Σ_j. Because of the preceding hypothesis, Σ_j is of dimension at most 2. *Suppose that all intersections of the varieties Σ_j, Σ_k $(j \neq k)$ are of dimension at most* 1.

The domain Δ constructed thus is pseudoconvex. We shall denote by Δ_1, Δ_2 and Δ_3 respectively the parts of Δ to the left of L_1, to the right of L_2, and between L_1 and L_2. Every connected component of any of the sets Δ_1, Δ_2, Δ_3 is a domain of holomorphy.

3. Let S be the set of points consisting of the part of L in Δ and its points of accumulation; S is contained in D_3 by hypothesis 1. Let σ be the boundary of S (considered as a set on L); σ is situated on the union of the Σ_j $(j=1, 2, \dots, \nu)$; the part of σ on Σ_j will be denoted by σ_j.

The σ_j are of dimension at most 2; the intersections of the varieties σ_j and σ_k $(j \neq k)$ are of dimension at most 1 because of hypothesis 6.

---

[2] These 3 hypotheses are based on the theorem of H. Cartan-P. Thullen that any finite domain of holomorphy is convex with respect to functions holomorphic on the domain. See: H. Cartan and P. Thullen, Math. Ann. 1932.

[3] There is a proposition regarding this hypothesis: Let D be a univalent and finite domain of holomorphy in the space (x, y). Given a positive number ε and a univalent, bounded domain D_0 contained, together with its boundary, in D, we can find a holomorphic function R of the variables x, y, x_0, y_0 in $(x, y) \in D_0$, $(x_0, y_0) \in D_0$, reducing to 1 for $x=x_0$, $y=y_0$, in such a way that we have the following: to any holomorphic function $f(x, y)$ on D, there corresponds a holomorphic function $\varphi(x, y)$ satisfying

$$|f-\varphi|<\varepsilon, \qquad (\varphi-\varphi_0)R=(x-x_0)P+(y-y_0)Q$$

on D_0, where φ_0 means $\varphi(x_0, y_0)$ and P and Q are holomorphic functions of the variables x, y, x_0, y_0. - One can prove this because of the theorem of H. Cartan-P. Thullen cited above, a theorem of A. Weil, and a theorem of the author. For these last 2 theorems, see: K. Oka, J. Sci. Hiroshima Univ., 1936, No. 4; and 1937, Theorem I.

Under these circumstances, let us consider the double integral taken over the two dimensional part of σ

$$I(x_0, y_0)=\frac{-1}{4\pi^2}\sum_j \int_{\sigma_j} \psi_j(x, y; x_0, y_0)\, \varphi(x, y)\, dx\, dy \qquad (j=1,2,\ldots,\nu),$$

$$\psi_j(x, y; x_0, y_0)=\frac{Q_j}{(x-x_0)(X_j-X_j^0)},$$ [4]

$\varphi(x, y)$ being an arbitrary holomorphic function on a certain open set containing S.

When $(x, y)\in\sigma_j$, ψ_j is holomorphic in $(x_0, y_0)\in\Delta_3$ except on L; the same is therefore true of $I(x_0, y_0)$. Let $I_1(x_0, y_0)$ be the part of $I(x_0, y_0)$ to the left of L, and I_2 that on the right. We find easily, because of hypothesis 6, that I_1 and I_2 can be analytically continued a little across L in Δ_3, in such a way that $I_1-I_2=\varphi$.

We are going to modify ψ_j. Let Σ_j' be the part of Σ_j on the set $|X_p(x, y)|\leqq 1$ $(p=1,2,\ldots,\nu; p\neq j)$; Σ_j' contains σ_j. We construct an open set V_j containing Σ_j', sufficiently close Σ_j', and such that all its connected components are domains of holomorphy. V_j certainly exists because Σ_j' is contained in D_3 because of hypothesis 1.

Since the first Cousin problem is always solvable on a univalent, finite, domain of holomorphy [5], we can find, because of hypothesis 2, *a meromorphic function* $\Phi_j(x, y; x_0, y_0)$ *in* $(x, y)\in V_j$, $(x_0, y_0)\in D_1$ *in such a way as to have the same poles as* ψ_j *when* $(x_0, y_0)\in D_3$ *and be holomorphic when* $(x_0, y_0)\notin D_3$.

$\Phi_j-\psi_j$ is holomorphic in $(x, y)\in V_j$, $(x_0, y_0)\in D_3$. D_3 being convex with respect to functions holomorphic on D_1, we can find, for a given positive number ε, *a holomorphic function* $\Psi_j(x, y; x_0, y_0)$ *in* $(x, y)\in V_j$, $(x_0, y_0)\in D_1$, *such that* $|\Phi_j-\psi_j-\Psi_j|<\varepsilon$ *on* $(x, y)\in V_j'$, $(x_0, y_0)\in D_3'$, where V_j' is an open set sufficiently close to V_j, contained, together with its boundary in V_j and given à priori, and D_3' is such a set relative to D_3.[6]

We have thus acquired functions Φ_j and Ψ_j relative to D_1. Let us set $A_j=\Phi_j-\Psi_j-\psi_j$. We construct similarly functions B_j for D_2, and consider the follow-

---

[4]) Let us explain some points concerning this integral. On any σ_j, $\partial X_j/\partial y$ does not vanish except perhaps at a finite number of points became of hypothesis 5. Let us take an arbitrary point P on σ_j outside the exceptional points. In the neighbourhood of P, we can represent the variety Σ_j, using real parameters u, v, in the form $x=x(u, v)$, $y=y(u, v)$, where the functions on the right hand side are power series in the variables u, v, and this, in such a way that (x, y) and (u, v) are in one-to-one correspondence; we are here considering (u, v) as a point in the plane by means of the usual rectangular axes. To the boundary of σ_j correspond a finite number of analytic arcs in the plane. Let x_2 be the imaginary part of x and θ_j the argument of $X_j(x, y)$; let us choose (u, v) so that $\partial(x_2, \theta_j)/\partial(u, v)>0$. We then have, by definition,

$$\int_{\sigma_j} \psi_j \varphi\, dx\, dy=\iint \psi_j \varphi \frac{\partial(x, y)}{\partial(u, v)}\, du\, dv$$

in the neighbourhood of the point P, the second integral being a double integral extended over the part of the (u, v) plane corresponding to the 2-dimensional part of σ_j.

[5]) See: K. Oka, J. Sci. Hiroshima Univ., 1937, No. 5.

[6]) See: K. Oka, J. Sci. Hiroshima Univ., 1937, No. 5.

ing integrals instead of $I(x_0, y_0)$:

$$J_1(x_0, y_0) = \frac{-1}{4\pi^2} \sum_j \int_{\sigma_j} (\psi_j + A_j)\, \varphi(x, y)\, dx\, dy$$

$$J_2(x_0, y_0) = \frac{-1}{4\pi^2} \sum_j \int_{\sigma_j} (\psi_j + B_j)\, \varphi(x, y)\, dx\, dy \qquad (j = 1, 2, \ldots, \nu).$$

$J_1(x_0, y_0)$ is holomorphic on Δ_1 since this is the case for $\psi_j + A_j$ when $(x, y) \in \sigma_j$; and similarly, $J_2(x_0, y_0)$ is holomorphic on Δ_2. These functions can be continued analytically a little across L in Δ, and there satisfy the relation

$$J_1(x_0, y_0) - J_2(x_0, y_0) = \varphi(x_0, y_0) - \frac{1}{4\pi^2} \sum_j \int_{\sigma_j} (A_j - B_j)\, \varphi(x, y)\, dx\, dy \qquad (j = 1, 2, \ldots, \nu).$$

Let us set $f = J_1 - J_2$; from this relation, it follows that f is holomorphic at every point of S.

We now consider f as given and φ as unknown in this relation, and we have a FREDHOLM integral equation of the second kind. Choosing ε sufficiently small, we therefore find the required $\varphi(x, y)$, holomorphic at every point of S. The following assertion results from this:

Under the conditions of No. 2, given a function $f(x, y)$ singlevalued and holomorphic at every point of the common boundary S of the sets Δ_1 and Δ_2, we can find 2 holomorphic functions $F_1(x, y)$ and $F_2(x, y)$ on Δ_1 and on Δ_2 respectively, such that we can continue them analytically a little across L in Δ, and such that we then have, identically, $F_1(x, y) - F_2(x, y) = f(x, y)$.

This gives us our starting point.

Bibliography

I. Domaines convexes par rapport aux fonctions rationelles
Journal of Science of the Hiroshima University **6** (1936), p. 245–255.

II. Domaines d'holomorphie
Journal of Science of the Hiroshima University **7** (1937), p. 115–130.

III. Deuxième problème de Cousin
Journal of Science of the Hiroshima University **9** (1939), p. 7–19.

IV. Domaines d'holomorphie et domaines rationnellement convexes
Japanese Journal of Mathematics **17** (1941), p. 517–521.

V. L'intégrale de Cauchy
Japanese Journal of Mathematics **17** (1941), p. 523–531.

VI. Domaines pseudoconvexes
Tôhoku Mathematical Journal **49** (1942), p. 15–52.

VII. Sur quelques notions arithmétiques
Bulletin de la Société Mathématique de France **78** (1950), p. 1–27.

VIII. Lemme fondamental
Journal of the Mathematical Society of Japan **3** (1951), p. 204–214; 259–278.

IX. Domaines finis sans point critique intérieur
Japanese Journal of Mathematics **27** (1953), p. 97–155.

X. Une mode nouvelle engendrant les domaines pseudoconvexes
Japanese Journal of Mathematics **32** (1962), p. 1–12.

Note sur les familles de fonctions analytiques multiformes etc.
Journal of Science of the Hiroshima University **4** (1934), p. 93–98.

Sur les domaines pseudoconvexes
Proceedings of the Imperial Academy, Tokyo (1941), p. 7–10.

Added in Print

In 1980 Toshio NISHINO from Kyushu University, Japan, started editing *Posthumous Papers of Kiyoshi Oka* with the help of Akira TAKEUCHI, Kyoto University. To date seven volumes have appeared. Most of these papers were written in Japanese, however there are also some papers in French which we would like to list here:

1. Fonctions algébriques permutables avec une fonction rationnelle nonlinéaire (vol. **6**; 87 pages; written about 1930)
2. Sur les ensembles de points à 4 dimensions engendrés analytiquement (vol. **7**; 146 pages; written about 1934)
3. Note sur les fonctions analytiques de plusieurs variables (vol. **3**; 7 pages; written about 1949)

H. Cartan

Œuvres – Collected Works

Volume 1–3
Editors: **R. Remmert, J-P. Serre**

1979. 3 portraits. XXXI, 1469 pages in French.
(In 3 volumes, not available separately)
ISBN 3-540-09189-0

From the preface: "We are happy to present the Collected Works of Henri Cartan.

There are three volumes. The first one contains a curriculum vitae, a Brève Analyse des Traveaux and a list of publications, including books and seminars. In addition the volume contains all papers of H. Cartan on analytic functions published before 1939. The other papers on analytic functions, e.g. those on Stein manifolds and coherent sheaves, make up the second volume. The third volume contains, with a few exceptions, all further papers of H. Cartan; among them is a reproduction of exposés 2 to 11 of his 1954/55 Seminar on Eilenberg-MacLane algebras. Each volume is arranged in chronological order.

The reader should be aware that these volumes do not fully reflect H. Cartan's work, a large part of which is also contained in his fifteen ENS-Seminars (1948–1964) and in his book Homological Algebra with S. Eilenberg.

Still, we trust that mathematicians throughout the world will welcome the availability of the **Œuvres** of a mathematician whose writing and teaching has had such an influence on our generation."

Springer-Verlag
Berlin
Heidelberg
New York
Tokyo

H. Grauert, R. Remmert

Theory of Stein Spaces

Translated from the German
by A. Huckleberry

1979. 5 figures, 1 table. XXI, 249 pages
(Grundlehren der mathematischen Wissenschaften, Band 236)
ISBN 3-540-90388-7

Contents: Sheaf Theory. – Cohomology Theory. – Coherence Theory for Finite Holomorphic Maps. – Differential Forms and Dolbeault Theory. – Theorems A and B for Compact Blocks in $\mathbb{C}^m$. – Stein Spaces. – Applications of Theorems A and B. – The Finiteness Theorem. – Compact Riemann Surfaces. – Bibliography. – Subject Index. – Table of Symbols.

The Theory of Functions of several variables is of importance not only in mathematics, but also in theoretical physics. Quantum field theory demands the construction of holomorphic hulls of special regions of the complex domain. These holomorphic hulls are Stein spaces.
This book contains a systematic presentation of the theory of these spaces, written by authors who have had a decisive part in the development of the theory. The proof of Theorems A and B from H. Cartan's seminar is the natural first step. Then, examples of applications are given and the finiteness theorems for compact complex spaces proved. This includes a treatment of compact Riemann surfaces from a higher point of view.

Springer-Verlag
Berlin
Heidelberg
New York
Tokyo

Printed in the United States
By Bookmasters